Second International Conference on Mechanical Engineering and Automation Science (ICMEAS 2016)

IOP Conference Series: Materials Science and Engineering
Volume 157

Singapore
13 – 15 October 2016

Editors:

Kyle Jiang
Ian McAndrew

ISBN: 978-1-5108-3273-2
ISSN: 1757-8981

Printed from e-media with permission by:

Curran Associates, Inc.
57 Morehouse Lane
Red Hook, NY 12571

Some format issues inherent in the e-media version may also appear in this print version.

Printed by Curran Associates, Inc. (2017)

For permission requests, please contact the Institute of Physics
at the address below.

Institute of Physics
Dirac House, Temple Back
Bristol BS1 6BE UK

Phone: 44 1 17 929 7481
Fax: 44 1 17 920 0979

techtracking@iop.org

Additional copies of this publication are available from:

Curran Associates, Inc.
57 Morehouse Lane
Red Hook, NY 12571 USA
Phone: 845-758-0400
Fax: 845-758-2633
Email: curran@proceedings.com
Web: www.proceedings.com

Table of contents

Volume 157

2016 Second International Conference on Mechanical Engineering and Automation Science (ICMEAS 2016)

13–15 October 2016, Singapore

Accepted papers received: 31 October 2016
Published online: 14 November 2016

Mechanical

General knowledge of mechnical engineering and automation science

IOP Conf. Series: Materials Science and Engineering **157** (2016) 011001 doi:10.1088/1757-899X/157/1/011001

PREFACE

ICMEAS 2016 was held in Singapore during October 13-15, 2016. ICMEAS 2016 was co-organized by Embry-Riddle Aeronautical University and Hong Kong Society of Mechanical Engineers (HKSME). The conference provides a useful and wide platform both for displaying the latest research and for exchanging research results and thoughts in broad Mechanical Engineering and Automation Science. The participants of the conference were from almost every part of the world, with a background in either academia or industry, even well-known enterprise. The success and prosperity of the conference is reflected in the high level of the papers received.

The proceedings are a compilation of the accepted papers and represent an interesting outcome of the conference. This book covers 3 chapters: Aeronautics; Mechanical; General knowledge of Mechanical engineering and automation science.

We would like to acknowledge all of those who supported ICMEAS 2016. Each individual and institution's help was very important for the success of this conference. Especially, we would like to thank the organizing committee for their valuable advice in the organization and helpful peer review of the papers.

We sincerely hope that ICMEAS 2016 will be a forum for excellent discussions that will put forward new ideas and promote collaborative researches. We are sure that the proceedings will serve as an important research source of reference and the knowledge, will lead to not only scientific and engineering progress but also other new products and processes.

Prof. Kyle Jiang, University of Birmingham, UK

Prof. Ian McAndrew, Embry Riddle Aeronautical University, UK

IOP Conf. Series: Materials Science and Engineering **157** (2016) 011002 doi:10.1088/1757-899X/157/1/011002

COMMITTEES

Advisory Committee Chairs

Prof. Kyle Jiang, University of Birmingham, UK
Prof. Harijono Djojodihardjo, Universiti Putra Malaysia, Malaysia

Conference Committee Chairs

Prof. Ian McAndrew, Embry Riddle Aeronautical University, UK (Co-Chair)
Prof. Du Ruxu, The Chinese University of Hong Kong, Hong Kong (Co-Chair)
Prof. Shinn-Liang Chang, National Formosa University, Taiwan

Program Committee Chairs

Prof. M. Chandrasekaran, Vels University, India
Prof. Yong Zhao, Nazarbayev University, Kazakhstan

International Technical Committee

Prof. Clint R. Balog, Embry-Riddle Aeronautical University, USA
Dr. Darwin Lau, the Chinese University of Hong Kong
Elena Navarro, Embry Riddle Aeronautical University, UK
Dr. Alex Ong Zhi Chao, University of Malaya, Malaysia
Dr. Li Wen, Beihang University, China
Prof. Tang Sai Hong, Universiti Putra Malaysia, Malaysia
Prof. C.V.Gopinath, Alliance University, Bangalore, India
Dr. EDMUND LIEW, Singapore Institute of Technology, Singapore
Prof. Spilios D. Fassois, University of Patras, Greece
Dr. Thomas Luft, Alexander-Universität Erlangen-Nürnberg, Germany
Dr. B. F. Yousif, University of Southern Queensland, Australia
Dr. Mohammadreza Behi, ClimateWell-KTH, Sweden
Dr. S. Rajakumar, Annamalai University, India
Dr. Balhassn S. M. Ali, Bursa Orhangazi University, Turkey
Dr. Mohamed Arezki MELLAL,UMBB, Algeria

Peer review statement

All papers published in this volume of *IOP Conference Series: Materials Science and Engineering* have been peer reviewed through processes administered by the proceedings Editors. Reviews were conducted by expert referees to the professional and scientific standards expected of a proceedings journal published by IOP Publishing.

Lift estimation of Half-Rotating Wing in hovering flight

X Y Wang, Y P Dong, Z Z Qiu, Y Q Zhang and J H Shan

School of Mechanical Engineering, Anhui University of Technology, Ma´anshan, 243002, P.R.China

wangxy@ahut.edu.cn, dyp704@163.com, qzzhlx@ahut.edu.cn, 15755503181@163.com, 2913@ahut.edu.cn.

Abstract. Half-Rotating Wing (HRW) is a new kind of flapping wing system with rotating flapping instead of oscillating flapping. Estimating approach of hovering lift which generated in hovering flight was important theoretical foundation to design aircraft using HRW. The working principle of HRW based on Half-Rotating Mechanism (HRM) was firstly introduced in this paper. Generating process of lift by HRW was also given. The calculating models of two lift mechanisms for HRW, including Lift of Flow Around Wing (LFAW) and Lift of Flow Dragging Wing (LFDW), were respectively established. The lift estimating model of HRW was further deduced, by which hovering lift for HRW with different angular velocity could be calculated. Case study using XFLOW software simulation indicates that the above estimating method was effective and feasible to predict roughly the hovering lift for a new HRW system.

1. Introduction

Flapping Wing Air Vehicle (FWAV) is widely applied in military and civil fields for its advantages. Some great results had been got in micro FWAV by lots of researchers in the world [1], [2]. However, the swing motion of flapping wing at high speed leads to large inertial force, by which the development of FWAV in large size had been further restricted. The movement forms of animal limbs are asymmetric swing in essence. Swing is a movement form in order to fit muscles for animal, while asymmetric motion is the basic condition to fly and walk. Based the above views, our work team had developed a rotating mechanism named Half-Rotating Mechanism (HRM) [3], which could not only generate asymmetrical motion but avoid swing. HRW flight mechanism was developed by improving HRM with lighter weight. It would make FWAV in large size possible for its simple structure, high efficiency and strong adaptability.

Lift prediction is a basic link to develop the bionic aircraft. Abundant research achievements [4-6] have been made on the lift of flapping flight. HRW aircraft as a new kind of flapping wing system has not yet researched about its lift. Therefore, the hovering lift of HRW aircraft was calculated in this paper based on LFAW and LFDW [7]. This research would provide primary quantitative analysis about the lift of HRW as well as guide the design and flight control for HRW aircraft.

2. Work principle of HRW and generating hovering lift by HRW

2.1. Motion and driving mechanism for HRW

The driving mechanism of HRW was a kind of simplified HRM which was suitable for flight. Figure 1 showed the driving mechanism which was consist of spindle, crank, blade of HRW, positioning slider and shaft with groove. The blade of HRW would turn half round anticlockwise when crank turned one round anticlockwise. Herein, point C was not only the center of shaft with groove but also the middle of blade of HRW. Point H was the center of positioning slider, which was the fixed point in HRM. To avoid discontinuous movement caused by the interference when point C moved to fixed point H, one edge of the crank was designed as shaft with groove. Furthermore, positioning slider was set at the location of point H to make sure the blade of HRW have accurate moving rule. The combination of positioning slider and shaft with groove could realize simplified design of HRM. HRW could produce continuous and certain motion when motor drove the crank to rotate.

As shown in figure 2, HRW rotated counterclockwise. The blade of HRW was divided into two parts by fixed point H. The part of HRW on fixed point H left moved downwards, which was equivalent to that wings of insect moved on downstroke. Thus this part could be called downstroke. Similarly, the other part of HRW on fixed point H right could be called upstroke.

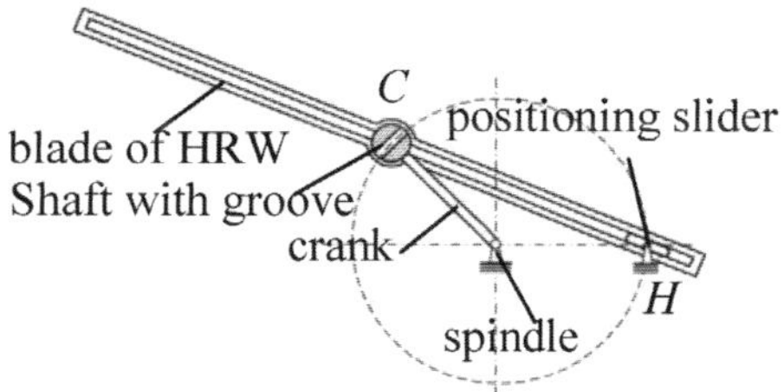

Figure 1. HRW driving mechanism

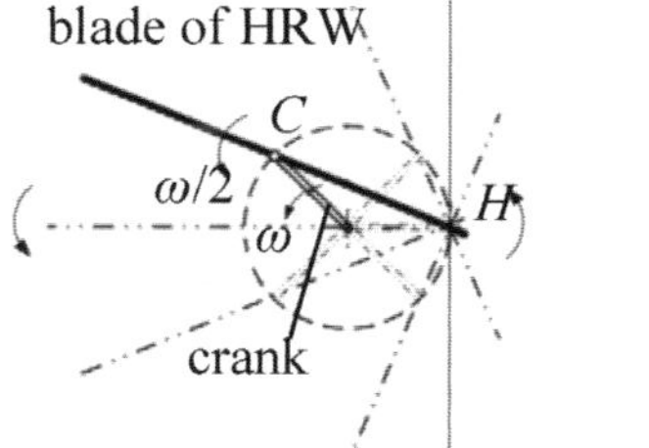

Figure 2. Motion interval division for HRW

2.2. *The spanwise airflow around HRW*

The effect of air on the blade was so complex during the whole motion of HRW that a simplified model that lift was generated by the relative motion between the air and wing was used. Generally, only chordwise airflow was around the flapping wing. Spanwise airflow herein arose from obvious telescopic movement of HRW along the spanwise direction, as shown in figure 3. The influence came from both chordwise and spanwise airflow on the lift estimation of the HRW should be considered. Typical flight states of aircraft included hovering flight and forward flight. The air relative to the HRW just produced spanwise airflow when the HRW aircraft was in the hovering flight. Not only the spanwise airflow but also chordwise airflow would be produced when HRW aircraft was in the forward flight. In this paper, the lift was studied only when HRW aircraft was in hovering flight, namely only spanwise airflow was around the wing.

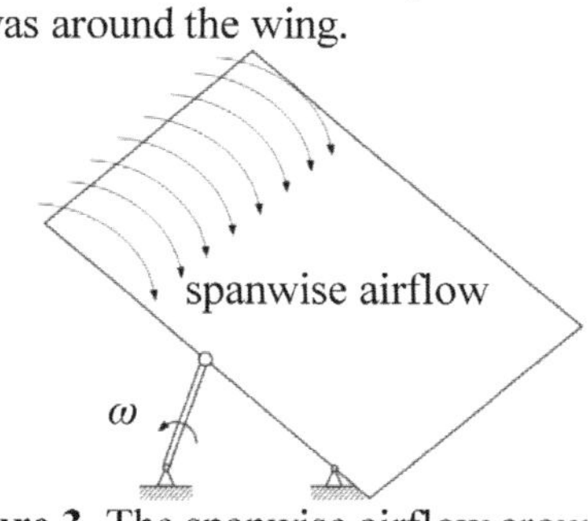

Figure 3. The spanwise airflow around HRW

2.3. *How to generate hovering lift by HRW*

The computation model of hovering lift by HRW was established. As shown in figure 4, AB and OC respectively represented the blade of HRW and crank, and the lengths of them were $2a$ and R. C was the shaft with groove, H denoted positioning slider. P was instantaneous center of velocity at this time. The angular velocity of HRW was $\omega/2$ when the angular velocity of crank was ω. φ represented the rotation angle of crank from original position where crank was in direction of horizontal-right. α denoted the angle of attack. It should be noted that all angles in counterclockwise direction were positive. The coordinate system which the origin was at point C and the positive X axis was along AB

was established. Then the range of downstroke and upstroke could be expressed by the bounds for x in $(-2R\sin\frac{\varphi}{2}, a]$ and $[-a, -2R\sin\frac{\varphi}{2}]$ respectively. The velocity of each point on the blade of HRW was so different that the relative velocity of airflow at corresponding points was different. For instance, the relative velocity of airflow at point D and velocity of point D were equal and opposite. Lift element was taken along the spanwise direction because the lift of each point on the blade of HRW was different, as shown in figure 5. Then angle of attack, LFAW and LFDW were calculated respectively. It should be noted that directions of lift generated by HRW in downstroke and upstroke were different.

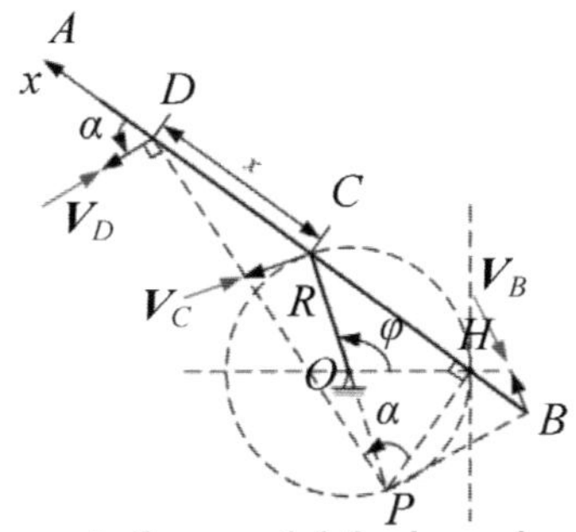

Figure 4. Computation model for hovering lift by HRW

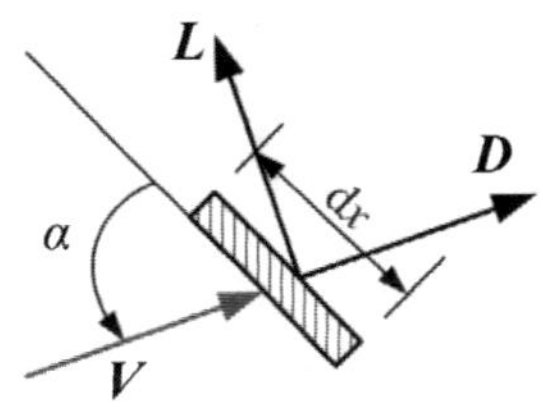

Figure 5. Lift element of wing

3. Lift estimation of HRW in hovering flight

3.1. Angle of attack for HRW flight

As the speed of each point on the wing was different when HRW rotated, its corresponding angle of attack was also different. The flow patterns of the airflow around the wing were different under the condition of small Angle of Attack (SAA) and Large Angle of Attack (LAA) when calculating the LFAW, and the lift estimation method was also different. Thus, the scope of angle of attack should be defined. The angle of attack of point C which was always $\varphi/2$ could be regarded as the decision value. Considering the cyclicality of blade motion, the stroke of the crank was set to one round in this paper in order to make calculation and description easier. When $\varphi/2<\pi/9$, namely $\varphi<2\pi/9$, the lift of any point on the wing was calculated by the method of SAA. When $\varphi/2>\pi/9$, namely $\varphi>2\pi/9$, the lift of any point on the wing was calculated by the method of LAA. Therefore, φ could be divided into four intervals that were $[0,\frac{2\pi}{9}),[\frac{2\pi}{9},\pi),[\pi,\frac{16\pi}{9}),[\frac{16\pi}{9},2\pi]$ .When $\varphi\in[0,\frac{2\pi}{9})$ or $\varphi\in[\frac{16\pi}{9},2\pi]$, the wing's angle of attack was SAA. When $\varphi\in[\frac{2\pi}{9},\frac{16\pi}{9})$, the wing's angle of attack was LAA.

Wherever the blade of HRW was, the angle of attack for any point on the wing could be got:

$$\alpha = \arcsin\frac{x+2R\sin\frac{\varphi}{2}}{\left(4R^2+x^2+4Rx\sin\frac{\varphi}{2}\right)^{1/2}} \tag{1}$$

Where R was the length of the crank, and φ was the rotation angle of crank.

3.2. LFAW estimation of HRW in hovering flight

3.2.1. LFAW estimation with SAA. Under the condition of SAA, air flowed around the wing could be regarded as air flowed around flat plate [8] then circulation around the wing which determined LFAW would be generated. Circulation at point C could be considered as the circulation of the whole wing, because the speed and circulation of each point on the wing was different.

The wing's angle of attack was SAA when $\varphi \in [0, \frac{2\pi}{9})$. According to the velocity direction of point *C*, it was known that clockwise circulation around the wing was as follows:

$$\Gamma = \pi v_C \cdot (2a) \sin \alpha_C \tag{2}$$

Where 2a was spanwise length of HRW, $v_C = \omega R$ was the velocity direction of point *C*, and $\alpha_C = \frac{\varphi}{2}$ was the angle of attack of point *C*.

The blade was composed of infinite point vortexes according to the thin wing theory [8]. If each point vortex was the same, circulation was equivalent to uniform distribution. Linear density of circulation was $\frac{\Gamma}{2a}$, namely the circulation around spanwise length element of wing dx was:

$$\mathrm{d}\Gamma = \pi v_C \sin \alpha_C \mathrm{d}x \tag{3}$$

Then the aerodynamic lift element was:

$$\mathrm{d}L_{\mathrm{S}} = \rho v \mathrm{d}\Gamma = \pi \rho h v v_c \sin \alpha_c \mathrm{d}x \tag{4}$$

Where ρ was air density, h was chord length of HRW, and v was the relative velocity of airflow.

LFAW elements on downstroke and upstroke were both:

$$\mathrm{d}F_{\mathrm{LAS}} = \mathrm{d}L_{\mathrm{S}} \sin(\frac{\varphi}{2} + \alpha) \tag{5}$$

LFAW element was integrated along the span of the wing to get the relationship between LFAW of whole wing and rotation angle φ:

$$F_{\mathrm{LAS}} = \int_{-a}^{a} \mathrm{d}F_{\mathrm{LAS}} = 2\pi a \rho h \omega^2 R^2 \sin\frac{\varphi}{2} \sin\varphi \tag{6}$$

Where ω was the angular velocity of crank.

Lift estimation method was the same as above when $\varphi \in [\frac{16\pi}{9}, 2\pi]$. The relationship between LFAW of whole wing and rotation angle φ was got:

$$F_{\mathrm{LAS}} = \int_{-a}^{a} \mathrm{d}F_{\mathrm{LAS}} = -2\pi a \rho h \omega^2 R^2 \sin\frac{\varphi}{2} \sin\varphi \tag{7}$$

3.2.2. LFAW estimation with LAA. Under the condition of LAA, the airflow was detached on the back of the wing. Thus, circulation around the wing could not be generated. The approximate calculation formulas of aerodynamic force originated from flowing around a flat at LAA [9] were used to estimate LFAW and LFDW with LAA. The pressure drag coefficient C_D measured by experiment was 4.6 in this paper.

The wing's angle of attack was LAA when $\varphi \in [\frac{2\pi}{9}, \pi)$. The estimation of aerodynamic lift element was in this form:

$$\mathrm{d}L_{\mathrm{L}} = \frac{1}{4} C_{\mathrm{D}} \rho h v^2 \sin 2\alpha \mathrm{d}x \tag{8}$$

Where C_D was pressure drag coefficient.

LFAW element on downstroke was:

$$\mathrm{d}F_{\mathrm{LAL1}} = \mathrm{d}L_{\mathrm{L}} \sin(\frac{\varphi}{2} + \alpha) \tag{9}$$

LFAW element on upstroke was:

$$\mathrm{d}F_{\mathrm{LAL2}} = -\mathrm{d}L_{\mathrm{L}} \sin(\frac{\varphi}{2} + \alpha) \tag{10}$$

The wing's angle of attack was LAA when $\varphi \in [\pi, \frac{16\pi}{9})$. The estimation of aerodynamic lift element was in this form:

$$\mathrm{d}L_{\mathrm{L}} = \frac{1}{4} C_{\mathrm{D}} \rho h v^2 \sin 2(\pi - \alpha) \mathrm{d}x = -\frac{1}{4} C_{\mathrm{D}} \rho v^2 \sin 2\alpha \mathrm{d}x \tag{11}$$

LFAW element on downstroke was:

$$\mathrm{d}F_{\mathrm{LAL1}} = -\mathrm{d}L_{\mathrm{L}} \sin(\frac{\varphi}{2} + \alpha) \tag{12}$$

LFAW element on upstroke was:

$$\mathrm{d}F_{\mathrm{LAL2}} = \mathrm{d}L_{\mathrm{L}} \sin(\frac{\varphi}{2} + \alpha) \tag{13}$$

LFAW element was integrated along the span of the wing to get LFAW of whole wing:

$$\begin{aligned} F_{\mathrm{LAL}} &= \int_{-2R\sin\frac{\varphi}{2}}^{a} \mathrm{d}F_{\mathrm{LAL1}} + \int_{-a}^{-2R\sin\frac{\varphi}{2}} \mathrm{d}F_{\mathrm{LAL2}} \\ &= \frac{1}{4} C_D \rho h \omega^2 R \left[\int_{-2R\sin\frac{\varphi}{2}}^{a} \frac{x^2 \cos^2\frac{\varphi}{2} + 3Rx\sin\varphi\cos\frac{\varphi}{2} + 2R^2\sin^2\varphi}{\left(4R^2 + x^2 + 4Rx\sin\frac{\varphi}{2}\right)^{1/2}} \mathrm{d}x \right. \\ &\left. - \int_{-a}^{-2R\sin\frac{\varphi}{2}} \frac{x^2 \cos^2\frac{\varphi}{2} + 3Rx\sin\varphi\cos\frac{\varphi}{2} + 2R^2\sin^2\varphi}{\left(4R^2 + x^2 + 4Rx\sin\frac{\varphi}{2}\right)^{1/2}} \mathrm{d}x \right] \end{aligned} \tag{14}$$

3.2.3. LFDW estimation of HRW in hovering flight. The lift was mainly derived from LFAW produced by circulation and LFDW could be ignored when the wing's angle of attack was SAA. However, LFDW made a significant contribution to lift in the range of LAA. The estimation of aerodynamic drag element was in this form:

$$\mathrm{d}D = \frac{1}{2} C_{\mathrm{D}} \rho h v^2 \sin^2 \alpha \mathrm{d}x \tag{15}$$

LFDW elements on downstroke and upstroke were both:

$$\mathrm{d}F_{\mathrm{LB}} = -\mathrm{d}D \cos(\frac{\varphi}{2} + \alpha) \tag{16}$$

LFDW element was integrated along the span of the wing to get LFDW of whole wing:

$$F_{\mathrm{LB}} = \int_{-a}^{a} \mathrm{d}F_{\mathrm{LB}} = -\frac{1}{8} C_D \rho h \omega^2 \int_{-a}^{a} \left(x + 2R\sin\frac{\varphi}{2} \right)^2 \cdot \frac{2R\cos\varphi - x\sin\frac{\varphi}{2}}{\left(4R^2 + x^2 + 4Rx\sin\frac{\varphi}{2} \right)^{1/2}} \mathrm{d}x \tag{17}$$

4. Simulation verification

4.1. Case study about lift calculation

To verify the results, the lift formulas obtained above were used to estimate the lift of HRW prototype which had been developed before. The dimensions of HRW were R=0.06m, a=0.141m, h=0.2m, and the air density ρ=1.225kg/m^3. The lift values of HRW at different positions within one round were calculated when ω was 6πrad/s, 10πrad/s, 20πrad/s, 30πrad/s respectively.

According to the dimensions of HRW prototype, the model was established in fluid simulation software XFLOW and the related parameters were set. Similarly, numerical calculation for the lift of HRW at four different angular velocities was also carried out.

4.2. Comparison between theoretical results and simulation results

Lift values from theoretical calculation and numerical calculation were processed by Matlab. When ω was 6πrad/s, 10πrad/s, 20πrad/s and 30πrad/s, the comparison results were shown respectively in figure 6.

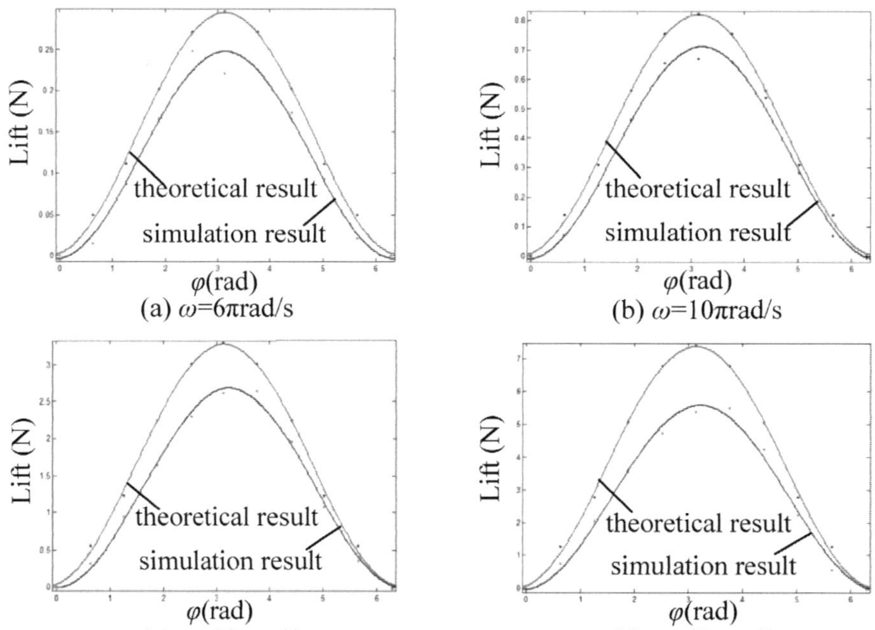

(a) ω=6πrad/s (b) ω=10πrad/s
(c) ω=20πrad/s (d) ω=30πrad/s

Figure6. Comparison between theoretical results and simulation results at different angular velocities

It could be seen from figure 6 that:

The changed trends of theoretical lift and simulation lift were consistent and corresponding lift values were also close, which could verify the validity of the theoretical calculation method about the lift of HRW in this paper.

The reasons for the difference between the two kinds of lift values:

1) The measuring accuracy of the pressure drag coefficient C_D in experiment.

2) Due to the complexity of airflow in HRW flight, the treatments of airflow were so different in theoretical calculation and numerical calculation that the calculation results might be different.

3) Flexible wing was used to measure the pressure drag coefficient C_D in experiment, while the rigid wing was used to calculate lift in simulation. The lift calculated in simulation was smaller than the lift obtained from theoretical calculation, for which the pressure drag coefficient of flexible wing was bigger than that of rigid wing.

5. Conclusions

1) According to motion characteristics of HRW driving mechanism, the blade of HRW was divided into downstroke and upstroke. It was illustrated that spanwise airflow was the main source of hovering lift. The theoretical estimation formulas of LFAW and LFDW were deduced respectively after establishing the lift estimating model for spanwise airflow, which were verified by XFLOW. The research in this paper provided the theory support for lift prediction and parameter design of HRW.

2) The velocity of each point along HRM spanwise direction was different. Hence, LFAW and LFDW were obtained after analyzing and integrating lift elements. The lift on downstroke and upstroke needed to be analyzed respectively because lift directions of them were different.

3) The lift of HRW was proportional to the square of angular velocity for the crank, and changed with the rotational angle of crank. The lift value could reach the maximum when the blade of HRW flapped down horizontally.

Acknowledgments

This work is supported by the National Natural Science Foundation of China (Grant No. 51375014 and No. 51405001)

References

[1] Pornsin-Sirirak T N, Lee S W, Nassef H and Grasmeyer J 2000 *13th IEEE Int. Conf. on Micro Electro Mechanical Systems (Miyazaki)* vol 1 (Piscataway: IEEE) pp 799-804

[2] Fearing R S, Chiang K H, Dickinson M H, Pick D L, Sitti M and Yan J 2000 *IEEE Int. Conf. on Robotics and Automation (San Francisco)* (Piscataway: Institute of Electrical and Electronics Engineers Inc) pp 1509-15

[3] Qiu Z Z 2011 *Half-Rotating Mechanism: Structure, Characteristics and Application* (Hefei: University of Science and Technology of China Press)

[4] Dickinson M H 1999 *Sci* **284** pp 1954-60

[5] Mazaheri K and Ebrahimi A 2011 *J. Fluids .Struct* **27**(4) pp 586-595

[6] Sun M andTang J 2002 *J. Exp.Biol* **205** pp 55-70

[7] Wang X Y, Dong Y P, Qiu Z Z,Wang H X and Chen F Q *Mech. Sci. Technol. Aero. Eng* (accepted)

[8] Wu W Y 1983*Fluid Mechanics (2)* (Beijing: Peking University Press) pp 80-1

[9] Jiang H B, Cao S L and Cheng Z Q 2011 *Chin. J. Appl.Mech* **28** pp 518-20

Analytical, Computational Fluid Dynamics and Flight Dynamics of Coandă MAV

H Djojodihardjo[1] and RI Ahmed[2]

[1]The Institute for the Advancement of Aerospace Science and Technology, Jl.Uranus V-17, Jakarta 15419, Indonesia; harijono.djojodihardjo@yahoo.com; [2]engriyadh64@gmail.com

[2]Graduate Student, Aerospace Engineering Department, Universiti Putra Malaysia, 43400 UPM Serdang. Selangor, Malaysia

Abstract. The paper establishes the basic working relationships among various relevant variables and parameters governing the aerodynamics forces and performance measures of Coandă MAV in hover and translatory motion. With such motivation, capitalizing on the basic fundamental principles, the Fluid Dynamics and Flight Mechanics of semi-spherical Coandă MAV configurations are revisited and analyzed as a baseline. To gain better understanding on the principle of Coandă MAV lift generation, a mathematical model for a spherical Coandă MAV is developed and analyzed from first physical principles. To gain further insight into the prevailing flow field around a Coandă MAV, as well as to verify the theoretical prediction presented in the work, a computational fluid dynamic CFD simulation for a Coandă MAV generic model are elaborated using commercial software FLUENT®. In addition, the equation of motion for translatory motion of Coandă MAV is elaborated. The mathematical model and derived performance measures are shown to be capable in describing the physical phenomena of the flow field of the semi-spherical Coandă MAV. The relationships between the relevant parameters of the mathematical model of the Coandă MAV to the forces acting on it are elaborated subsequently.

1. Introduction

Coandă effect and Coandă jet have found many applications in engineering, among others in aircraft and vehicle technology [1-5], as well as wind-turbines [6-9]. In particular, Coandă MAV' share in active developments, and various configurations have been proposed and developed. For the purpose of designing a Coandă MAV which could meet the desired mission and design requirements, it is mandatory to establish the basic working relationships among various relevant variables and parameters governing the aerodynamics forces. To assist the analysis, design and developments of Coandă MAV's, several tools can be resorted to. The first is the analytical tool, which capitalizes on the basic fundamental principles. The second is the utilization of Computational Fluid Dynamics, (CFD) which has the advantage of providing visualization for significant insight and identification to the problem at hand, which then can be utilized in enhancing the analysis and identifying specific details. Experimental tools can benefit from the insight gained by analysis, CFD computational and visualization studies, in the preliminary study stage by designing specific experiments as well as in the conceptual and prototype design stages. Various CFD studies have also indicated that there are still significant discrepancies between CFD and experimental studies [6], which necessitate the existence of specific baseline configurations for validation purposes. Coandă MAV is expected to be capable of manoeuvring, and the analysis carried out here is associated with hovering and cruising conditions, which should give further insight on, and can be further elaborated into its manoeuvring performance.

In the study of Coandă MAV, it is desired that the Coandă propulsion system will be able to provide both an efficient cruise phase as well as hover capability, by using Coandă jet blanket for semi-spherical Coandă MAV, which is deflected downward using a curved surface. Thus lift will be generated, first to hover and later for propulsion as well. Such Coandă jet has been utilized for circulation enhancement [3-9] for fixed wing aircraft and turbine blades in forward flight and movement. The principle of Coandă MAV lift generation as well as the equation of motion for its translatory motion will be derived and elaborated.

The present analysis is carried out on Coandă MAV spherical configuration as a baseline, which utilizes some results from the authors' previous analysis. The equations can be easily modified for other configurations and can be compared to other investigators' results for further assessments. CFD visualization studies will also be utilized to gain further understanding in developing the physical model in the analysis.

Some basic results drawn from the physical and mathematical model has been obtained to describe the physical phenomena of the flow field related to the relevant surfaces influenced by the Coandă effect jet sheets, and rewritten in a more convenient form as necessary to gain better insight on the relationship among the relevant parameters.

2. System of Coordinates

For the derivation of the equation of motion of the Coandă MAV, the system coordinates that are required for setting the equation will be defined. Following the convention in flight dynamics, four main reference frames are identified, namely the inertial coordinate system, the local horizon reference frame, the body reference frame, and the wind reference frame, as required in the analysis, and are depicted in Figure1a. Without loss of geerality, for simplicity and instructiveness, only two-dimensional coordinate systems in the plane of symmetry of the Coandă MAV will be elaborated. These four coordinates are:

a) ***Inertial coordinate system***, which is used for defining and the application of the Newton's law of motion. In two dimensions, this coordinate system is depicted by the $O_E xyz$.
b) ***The local horizon coordinate system*** $\boldsymbol{Ox_h y_h z_h}$, which is fixed at the center of mass of the vehicle, and is parallel with the earth Inertial Coordinate system.

(a) 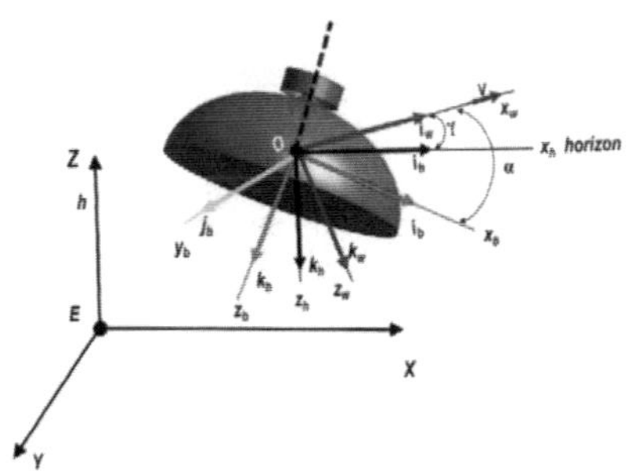
(b) 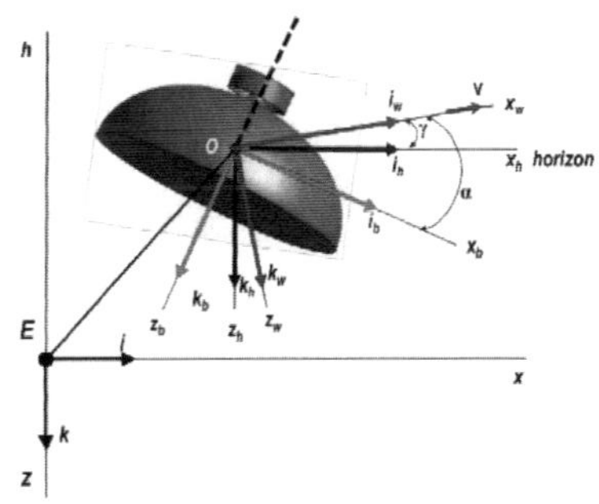

Figure 1. (a) The details of the four coordinate systems for Coandă MAV flight vehicle; (b) Axes systems in the vertical plane perpendicular to the Earth's Surface

c) ***The body coordinate system*** $\boldsymbol{Ox_b y_b z_b}$, which is fixed to the vehicle, and follows a conveniently chosen axis of the vehicle.
d) ***The wind axes system*** $\boldsymbol{Ox_w y_w z_w}$, which moves with the vehicle and the x_w axis is coincident with the velocity vector (flight path of the vehicle).

In the figure, the wind axes are oriented to the flight path angle γ relative the local horizon axes and by the angle of attack α relative to the body axes. The two dimensional configuration in the vertical plane perpendicular to the earth surface is shown in Figure1 (b).

3. Mathematical Model and Fluid Dynamic Analysis for hovering Semi-spherical Coandă MAV as a Baseline Configuration

The mathematical model of Coandă MAV elaborated here follows closely that of Ahmed et al [10-11] and Djojodihardjo and Ahmed [12-13], which was based on first principles and are summarized and reproduced here as a baseline reference for other development. For Coandă MAV with the configuration depicted in Figure2, an actuator rotor can be visualized to be located at the center of the body. For conceptual development, the dimension of the rotor can be set to be small. Part of the flow being drawn by the actuator can be utilized for lift (like in a helicopter), and part of it will be utilized for establishing radial flow for Coandă jet blanket on the surface of the body. Momentum analysis is carried out in the analysis of Coandă effect related to finding the relevant aerodynamic forces and defining the Coandă MAV performance parameters. The latter is developed for a baseline and simplified semi-spherical Coandă MAV as depicted in Figure 3. In the following, a rigorous analysis is carried out to elaborate how Coandă effect contributes to the generation of lift capability of Coandă MAV.

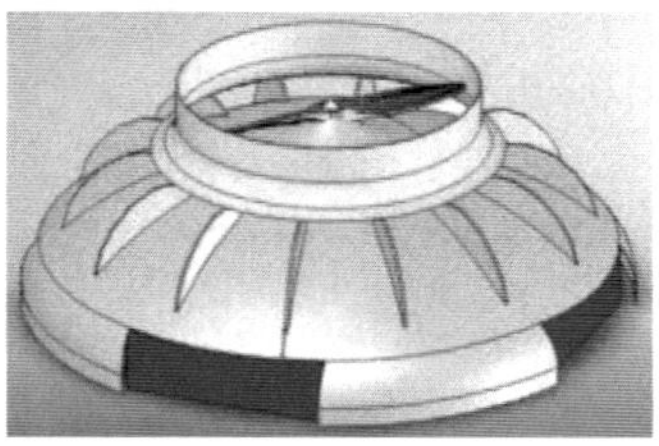

Figure 2. A semi-spherical Coandă MAV indicating the development of Coandă jet over the upper surface of the Coandă MAV.

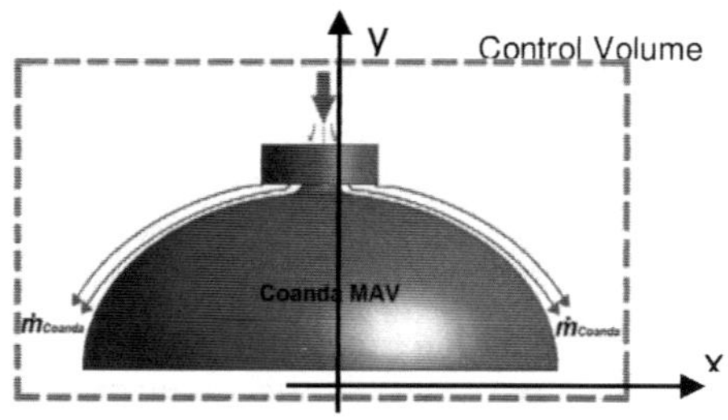

Figure 3. An equivalent Semi-Spherical Coandă MAV for generic analysis.

First, the analysis will address the Coandă effect for lift by applying the fundamental conservation analysis on the control volume CV (red dashed rectangle in Figure3.

Applying the continuity equation on the control volume, which reduces to the Coandă Blanket as depicted in Figure3b, there is obtained

$$\rho \cdot V_{j\text{-}R} \cdot 2\pi R h = \rho \cdot V_{j\text{-}in} \cdot 2\pi R_i h_i = \rho \cdot V_{j\text{-}out} \cdot 2\pi R_o h_o \tag{1}$$

and

$$\begin{aligned} V_{j\text{-}R} &= \frac{R_i h_i}{R h_R} V_{j\text{-}in} = \frac{\dot{m}}{2\pi \rho R h_R} \\ V_{j\text{-}out} &= \frac{R_i h_i}{R_o h_o} V_{j\text{-}in} = \frac{\dot{m}}{2\pi \rho R_o h_o} \end{aligned} \tag{2}$$

where $V_{j\text{-}R}$ is the jet flow velocity, R is the vehicle body radius, h is the jet slot thickness and $\dot{m}$ is the jet mass flow rate. The momentum equation applied to the control volume in the y (vertical) direction can be differentiated into the contribution due to the Coandă Blanket momentum and the pressure difference on the body due to Coandă Blanket:

$$\begin{pmatrix} \text{Total Lift force} \\ \text{due to Coandă Blanket} \end{pmatrix} = \begin{pmatrix} \text{Vertical component of momentum} \\ \text{balance due to Coandă Blanket} \end{pmatrix} + \begin{pmatrix} \text{Pressure difference on the body} \\ \text{of MAVsubject to Coandă Blanket} \end{pmatrix} \tag{3}$$

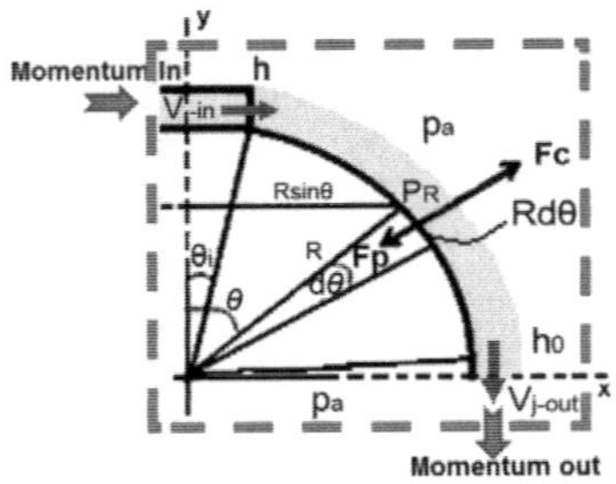

Figure 4. Schematic of Coandă jet (Blanket) for a Spherical (doughnut shaped) MAV.

The contribution of lift from the momentum flux through the control volume CV in the y (vertical) direction is given by:

$$\begin{pmatrix}\text{Force in the}\\ \text{y-direction}\end{pmatrix}=\begin{pmatrix}\text{Momentum Out}\\ \text{in the y-direction}\end{pmatrix}-\begin{pmatrix}\text{Momentum In}\\ \text{in the y-direction}\end{pmatrix} \quad (4)$$

Since the Momentum in the radial direction does not contribute to lift, then the momentum equation in the y-direction for the control volume CV depicted in Figure4 is:

$$F_{Coanda\ jet\ Blanket} = \dot{m}V_{j\text{-}out} = 2\pi R_i h_i\ \rho \cdot V_{j\text{-}in} \cdot V_{j\text{-}out} \quad (5)$$

The contribution of lift from the pressure difference between the upper part (curved MAV body covered by Coandă Blanket) and the lower part of the MAV body can be found by considering various fundamental physical approach. The detail of the elaboration can be found in Ahmed and Djojodihardjo [10-11] and yields:

$$F_{\substack{Inducedpressure\\ difference}} = F_{lower-surface} - F_{upper-surface} = \pi\left(R_0^2 - R_i^2\right)p_a - \int_{\theta_i}^{\theta_0}\left(p_a - \frac{h}{R}\rho\left(\frac{\dot{m}}{2\pi\rho Rh}\right)^2_{j-R}\right)\pi R^2 \sin 2\theta d\theta \quad (6)$$

where θ_i and θ_0 are the turning angles at which the jet flow injected and separated from the Coandă surface respectively, as depicted in Figure4. The jet flow assumed to be uniform outflow separating at the sharp edge of the Coandă MAV curved surface. Then the contribution for lift due to the pressure difference across the surfaces of the MAV body, for the latter approach is given by (as elaborated by Djojodihardjo and Ahmed [12-13]), for the significant value of θ_i):

$$F_{Induced\ pressure\ difference} = \left(\frac{\dot{m}^2}{4\pi\rho h_R R}\right)_{j-R} \quad (7)$$

Hence the total lift due to Coandă jet blanket momentum and Coandă jet blanket induced pressure difference is given by

$$Lift_{\substack{Spherical-Coanda\ jet\ Blanket+\\ induced\ pressure\ difference}} = 2\pi R_i h_i\ \rho V_{j\text{-}in} \cdot V_{j\text{-}out} + \left(\frac{\dot{m}^2}{4\pi\rho Rh}\right)_{j-out} = \dot{m}\cdot V_{j\text{-}out} + \frac{1}{2}\dot{m}V_{j\text{-}out} \quad (8)$$

Hence, due to the presence of the Coandă blanket, the Coandă MAV has additional lift given in (8). It should be noted that a more exact approach requires additional information, i.e. the energy conservation equation, which is then utilized as the fourth equation. This is elaborated subsequently.

Applying the energy conservation equation on the same control volume, which has been redrawn for convenience in Figure5, assuming uniform properties across the sectional areas at the input and output of the Coandă jet blanket and ignoring the entrainment energy exchange between the ambient air and the Coandă jet blanket, the energy equation can be written as:

$$\left(\frac{p}{\rho}\right)_{in} + \frac{1}{2}\left(V_{j\text{-}in}^{\ 2}\right) = \left(\frac{p}{\rho}\right)_{out} + \frac{1}{2}\left(V_{j\text{-}out}^{\ 2}\right) \quad (9)$$

which is essentially the Bernoulli equation along any streamline between the inlet and the outlet sections. The contribution of the entrainment energy exchange along the Coandă jet blanket can later be incorporated, such as by adopting certain assumptions as a heuristic approach or, applying the complete viscous equation by numerical simulation or CFD approach.

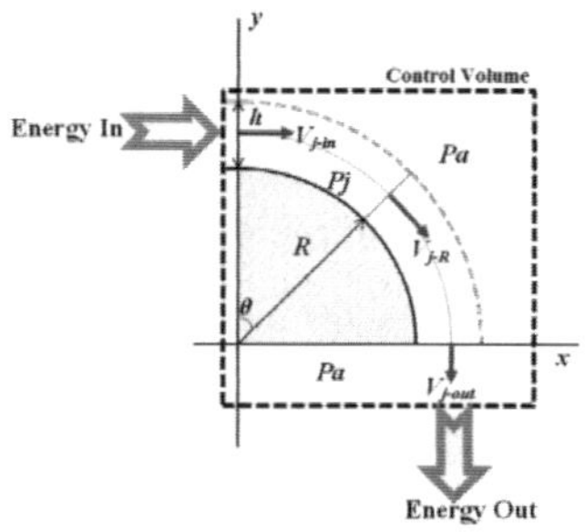

Figure 5. Schematic of the control volume around the MAV for the application of the energy.

The latter should be carried out for more meticulous approach, using the present simplified analytical approach as guidelines. At the same time, the CFD results, by appropriate considerations of the relevant parameters utilized there can be used in developing the heuristic approach as well as for validating and assessing the merit of the analytical approach. Noting that from the outset, the flow is considered to be incompressible, equation (9) reduces to

$$\frac{1}{2}\rho V_{\text{j-in}}^{2} = \frac{1}{2}\rho V_{\text{j-out}}^{2} \tag{10}$$

Hence, by the application of the conservation energy principle for incompressible flow, and assuming ambient pressure at the inlet and outlet of the Coandă jet blanket and neglecting the energy exchange between the ambient air and the Coandă jet blanket, the following velocity relationship between the inlet and outlet sections of the Coandă jet blanket is given by

$$V_{j\text{-}out} = V_{j\text{-}in} \tag{11}$$

This should simplify the solution given in equation (8) and reduce the use of idealization or the number of assumption since fewer unknowns are included in the equations. Combining equations (8) and (9) yields:

$$Lift_{\substack{Coanda-\,jet\,Blanket+\\ induced\,pressure\,difference}} = \frac{3}{2}\dot{m}V_{j-in} \tag{12}$$

The results will be assessed a posteriori by using more structured CFD simulations, although the latter also contain uncertainties and inaccuracies.

Mathematical Model and Flight Dynamic Analysis for Semi-spherical Coandă MAV in Translatory Motion

In the development of the flight dynamics of Coandă MAV in Translatory Motion, without loss of generalities, the equation of motion is developed in the vertical plane perpendicular to the earth motion, rendering two-dimensional planar motion. The hovering state, as depicted in Figure6, will be utilized as a reference.

Balance of forces in the free-body diagram exhibited in Figure6 and the use of the Coandă lift from equation (12) leads to:

$$L_{Hover} \equiv T_{Coandă-MAV} = \frac{3}{2}\dot{m} \cdot V_{j\text{-}in} = 3\pi\rho R_i h_i V_{j\text{-}in}^2 = W_{Coandă-MAV-Take-off} \tag{13}$$

In the development of the equation of motion during translatory motion, a further simplifying assumptions will be made, which could be refined at later stage to incorporate more realistic ones. These are:

1. The thrust of the Coandă MAV will not be affected by the attitude of the Coandă MAV configuration during hover. This assumption implies that Coandă jet flow relative to the Coandă MAV does not change during maneuver. More elaborate analytical model should be developed assisted by CFD simulation.
2. During the translatory motion, the thrust of the Coandă MAV is assumed to be:

$$T_{Coandă-MAV} = \frac{3}{2}\dot{m} \cdot V_{j\text{-}in}\Big|_{hover-equivalent} = 3\pi\rho R_i h_i V_{j\text{-}in}^2\Big|_{hover-equivalent} \tag{14}$$

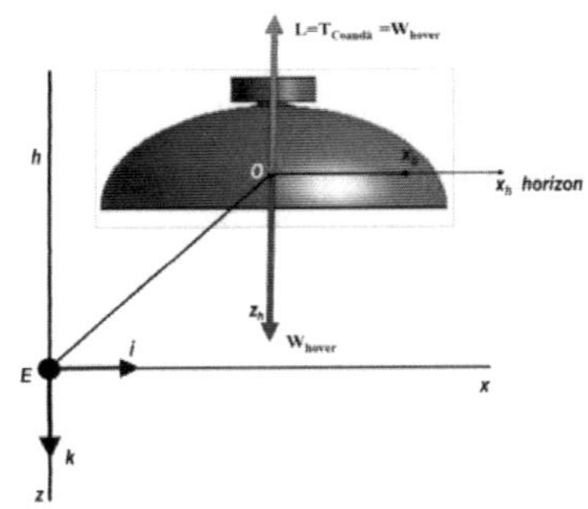

Figure 6. Coordinate systems for Coandă MAV in hover, in two-dimensional configuration (at the vertical plane of symmetry and motion)

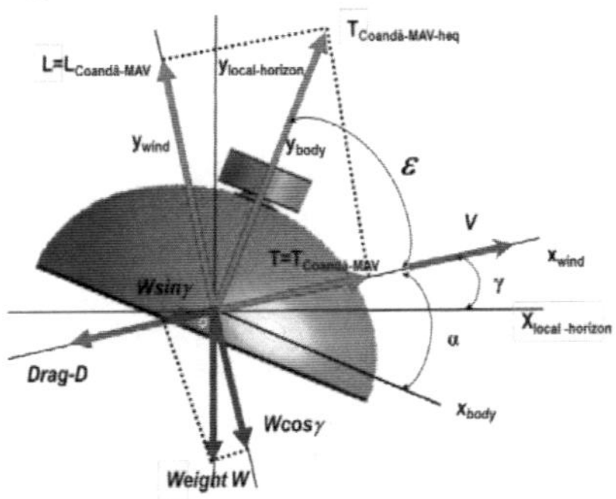

Figure 7. Schematic of Coandă MAV in two-dimensional tranlatory motion in the vertical plane (at the vertical plane of symmetry and motion)

The equation of motion of the Coandă MAV in translational motion in the two-dimensional vertical plane can be written by referring to Figure7 using the basic flight dynamic equation for flight vehicle:
Kinematic:

$$\dot{x} = V\cos\gamma \qquad \left[LT^{-1}\right] \tag{15}$$

$$\dot{h} \equiv \dot{y} = V\sin\gamma \qquad \left[LT^{-1}\right] \tag{16}$$

$$\varepsilon = \tan^{-1}\frac{\ddot{y}_{wind}}{\dot{V}} = \tan^{-1}\frac{g\left(L - \mathrm{W}\cos\gamma\right)}{W\dot{V}} \qquad [-] \tag{17}$$

$$\dot{\varepsilon} = \tan^{-1}\frac{\ddot{y}_{wind}}{V} = \tan^{-1}\frac{g\left(L - \mathrm{W}\cos\gamma\right)}{WV} \qquad \left[T^{-1}\right] \tag{18}$$

Coandă MAV Fluid Dynamics:

$$T_{Coandă-MAV-heq} \equiv T_{Coandă-MAV-hover-equivalent} = \frac{3}{2}\dot{m} \cdot V_{j\text{-}in} = 3\pi\rho R_i h_i V_{j\text{-}in}^2 \qquad \left[MLT^{-2}\right] \tag{19}$$

Coandă MAV Flight Dynamics:

$$\dot{V} = (g/W)\left(T_{Coandă\ MAV-heq}\cos\varepsilon - D - W\sin\gamma\right) \qquad \left[LT^{-2}\right] \quad (20)$$

$$\ddot{y} = (g/W)\left(T_{Coandă\ MAV-heq}\sin(\varepsilon+\gamma) - W - D\sin\gamma\right) \qquad \left[LT^{-2}\right] \quad (21)$$

Weight-Fuel Consumption:

$$\dot{W} = -CT \qquad \left[MT^{-1}\right] \quad (22)$$

It should be noted, that the above relationships are derived for balance of forces and Newton's equation for semi-spherical Coandă MAV treated as a point mass moving in a plane perpendicular to the planar Earth.

5. Performance Measure

5.1 Performance Measure during Hover

The feasibility of using Coandă techniques to enhance aerodynamic performance of Coandă MAV can be assessed using some non-dimensional quantities of performance measure. The most logical measure that has commonly been utilized is the Coandă jet momentum coefficient, C_μ (Mamou and Khalid [14], 2007; Djojodihardjo [9]),, which are defined as:

$$C_\mu = \frac{\dot{m}V_{Coandă-jet}}{\frac{1}{2}\rho_j V_\infty^2 A_{ref}} \quad (23)$$

While (23) may be appropriate for any Coandă activated vehicle during hover and lift-off, one can define other performance measures to evaluate the aerodynamic performance of the spherical Coandă MAV considered here, based on the rate of the energy of the Coandă jet ejected by Coandă MAV at its ideal inlet velocity $V_{j\text{-}in}$ at the Coandă jet injection (inlet) or $V_{j\text{-}out}$ at its peripheral outlet compared to the rate of momentum influx of the Coandă jet times its inlet velocity $V_{j\text{-}in}$ (non-dimensional):

Further realistic assumption can be drawn from CFD visualization, which also added further insight to the problem. Similar performance measures can be defined for Cylindrically Shaped Coandă MAV.

Taking note that Mirkov and Rasuo [15-16] considered the inlet velocity to thrust ratio for a given inlet size could be the most important output value resulting either from experiment on UAV's or by numerical simulation, another performance measure could be defined to be the lift to inlet velocity, which is a dimensional quantity. Substituting equation (13) for the lift, it gives;

$$PM1_{\substack{Spherical\\Coandă-MAV}} = \frac{\text{Lift produced by Coandă MAV} \cdot V_{j-out} \text{ at its peripheral outlet}}{\text{the rate of momentum influx of the Coandă jet times its inlet velocity } V_{j-in}}$$
$$= \frac{Lift.V_{j-out}}{\dot{m}V_{j\text{-}in}^{\ 2}} = \frac{\frac{3}{2}\dot{m}V_{j\text{-}in}^{\ 2}}{\dot{m}V_{j\text{-}in}^{\ 2}} \quad (24)$$

$$PM2_{\substack{Spherical\\Coandă-MAV}} = \frac{Lift_{Coanda-jet\ Blanket+induced\ pressure\ difference}}{Coandă\ jet\ inlet\ Velocity} = \frac{\dot{m}\cdot V_{j\text{-}in} + \frac{1}{2}\dot{m}V_{j-in}}{V_{j\text{-}in}} = \frac{3}{2}\dot{m} \quad (25)$$

A third performance measure, which is also a dimensional quantity, may be defined based on the Coandă jet rate of momentum influx per unit Coandă jet kinetic energy input [10-11]:

$$PM3_{\substack{Spherical\\Coandă-MAV}} = \frac{Lift_{Coanda-jet\ Blanket+induced\ pressure\ difference}}{Jet\ Inlet\ velocity^2} = \frac{3}{2}\frac{\dot{m}}{V_{j\text{-}in}} \quad (26)$$

5.2 Performance Measure for Translation

Performance measures during translation that are elaborated are:

a. For level flight:

$$\begin{aligned} PM_{\substack{Coand\breve{a}-MAV \\ LevelFlight}} &= \frac{Power\ required\ to\ overcome\ Drag}{Power\ Input\ by\ Coand\breve{a}-jet-action} \\ &= \frac{Drag * Velocity}{Coand\breve{a}\ Jet\ Momentum\ Gain * Coand\breve{a}\ Jet\ Velocity\ In} \\ &= \frac{D \cdot V\cos\gamma^2}{\dot{m}V_{j\text{-}in}{}^2} \end{aligned} \tag{27}$$

For purely level flight, as depicted in Figure8, γ=0, and Equation (28) reduces to:

$$PM_{\substack{Coand\breve{a}-MAV \\ LevelFlight}} \approx \frac{1}{4} C_{D\to x} \frac{\pi\rho R_o{}^2 \sin\alpha}{\dot{m}h_i} \frac{V^3}{V_{j\text{-}in}{}^2} \tag{28}$$

The Performance Measure $PM_{Coand\breve{a}\text{-}MAV\ Level\ Flight}$ can be expressed as a function of C_D, $\dot{m}$, R_o, h_i, α, V, $V_{j\text{-}in}$

b. For climbing:

$$\begin{aligned} PM_{\substack{Coand\breve{a}-MAV \\ Climbing}} &= \frac{Power\ required\ to\ Climb}{Power\ Input\ by\ Coand\breve{a}-jet-action} \\ &= \frac{\mathbf{T}_{Coand\breve{a}-MAV-heq} \bullet \dot{\mathbf{y}} - \mathbf{D} \bullet \dot{\mathbf{y}} - \mathbf{W} \bullet \dot{\mathbf{y}}}{Coand\breve{a}\ Jet\ \text{Momentum Gain} * Coand\breve{a}\ Jet\ \text{Velocity In}} \\ &= \frac{3}{2} \frac{\cos(\alpha-\gamma) V \sin\gamma}{V_{j\text{-}in}} - \frac{(D\sin\gamma + W) V \sin\gamma}{\dot{m}V_{j\text{-}in}{}^2} \end{aligned} \tag{29}$$

For purely climbing flight, γ=π/2, and Equation (29) reduces to

$$PM_{\substack{Coand\breve{a}-MAV \\ Climbing}} = \frac{\frac{3}{2}\dot{m} \cdot V_{j\text{-}in} V - \left(\frac{1}{2}\rho C_{D\to y} \pi R_o{}^2 V^2 + W\right) V}{\dot{m}V_{j\text{-}in}{}^2} \tag{30}$$

The Performance Measure $PM_{Coand\breve{a}\text{-}MAV\ Climbing}$ as function of C_D, R_o, $\dot{m}$, V, $V_{j\text{-}in}$

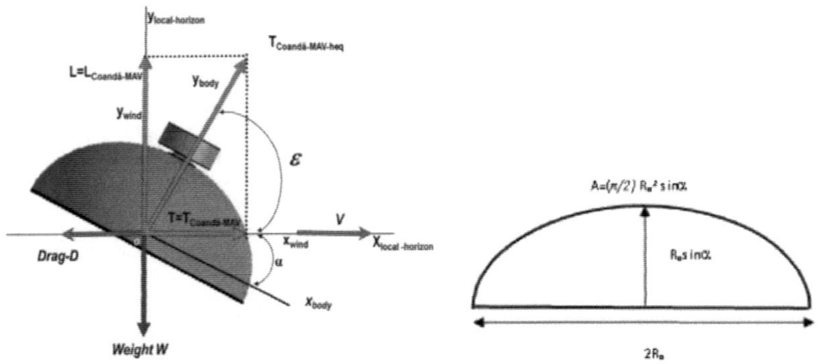

Figure 8. Schematic of Coandă MAV in two-dimensional cruising motion in the vertical plane of symmetry and motion. Inset: Cross sectional area perpendicular to direction of motion.

These are the basic relationships that can further be modified and elaborated for other flight conditions.

6. Some Results and Examples

To assess the merit and plausibility of the theoretical analysis using first principles carried out here, a comparison of the theoretical prediction for the performance measures PM1, PM2 and PM3 given by (25) - (26) are compared with CFD simulation carried out using ANSYS FLUENT® for Spherically Shaped Coandă MAV. The results are depicted in Figs. 9 and 10. The simulation was performed using steady RANS with two equations k-ω Shear Stress Transport (SST) turbulence model. Simple pressure-velocity coupling scheme with the least squares cell base as discretization gradient was applied in the solution method together with second order upwind for the momentum equation and the turbulent kinetic energy.

The performance *PM2* as depicted in Figure10 shows the influence of the ratio of the jet slot thickness h to the reference radius R_o. The influence of the jet inlet velocity on the air vehicle performance "lift" may also be of interest from designer's point of view, which enables them to befittingly select the right propulsion system for such MAV. The jet inlet velocity influence on the performance of the Spherical Coandă MAV at constant jet thickness (h=50mm) is depicted in Figure11. For comparative and validation purposes, Figs. 12 (a) and (b) are produced to assess the present theoretical prediction with CFD simulation results for the case considered by Schroijen and van Tooren [15] for cylindrical Coandă MAV, for h/R=0.075 at two different jet inlet velocities.

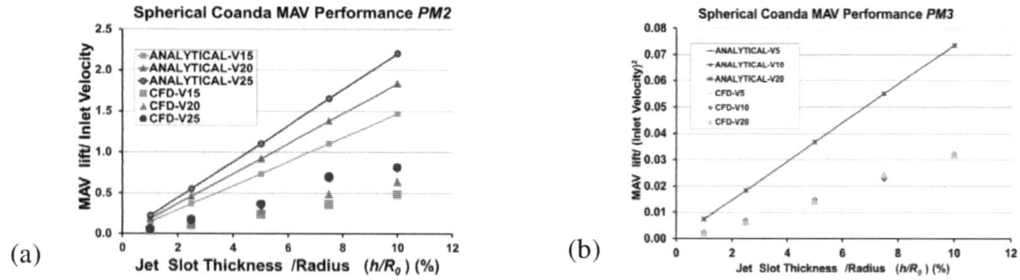

Figure 9. Qualitative comparison of analytical (non-viscous) and CFD (viscous) computed Performance Measure for spherical Coandă MAV with various jet slot velocity as function of jet slot thickness / outer Radius R_o; (a) *PM2*; (b) *PM3*.

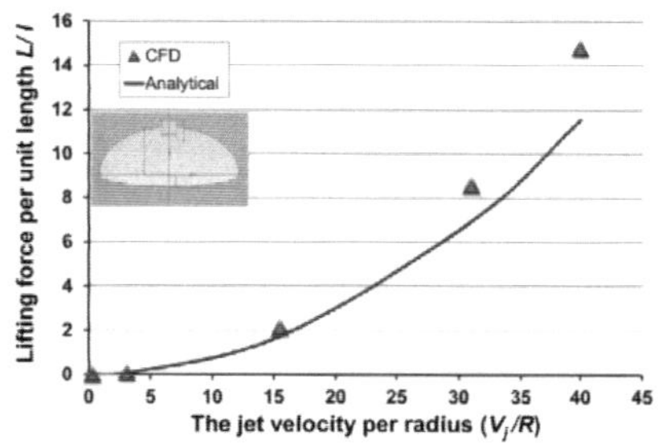

Figure 10. Analytical and CFD models comparison for jet inlet velocity influence on performance of Spherical Coandă MAV at constant jet thickness (h=50mm).

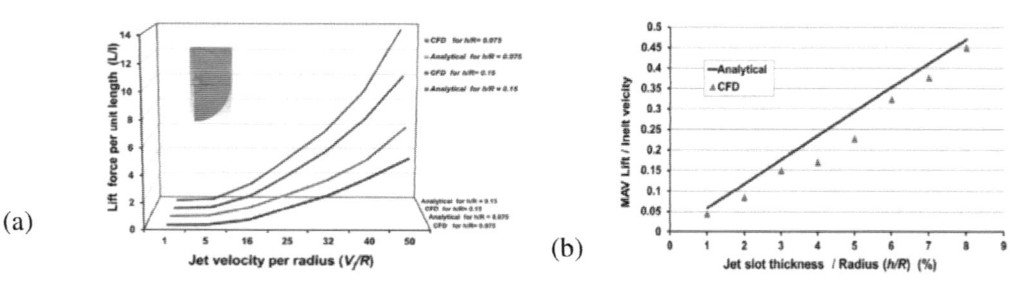

Figure 11. Performance measure comparison between mathematical and CFD models for Cylindrical Coandă MAV for various (a) Inlet velocity and (b) Jet slot thickness

Relatively good qualitative agreements between the theoretical results and CFD simulations shown there lend support to the present analysis, noting that in the analytical study many simplification have been introduced, while in the CFD simulation, the full Navier-Stokes equation for incompressible fluid was used.

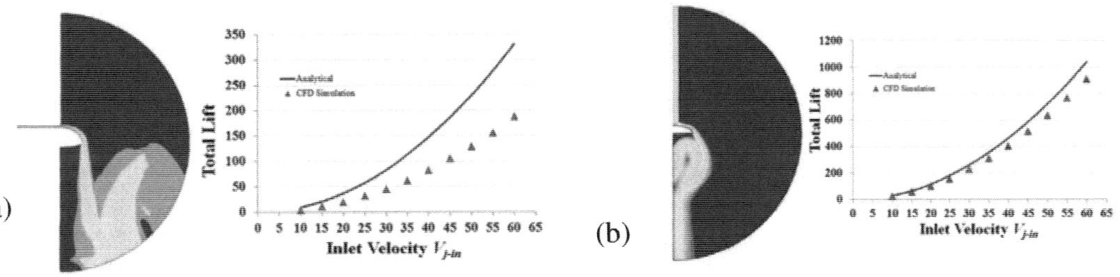

Figure 12. Lift versus input jet velocity for similar configuration considered in (a) Mirkov and Rasuo [16], (b) Ghozali [18]

The effect of viscosity that can be revealed by CFD simulation as exemplified by Figure12 for two different locations of the Coandă jet introduction. The CFD images exhibited in Figure12 (a) and (b) show the velocity magnitude contours for the given configuration and inlet conditions for cases considered by Mirkov and Rasuo [16-17] and Ghozali [18].

Meticulous attempts and grid sensitivity studies on the size of the mesh cells have been performed to enable plausible visualizations of the flow around the whole body with best details.

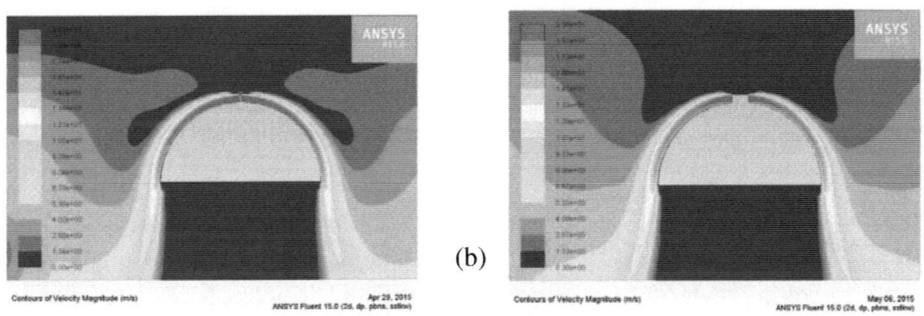

Figure13. Symmetrical Coandă effect - velocity magnitude contour with different inlet radial position; a) Ri=5 mm; b) Ri=50mm

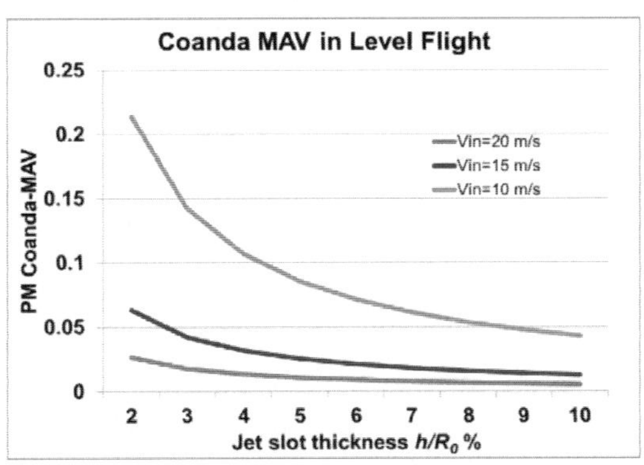

Figure 14. Coandă MAV Performance Measure as a function of jet slot thickness for various cruising velocity, calculated using Eq. (29) and CFD simulation for a baseline Coandă MAV configuration.

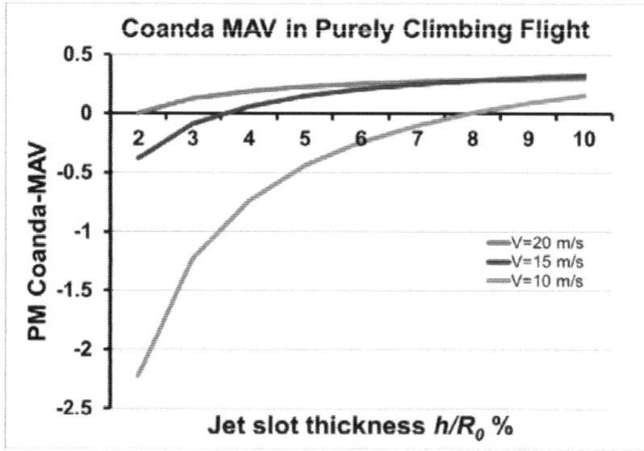

Figure15. Coandă MAV Performance Measure as a function of jet slot thickness for various climbing rate (velocity), calculated using Eq. (31) and CFD simulation for a baseline Coandă MAV configuration.

The numerical study was performed with a moderately small Reynolds number, Re= 68458, based on mean velocity and jet inlet height. The velocity across the inlet, which has a thickness of 0.05 m, is uniform ($Vj\text{-}in$ = 20 m/s). The number of the mesh cells was 52830, and the mesh quality was found to be acceptable. The orthogonality quality was 4.96943e-01, which is acceptable from the ANSYS orthogonality quality requirements. Results obtained on the influence of the inlet jet radius on the air vehicle performance (lift) investigated using CFD simulation for two inlet jet radius, Ri=5 mm and Ri=50mm are presented as velocity magnitude contours shown in Figure13.

For translatory flight, some examples of Performance Measures simulation studies for Coandă MAV in purely level and climbing flights are exhibited by Figs. 14 and 15. The figures are obtained using equations (29) and (31) and CFD simulation for level and climbing flights, respectively.

6. Conclusions

A comprehensive effort has been made to analyze and describe the governing equations applicable to Coandă MAV in hover and translatory motion utilizing first principles as articulated in fluid dynamics and flight mechanic, using conservation principles for a control volume and free-body diagram for the application of Newton's law of motion. The paper establishes the basic working relationships among various relevant variables and parameters governing the aerodynamics forces and performance measures of Coandă MAV in hover and translatory motion. CFD simulations for a Coandă MAV generic model are carried out to gain insight into the flow situation, establishing the analytical model and validating it subject to the theoretical assumptions. The mathematical model and derived performance measures are shown to be capable in describing the physical phenomena of the flow field of the semi-spherical Coandă MAV. Comparison with CFD simulation serves also to assess the uncertainties and accuracy of the theoretical approach. In addition, the CFD computational and visualization studies, can provide further insight in revealing other characteristics of the flow field, and can assist further in-depth analysis, as well as in the design of experiments to that end. For example, to find out, under what conditions the effect of viscosity may cause the Coandă jet to separate. CFD visualization has certainly assisted the analytical work in assessing the application of fundamental conservation principle in continuum mechanics as well as elaborating the equation of motion for the flight dynamics of the Coandă MAV.

References

[1] Gad-el-Hak M. *Flow Control - Passive, Active and Reactive Flow Management*. Cambridge University Press, London, United Kingdom, 448 pages 2000
[2] Yi Liu, Lakshmi N. Sankar, Robert J. Englar, Krishan K. Ahuja. Numerical simulations of the steady and unsteady aerodynamic characteristics of a circulation control wing airfoil. *AIAA* Paper

2001-0704

[3] Kweder J., Panther C.C., Smith J.E. Applications of circulation control, yesterday and today, *International Journal of Engineering, (IJE)* 195;, vol.4 (5), pp. 411-429, 2011.

[4] Djojodihardjo H. and Thangarajah N. Research, development and recent patents on aerodynamic surface circulation control-A critical review. *Recent Patents on Mechanical Engineering* .vol. 7, pp. 1-37, 2014.

[5] Djojodihardjo H. Progress and development of Coandă jet and vortex cell for aerodynamic surface circulation control–An overview. *The SIJ transactions on advances in space research & earth exploration (ASREE)*, Vol.1, No.1, September-October 2013

[6] Rumsey CL, Nishino T. Numerical study comparing RANS and LES approaches on a circulation control airfoil. *International Journal of Heat and Fluid Flow* 32: 847-864, 2011.

[7] Tongchitpakdee C., Benjanirat S., Sankar L.N. Numerical studies of the effects of active and passive circulation enhancement concepts on Wind Turbine Performance. *Journal of solar energy engineering transactions of the ASME*; 128: 432 – 444, 2006.

[8] Hamid, MFAH, Djojodihardjo, H., Suzuki, S. and Mustapha, F. Numerical assessment of Coanda Effect as airfoil lift enhancer in wind-turbine configuration. *Regional conference on mechanical and aerospace technology* Bali, February 9 – 10, 2010

[9] Djojodihardjo H., Abdul-Hamid M. F., Jaafar A. A., Basri S., Romli F. I., Mustapha F., Mohd-Rafie A. S. and Abdul-Majid D. L. A. Computational study on the aerodynamic performance of wind turbine airfoil fitted with Coandă Jet. *Journal of renewable energy*, volume 2013, Article ID 839319, 17 pages, 2013

[10] Ahmed R.I., Djojodihardjo H., Abu Talib A.R., Abd Hamid M.F. Application of Coandă jet for generating lift for micro air vehicles – Preliminary design considerations, *Applied Mechanics and Materials* Vol. 629,pp. 139-144, 2014. URL: http://www.scientfic.net/ AMM.629.139

[11] Ahmed R.I., Djojodihardjo H., Abu Talib A.R., Mohd-Rafie A. S., First Principle Analysis of Coandă Micro Air Vehicle Aerodynamic Forces for Preliminary Sizing" , Aircraft Engineering and Aerospace Technology, DOI: 10.1108/AEAT-03-2015-0080.R2 ,2015.

[12] Djojodihardjo H. Ahmed R.I. An analysis on the lift generation for Coandă Micro Air Vehicles. *IEEE, ICARES conference proceeding,* Yogyakarta, Nov. 2014

[13] Djojodihardjo, H., Ahmed, R.I., Thalib, A.R.A. and Rafie, M.A.S. Analytical and CFD visualization studies of Coandă MAV, *Proceedings,* The 13th Asian Symposium on Visualization, Novosibirsk, Russia, June 22–26, 2015,

[14] Mamou, M. and Khalid, M. Steady and unsteady flow simulation of a combined jet flap and Coandă jet effects on a 2D airfoil aerodynamic performance. *Revue des Energies Renouvelables CER'07* Oujda 55 – 60, 2007

[15] Schroijen, M. Van Tooren, M. MAV propulsion system using the Coandă effect. *45th AIAA/ASME /SAE/ASEE Joint Propulsion Conference & Exhibit* Denver, 2009.

[16] Mirkov, N. and Rasuo, B. Numerical simulation of air jet attachment to convex walls and applications, *ICAS* 2010-621

[17] Mirkov, N. and Rasuo, B. Maneuverability of a UAV with Coandă Effect based lift production, *ICAS* 2012-617

[18] Ghozali, D. Analysis of Coandă Effect using computational-fluid-dynamic, Thesis, *Universitas Gajah Mada*, Indonesia, 2013.

Acknowledgments

The authors would like to thank Universiti Putra Malaysia (UPM) for granting Research University Grant Scheme (RUGS) No.9378200, and the ministry of higher education ERGS: 5527088 ; FRGS:5524250 under which the present research is carried out.

Crack Diagnosis of Wind Turbine Blades Based on EMD Method

CUI Hong-yu[1], DING Ning[1,2], HONG Ming[1]

[1] School of Naval Architecture, Dalian University of Technology, Dalian 116024, China
[2] COSCO-SHIPYARD, COSCO Ship DESIGN CENTER, Dalian 116600, China

E-mail address: cuihongyu@dlut.edu.cn

Abstract. Wind turbine blades are both the source of power and the core technology of wind generators. After long periods of time or in some extreme conditions, cracks or damage can occur on the surface of the blades. If the wind generators continue to work at this time, the crack will expand until the blade breaks, which can lead to incalculable losses. Therefore, a crack diagnosis method based on EMD for wind turbine blades is proposed in this paper. Based on aerodynamics and fluid-structure coupling theory, an aero-elastic analysis on wind turbine blades model is first made in ANSYS Workbench. Second, based on the aero-elastic analysis and EMD method, the blade cracks are diagnosed and identified in the time and frequency domains, respectively. Finally, the blade model, strain gauge, dynamic signal acquisition and other equipment are used in an experimental study of the aero-elastic analysis and crack damage diagnosis of wind turbine blades to verify the crack diagnosis method proposed in this paper.

1. Introduction

Wind turbine blades are the key component for energy conversion in a wind power generation system[1,2]. It is possible for cracks to occur on the blade surface under poor service conditions or following long running times, and fatigue damage will accumulate until the blade breaks, which leads to incalculable losses[2,3]. Therefore, accurately monitoring the blade surface condition before crack growth is a prerequisite and is necessary for wind power generation systems[3,4].

Blades operate under the coupling of gravity, aerodynamic and centrifugal forces, and the vibration signal is characterized as nonlinear, stochastic and coupling[5]. Presently, there are many blade structure analysis methods for analyzing the signal characteristics during blade operation in the time or frequency domains. Empirical mode decomposition (EMD) is a signal analysis method in the time-frequency domain proposed by Norden E. Huang in 1999. EMD can decompose a signal into a series of intrinsic mode functions (IMF) with different time characteristics and smooth processing[6]. Linear and nonlinear signals can be analyzed by the EMD method due to its adaptability. Therefore, EMD is advantageous for analyzing the vibration signals of rotating machinery and other structural characteristics signals.

Once cracks occur on a blade, the structure stress near the cracks will change. The structural stress signal is obtained from numerical simulations without any noise pollution, but the actual signal that is acquired cannot be used in crack recognition because of the pollution. Therefore, we propose a crack damage diagnosis method in which collected signals are decomposed based on the EMD method and an energy method is applied to recognize crack damage.

2. Wind turbine blade aero-elastic analysis

In this paper, a fluid-structure coupling method is used for the aero-elastic analysis of blades. Computational fluid dynamic and structural dynamic theory are used to calculate the flow domain and structure, respectively[7,8].
The structure dynamic equation of wind turbine blades is as follows:

$$\boldsymbol{M}\ddot{\boldsymbol{u}} + \boldsymbol{C}\dot{\boldsymbol{u}} + \boldsymbol{K}\boldsymbol{u} = \boldsymbol{F}_s + \boldsymbol{F}_a \tag{1}$$

where $\boldsymbol{M}$ is the mass matrix; $\boldsymbol{C}$ is the damping matrix; $\boldsymbol{K}$ is the stiffness matrix; $\boldsymbol{u}$, $\dot{\boldsymbol{u}}$, and $\ddot{\boldsymbol{u}}$ are displacement, velocity and acceleration, respectively; $\boldsymbol{F}_s$ is the loading on the blades; and $\boldsymbol{F}_a$ is the fluid loading, which is a function of displacement, velocity and acceleration.

Computational Fluid Dynamic (CFD) is applied to simulate the flow phenomena of fluid. The equations of fluid mechanics are continuity equation, momentum equation and energy equation.
According to the law of conservation of mass, the continuity equation of fluid is obtained as follows:

$$\frac{\partial \rho}{\partial t} + \nabla \cdot (\rho \boldsymbol{V}) = 0 \tag{2}$$

where $\nabla = \frac{\partial^2}{\partial x^2} + \frac{\partial^2}{\partial y^2} + \frac{\partial^2}{\partial z^2}$ is the Hamilton operator; ρ is the fluid density; and $\boldsymbol{V}$ is the fluid velocity vector.

According to the law of conservation of momentum, the momentum equation of fluid or Navier-Stokes (NS) equation, is obtained as follows:

$$\begin{cases} \rho \dfrac{d\boldsymbol{u}}{dt} = \rho \boldsymbol{f}_x + \dfrac{\partial \boldsymbol{p}_{xx}}{\partial x} + \dfrac{\partial \boldsymbol{p}_{yx}}{\partial y} + \dfrac{\partial \boldsymbol{p}_{zx}}{\partial z} \\ \rho \dfrac{d\boldsymbol{v}}{dt} = \rho \boldsymbol{f}_y + \dfrac{\partial \boldsymbol{p}_{xy}}{\partial x} + \dfrac{\partial \boldsymbol{p}_{yy}}{\partial y} + \dfrac{\partial \boldsymbol{p}_{zy}}{\partial z} \\ \rho \dfrac{dw}{dt} = \rho \boldsymbol{f}_z + \dfrac{\partial \boldsymbol{p}_{xz}}{\partial x} + \dfrac{\partial \boldsymbol{p}_{yz}}{\partial y} + \dfrac{\partial \boldsymbol{p}_{zz}}{\partial z} \end{cases} \tag{3}$$

where $\boldsymbol{f}_x$, $\boldsymbol{f}_y$ and $\boldsymbol{f}_z$ are the components of fluid mass force in the normal direction and $\boldsymbol{p}$ is the fluid stress tensor.

According to the law of conservation of energy, the energy derivative with respect to time is equal to the sum of the net inflow of heat per unit time and the acting of mass force and surface forces on the fluid infinitesimal. The integral form of the energy equation is as follows:

$$\rho \frac{d}{dt}\left(\boldsymbol{e} + \frac{\boldsymbol{V} \cdot \boldsymbol{V}}{2} \right) dV = \nabla \cdot (\sigma \cdot \boldsymbol{V}) + \boldsymbol{V} \cdot \rho \boldsymbol{f} + \nabla \cdot \boldsymbol{q} \tag{4}$$

where $\boldsymbol{e}$ is the internal energy of fluid infinitesimal; σ is the surface force of fluid infinitesimal; $\boldsymbol{f}$ is the mass force of fluid infinitesimal; and $\boldsymbol{q}$ is the thermal conductivity.

In this paper, a fluid-structure coupling calculation method is applied in aero-elastic analysis of wind turbine blades. Computational fluid dynamics and structural dynamics theory are used to calculate the flow domain and structure, respectively, and thereby generate two sets of finite elements and boundary conditions, including coupling parameters that are specially defined. In the calculation, the coupling amount will be exchanged through the middle platform, and the computing parameters will be transferred through the fluid-solid coupling boundary. Specifically, during the fluid-structure coupling iterative calculation, the structure and fluid domains are calculated once at each time step, and their results are exchanged for another calculation until the results converge to achieve the fluid-structure coupling calculation analysis.

3. Numerical simulations of blades aero-elastic analysis based on ANSYS Workbench

ANSYS Workbench is applied to analyze the aero-elasticity of blades. The geometry model of blade, which is based on the initial blade profile design, is built in ANSYS Workbench and meshed into finite elements. Then, the FE model is imported into ANSYS Workbench to obtain the response in the time domain based on bi-directional flow-solid coupling method to achieve a numerical simulation of the blades' aero-elastic analysis.

The basic parameters of a wind turbine blade are designed by using Bates theory and blade element theory[9]. The basic parameters are shown in Table 1.

Table 1. Basic parameters of wind turbine blade design

Basic design parameters	Value
rated power P	1200kW
designed wind speed V	15 m/s
wind energy utilization factor C_p	0.4
generator efficiency η_1	0.9
mechanical transmission efficiency η_2	0.8
air density ρ	1.225 kg/m^3

Based on the parameters in Table 1, the chord and twist angle for each location are obtained as shown in Table 2.

Table 2. Chord length and torsion angle at each cross-section of wind turbine blade

Location	l_i (m)	Twist angle γ_i (°)	Location	l_i (m)	Twist angle γ_i (°)
0.1R	6.2	55.2	0.6R	1.3	87.1
0.2R	3.7	72.4	0.7R	1.1	88.2
0.3R	2.6	79.5	0.8R	1.0	89.1
0.4R	2.0	83.2	0.9R	0.9	89.7
0.5R	1.6	85.5	1.0R	0.8	90.0

Based on the basic parameters of a designed wind turbine blade, PROFILI is applied to design blade profiles. Then, the profiles are imported into ANSYS to create a geometry model, as shown in Figure 1.

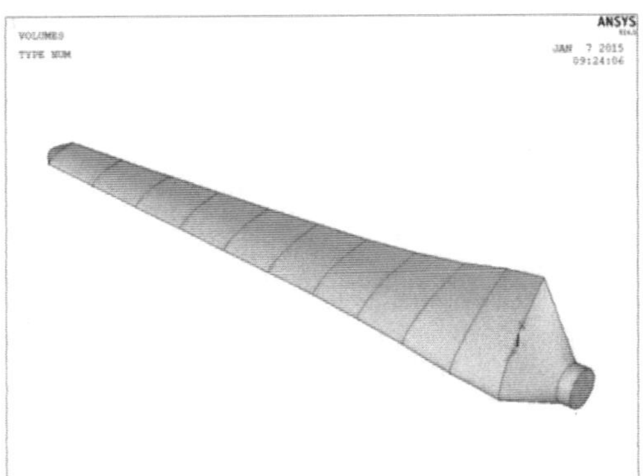

Figure 1. Three-dimensional geometric model of wind turbine blade

Because CFX was the first fluid dynamics analysis software in the world, ANSYS+CFX is applied to analyze the aero-elasticity of the designed blade. The calculation process is as follows:

a. Import the created blade model and create flow model.

b. The blade and fluid model parameters are set, and the models are meshed, as shown in Figure 2.

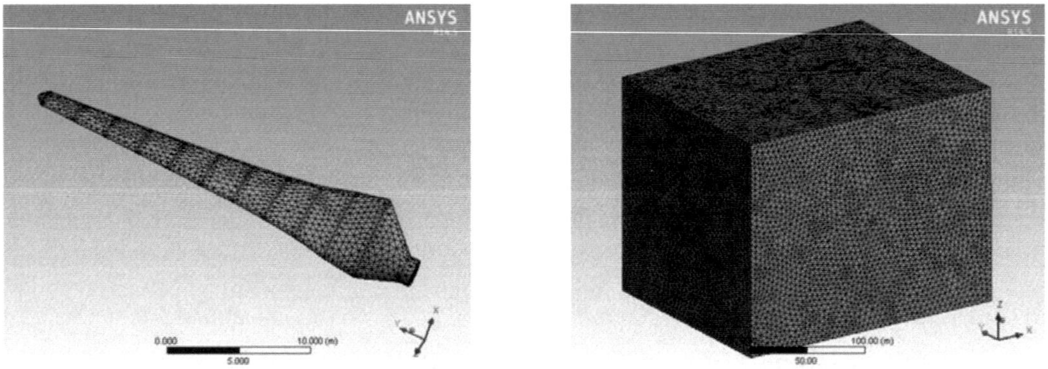

Figure 2. Mesh for wind turbine blade and fluid domain

c. The elements of the blade and fluid are imported into CFX to make the bi-directional flow-solid coupling calculation and obtain the pressure on the blade surface in the time domain. The distribution of the pressure on the blade surface and the air speed in the longitudinal profile of the fluid domain at 3 s are shown in Figure 3.

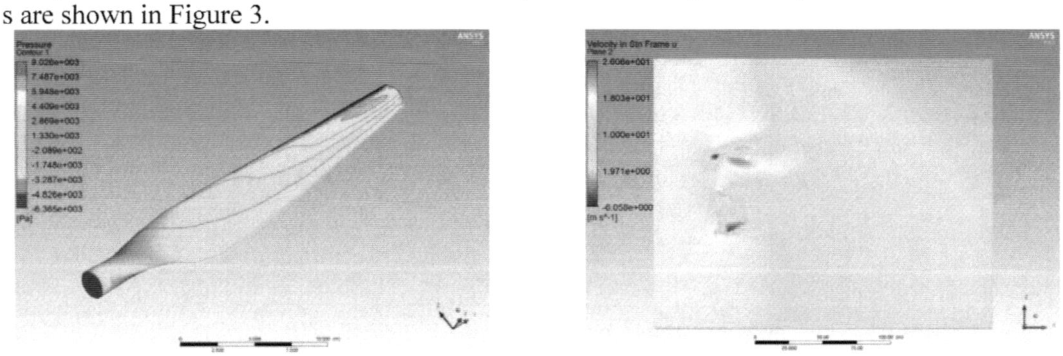

Figure 3. Surface pressure distribution and air flow rate in longitudinal section diagram at 3 s

d. The pressure calculated in CFX is used as the load on the blade surface, and the structural dynamics are solved to obtain the response in the time domain to achieve an aero-elastic analysis of blades. The stress and displacement response at the 3 s moment are shown in Figure 4.

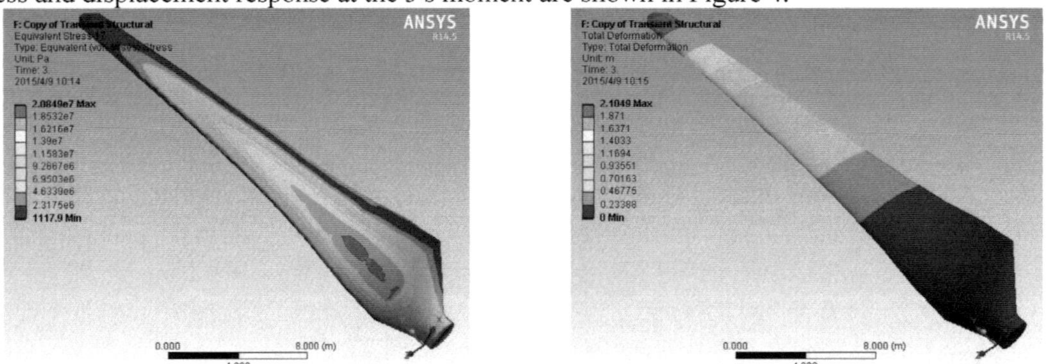

Figure 4. Stress and displacement response of blade at 3 s

4. Crack diagnosis method of wind turbine blade based on numerical simulation

In this paper, a crack diagnosis method based on a numerical simulation of the blade aero-elastic analysis is proposed. A crack is created 10 m from the blade root on the blade geometry model. The length of the crack is fixed at 0.1 m. The width of the crack is fixed at 0.01 m. The depth of the crack is designed as 0.004 m, 0.005 m, 0.01 m, and 0.02 m. The aero-elasticity of the blade containing a crack is analyzed, and the responses of selected points as a characteristic signal of crack diagnosis are obtained. The crack location and distribution of feature points are shown in Figure 5.

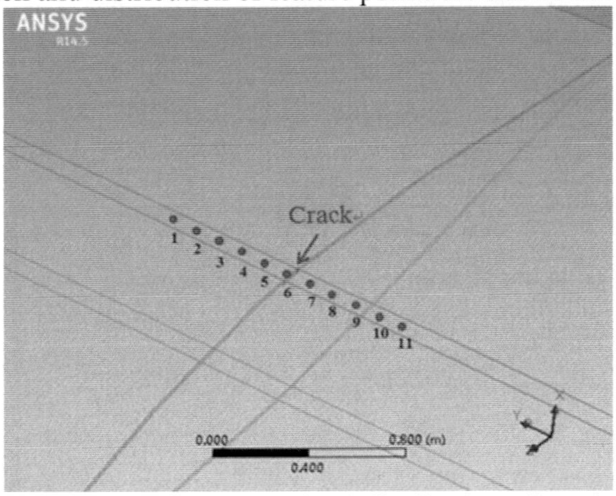

Figure 5. Extracted feature points' distribution of crack signal

In this paper, the energy method, which is widely applied in the extraction of signal characteristics, is used to identify the crack of a wind turbine blade. The response of feature point i is collected first. Then, the stress response energy of point i is calculated based on formula (5) as follows:

$$E_i = \sum_{k=1}^{n} x_{ik}^2 \qquad (k = 1, 2, 3\text{L } n) \tag{5}$$

where k is number of sampling points. The energy E_i of feature point i is used as the eigenvalue to identify the crack in the wind turbine blade. The variation in the energy values for each feature point with varying depths of the crack is shown in Figure 6, where "D" represents the crack depth and "B" represents the crack width.

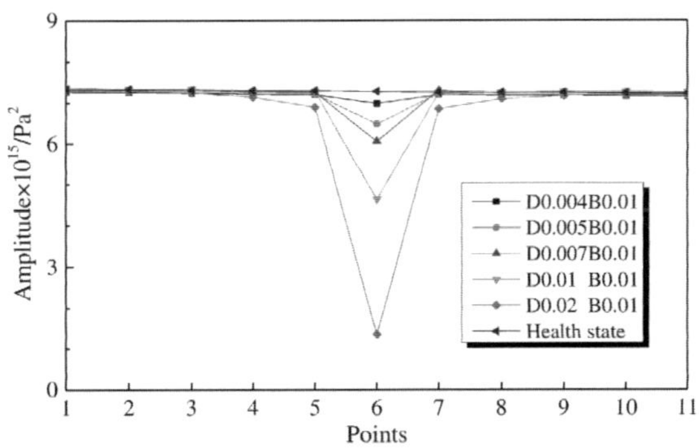

Figure 6. Energy feature point of crack variation with depth

It can be seen from Figure 6 that the energy eigenvalue of the feature point at the crack is much lower than the energy of the same point in healthy conditions. Further, the energy eigenvalue of a feature point reduces gradually with increases in the crack depth.

The EMD method is very effective in handling both non-stationary and nonlinear signals. Signals are decomposed into several IMFs from high to low frequency, and a residual by EMD method. The collected signal $x(t)$ is decomposed by EMD as follows:

$$x(t) = \sum_{i=1}^{N} c_i(t) + r(t) \tag{6}$$

where $c_i(t)$ are IMF components and $r(t)$ is the residual.

In this paper, signals at different locations are decomposed by the EMD method to obtain the first two IMF components, and the energy eigenvalue of each IMF is calculated. The energy eigenvalues of the first two IMF components in different crack states are shown in Figure 7.

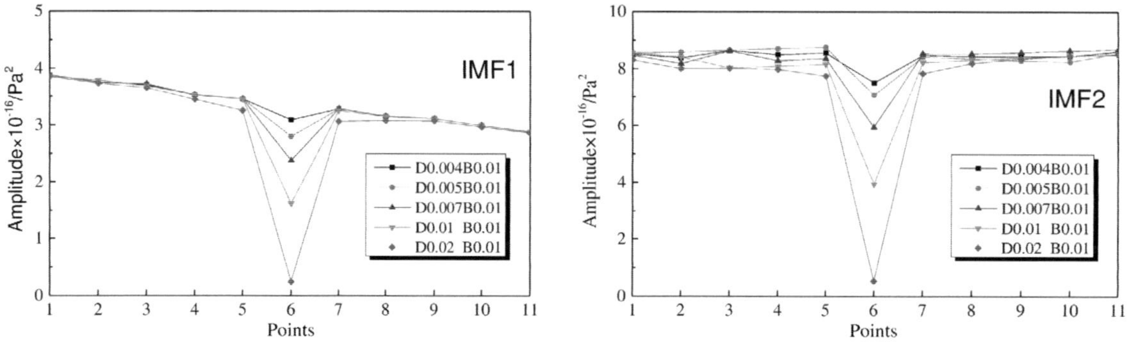

Figure 7. Energy feature point of crack variation with breadth of IMF1 and IMF2

Figure 7 shows, for the first two IMF components decomposed by EMD, that the energy eigenvalue of the feature point at the crack is much lower than that of the same point in healthy conditions. Further, the energy eigenvalue of a feature point reduces gradually with increasing crack depths.

5. Experiment of crack diagnosis of wind turbine blade

The structural stress signals obtained from numerical simulation calculation are ideal, whereas the actual collected signals are difficult to use in the diagnosis of cracks due to noise pollution of the signals. Crack diagnosis experiments are performed to prove the validity, efficiency and feasibility of the method proposed in this paper. Variable wind power, as surface loading, is controlled by adjusting an electric fan during the experiment, and the stress signals at different positions are collected by strain gauges connected to signal acquisition instruments. The wind turbine blade, crack and strain gauges are shown in Figure 8.

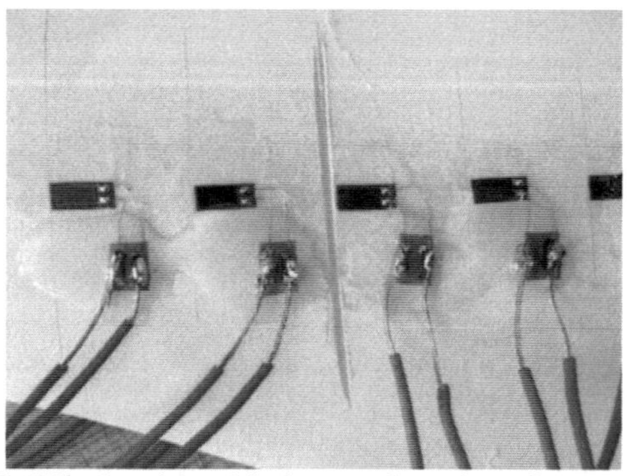

Figure 8. Wind turbine blade, crack and strain gauges

The collected blade stress signal in the time domain is shown in Figure 9.

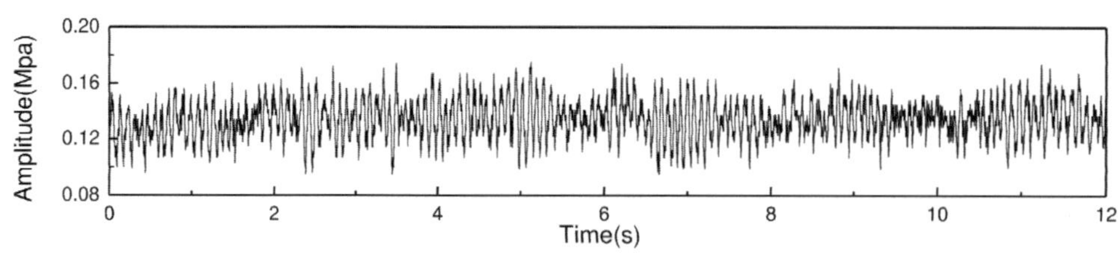

Figure. 9 Time domain distribution of stress signal at point 1 of crack 1

Each feature extraction was obtained using the energy method to obtain the stress crack state at different measuring points. To compare the energy characteristics of each measuring point effectively, the energy eigenvalue of each measuring point is normalized. The energy eigenvalues of cracks at different depths after the normalization process are shown in Figure 10.

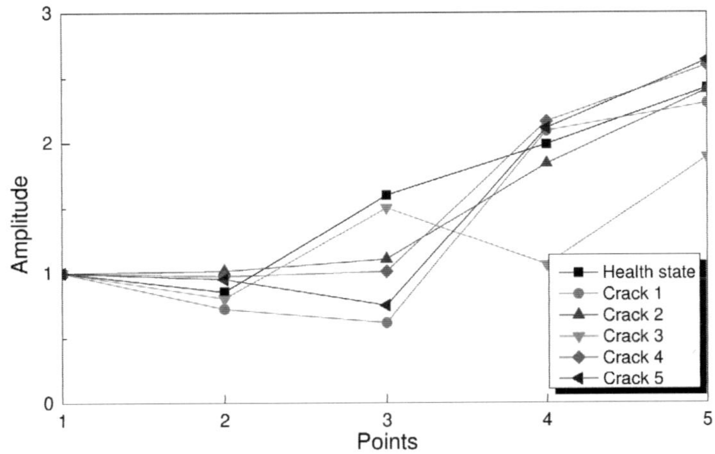

Figure 10. Energy feature point of crack variation with width

Figure 10 shows that the energy eigenvalues of the feature point at the crack are much lower compared with the energy of the same point in healthy conditions, which is same as the conclusion obtained from numerical simulations. The variation in the energy eigenvalues with increasing crack depth is not significant enough, which is in large contrast to the conclusion obtained from numerical simulations.

The EMD-based energy method is used to diagnose the crack damage of wind turbine blades. The variation in the energy eigenvalues with increasing crack depth for the IMF1 and IMF2 components is shown in Figure 11.

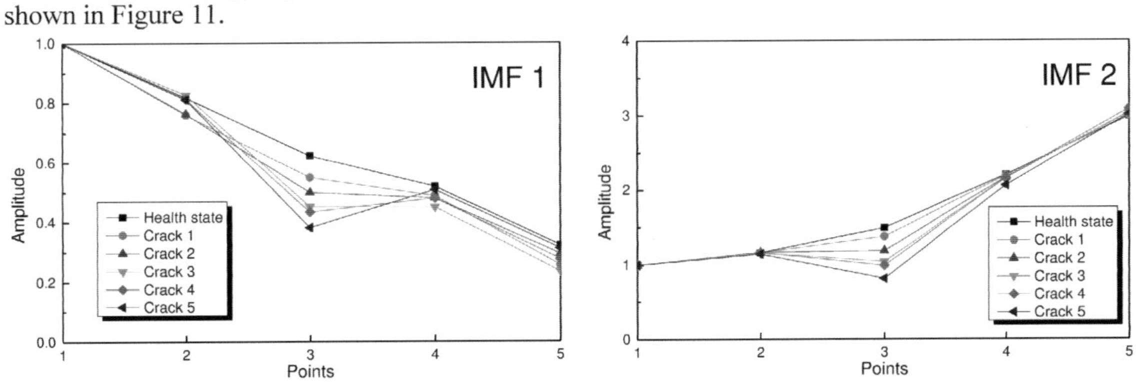

Figure 11. Energy feature point of IMF1 and IMF2 with variations in crack width

Figure 8 shows that the stress signal obtained from the experiments can be decomposed into several IMFs based on EMD method; thus, the first two IMF components are taken as the original signals in the diagnosis. The energy eigenvalues of the feature point at the crack are much lower than the energy of the same point in healthy conditions, and the energy eigenvalue of feature points reduces gradually with increasing crack depth.

The experimental conclusions are similar to the results of numerical simulations that used the crack diagnosis method based on EMD, as proposed in this paper. This proves the validity, efficiency and feasibility of the method.

6. Conclusions

In this paper, according to the basic theory of fluid-solid coupling, ANSYS Workbench is applied to analyse aero-elasticity, and an EMD-based energy method is used to diagnose the crack damage of wind turbine blades. The results show that the energy eigenvalues of a feature point at the crack are much lower than the energy of the same point in healthy conditions and that the energy eigenvalues of feature points decrease gradually with increasing crack depth. Ideal diagnosis results can be obtained in either numerical simulation research or experimental research to verify the correctness, validity and feasibility of the method proposed in this paper.

Acknowledgments
This work was funded by the Natural Science Foundation of China (Grant No. 51109034)

References
[1] Ying Wang, Xiaojing Sun, Xiaohua Dong, Bing Zhu, Diangui Huang and Zhongquan Zheng 2016 Energy Conversion and Management vol 108, Numerical investigation on aerodynamic performance of a novel vertical axis wind turbine with adaptive blades pp 275-286.
[2] Ruizhen Yang, Yunze He and Hong Zhang 2016 Renewable and Sustainable Energy Reviews vol 60,Progress and trends in nondestructive testing and evaluation for wind turbine composite blade pp 1225-1250.
[3] N. Dervilis, M. Choi, S.G. Taylor, R.J. Barthorpe, G. Park, C.R. Farrar and K. Worden 2014 Journal of Sound and Vibration vol 333, On damage diagnosis for a wind turbine blade using pattern recognition pp1833-1850.
[4] Julián Sierra-Pérez, Miguel Angel Torres-Arredondo and Alfredo Güemes 2016 Composite Structures vol 135, Damage and nonlinearities detection in wind turbine blades based on strain field pattern recognition. FBGs, OBR and strain gauges comparison pp 156-166.
[5] H. Hamdi, C. Mrad, A. Hamdi and R. Nasri 2014 Applied Acoustics vol 86, Dynamic response of a horizontal axis wind turbine blade under aero-dynamic, gravity and gyroscopic effects pp 154-164.
[6] Huang NE, Shen Z, Long SR, Wu MC, Shih HH, Zheng Q, Yen NC, Tung CC and Liu HH 1998 Proceed-ings of the Royal Society A: Mathematical Physical and Engineering Sciences vol 454, The em-pirical mode decomposition and the Hilbert spectrum for nonlinear and non-stationary time series analysis pp 903-995.
[7] Cui Peng and Jinglong Han 2011 Journal of Fluids and Structures vol 27, Numerical investigation of the effects of structural geometric and material nonlinearities on limit-cycle oscillation of a cropped delta wing pp 611-622.
[8] Dong Ok Yu and Oh Joon Kwon 2014 Renewable Energy vol 70, Predicting wind turbine blade loads and aeroelastic responseusing a coupled CFD-CSD method pp184-196.
[9] SM Habali, IA Saleh and Local design 2000 Energy Conversion & management vol 4, testing and manufacturing of small mixed airfoil wind turbine blades of glass fiber rein-forced plastics, Part I: Design of the blade and root pp 249-280.

Research on Modelling of Aviation Piston Engine for the Hardware-in-the-loop Simulation

Bing Yu[1 2], **Wenjun Shu**[1 2] **and Wenchao Bian**[1 2 3]

[1]Jiangsu Province Key Laboratory of Aerospace Power Systems ,College of Energy and Power Engineering, Nanjing University of Aeronautics & Astronautics, Nanjing 210016, China;
[2] Advanced Aero Engine Collaborative Innovation Center, Nanjing University of Aeronautics & Astronautics, Nanjing 210016, China;
[3] College of Aeronautics, Shandong Jiaotong University, Jinan 250357, China

Corresponding author:Bing Yu (Yb203@nuaa.edu.cn)

Abstract. In order to build the aero piston engine model which is real-time and accurate enough to operating conditions of the real engine for hardware in the loop simulation, the mean value model is studied. Firstly, the air-inlet model, the fuel model and the power-output model are established separately. Then, these sub models are combined and verified in MATLAB/SIMULINK. The results show that the model could reflect the steady-state and dynamic performance of aero engine, the errors between the simulation results and the bench test data are within the acceptable range. The model could be applied to verify the logic performance and control strategy of controller in the hardware-in-the-loop (HIL) simulation.

1. Introduction

Developing a controller for aircraft engine is a very long term work, which will cost a lot of manpower and material resources. In order to improve design efficiency and reduce the risk, the controller is usually applied to the HIL (hardware in the loop) platform before controlling the real engine [1]. Therefore, establishing a general HIL platform is important for an aircraft engine controller. In this article, a piston engine model for aircraft is built based on the mean value engine modelling approach. At the same time the built model was verified in MATLAB/SIMULINK, and which could provide a fast modelling approach for real-time engine model based on the code generation technology. Considering the control period of aero engine is generally less than 20ms, the complex aero engine model with a long computing period is difficult to satisfy the requirement of real time. [2-3] According to the demand of the HIL system, the engine model should meet the requirements as followings: (1) The engine model can reflect all operating conditions; (2) Its accuracy could meet the requirements on verifying the logic and performance of controller [4]; (3) Both steady-state simulation and dynamic simulation can be carried out [5]; (4) It has high universality and could be applied to the different engines after slight modification.

Engine dynamic model mainly includes mean value engine model (MVEM) and each cylinder respectively controlled engine model (CCEM) at present. Compared to each cylinder respectively controlled engine model, mean value engine model has the advantages of simple structure, lower computing iteration amount, the overall accuracy can meet the requirements of simulation, the model can describe the steady and dynamic characteristics of engine, and it could be applied to the electronic control system design and analysis of engine modelling [6]. Therefore, mean value engine modelling method is selected to establish aviation piston engine model in the article.

2. Introduction of MVEM model

MVEM model was first proposed by Rasmussen, however, its structure and general expression form were proposed by Elbert Hendricks [7]. MVEM model focuses on the dynamic characteristics of the engine, and ignores the difference of each cylinder working state in different crank shaft angle during a work cycle, and the difference of each cylinder is processed through the average value method [8-10]. According to the engine operating principles, the analysis and simulation of the working process of engine are started from the energy conversion and working medium flow. In the process of modelling, differential or algebraic equations are used to describe the simple and clear physical processes. Empirical formula is instead of those parts which are difficult to describe with mathematical models or complex physical processes. According to modular design method, the established model includes three sub-models of intake system, fuel system and power train system. The schematic diagram of the engine model is shown in figure 1. The main purpose of the hardware-in-the-loop simulation experiment is aimed at validating the electronic controller interface circuit and true-false of program logic, and in order to realize the tuning of system parameters, the engine model with fewer parameters and lower order is usually utilized to meet the requirements of the controller verifying [11-12].

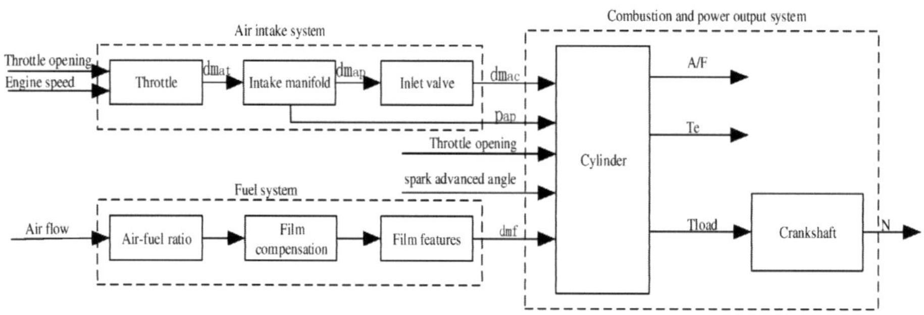

Figure 1. Schematic diagram of the engine model

3. Establishment of air intake system model

3.1. The throttle air mass flow model

Throttle plays an important role in adjusting the air flow into the cylinder, so it is significant to design a model of air mass flow at throttle reflecting actual physical process for predicting the state of the air inside the intake manifold and measuring the actual air quantity entering into the cylinder each cycle.

This article adopts the model of air mass flow at throttle proposed by Hendricks, the model divides the air flow at throttle disc into two parallel master-slave isentropic and master-slave air flows, and which converge in the back of the throttle body [13-14]. The expressions of the model are as follows:

$$\dot{m}_{at} = \frac{p_{amb}}{\sqrt{T_{amb}}} \beta_1(\alpha) \cdot \beta_2(p_r) \tag{1}$$

$$\beta_1(\alpha) = 1 - \varphi_1 \cos(\alpha) + \varphi_2 \cos^2(\alpha) \tag{2}$$

$$\beta_2(p_r) = \begin{cases} \sqrt{p_r^{p_1} - p_r^{p_2}} / p_n, \text{if } p_r \geq p_c \\ 1 \quad , \text{if } p_r < p_c \end{cases} \left(p_c = \left(\frac{p_1}{p_2}\right)^{\frac{1}{p_2 - p_1}}, \quad p_n = \sqrt{p_c^{p_1} - p_c^{p2}} \right) \tag{3}$$

Where P_1, P_2, φ_1, φ_1, P_c and P_n are constants used in the simplified process of the throttle valve air mass flow model. These constants are independent of the specific throttle structure, therefore, Hendricks can obtain these common model parameters after fitting test data form different engine throttle bench:

$$\begin{cases} p_1 = 0.4404 \quad p_2 = 2.3143 \\ \varphi_1 = 1.4073 \quad \varphi_2 = 0.40877 \\ p_n = 0.7404 \quad p_c = 0.4125 \end{cases} \tag{4}$$

Where $\dot{m}_{at}$ is air mass flow at throttle; P_{amb} and T_{amb} are the throttle inlet air pressure and temperature, respectively; $\beta_2(\alpha)$ is the throttle efficient coefficient of flow area; $\beta_2(P_r)$ is the inlet air corrected value depended on the state of air at throttle; α is the throttle opening degree; P_r is the ratio between the throttle inlet and outlet pressure, that is P_{ap}/P_{amb}; P_{ap} is pressure inside the intake manifold.

3.2. *Intake manifold pressure model*

Assuming that the intake manifold is a control body, $\dot{m}_{at}$ and $\dot{m}_{ac}$ respectively are air flow into the intake manifold and air flow out of the intake manifold. According to the conservation law of mass, air flow $\dot{m}_{ap}$ inside the intake manifold can be obtained as follows:

$$\dot{m}_{ap} = \dot{m}_{at} - \dot{m}_{ac} \tag{5}$$

The ideal gas state equation is $m = PV(RT)^{-1}$, and derivation on both sides of the equation:

$$\dot{m}_{ap} = \frac{V_{ap}}{RT_{ap}} \dot{p}_{ap} - \frac{p_{ap} - V_{ap}}{RT_{ap}^2} \dot{T}_{ap} \tag{6}$$

Since the gas temperature inside the intake manifold changes little, compared with the gas pressure change inside intake manifold, it can be neglected and $\dot{T}_{ap} \approx 0$ could be obtained. Substituting equation (5) into equation (6), an intake manifold pressure model expression can be deduced as follows:

$$\dot{p}_{ap} = \frac{RT_{ap}}{V_{ap}} (\dot{m}_{at} - \dot{m}_{ac}) \tag{7}$$

Where T_{ap}, m_{ap} and P_{ap} are respectively air absolute temperature, mass and pressure inside the intake manifold; V_{ap} is the intake manifold volume; R is the air gas constant, which is generally taken 287 J/(kg • K).

3.3. *The intake valve air mass flow model*

Air mass flow at intake valve has a direct influence on fuel quantity into the cylinder, thereby affecting the engine speed and torque, therefore, it is important to build an intake valve air flow model [15]. However, the model is a difficulty in the mean-value modelling, it is determined initially by experimental data identification and data regression analysis method. Although the model built by this method is applicable for a specific intake system, it can't be applied to other types of engine air intake system effectively. In order to construct an intake valve air flow model with good portability, and avoid mass collection of experimental data to identify and analysis for different engines, this article applies a generic air flow model of the intake valve, and utilizes the speed-density method to establish air mass flow model at intake valve for four-stroke engine. The expression is as follows:

$$\dot{m}_{ac} = \frac{NV_s p_{ap}}{120RT_{ap}} \phi_v \tag{8}$$

Where N is the speed of the engine crankshaft; V_s is the engine displacement; $\emptyset_v$ is the engine volumetric efficiency.

Integration on equation (8), intake air mass of the engine single cylinder in each work cycle can be obtained:

$$m_{ac} = \int_{i_o}^{i_c} \dot{m}_{ap} dt \approx \frac{NV_s p_{ap} \phi_v}{120RT_{ac}} \cdot \frac{30}{N} = \frac{V_s}{4RT_{ac}} (s_i p_{ap} - y_i) \tag{9}$$

Where s_i and y_i respectively are the slope and intercept air temperature inside the intake manifold, because both s_i of y_i are nearly same with different the engine speeds, they could be considered as constants; i_o and i_c respectively are the moment to open and close intake valve; $30/N$ is intake time of the per cylinder each cycle.

4. Establishment of fuel system model

Fuel ejected into cylinders by injector is difficult to mix with air absolutely. Some fuels attach to wall surface of the intake manifold, intake or the back of intake valve, these fuels will usually form film. The film is evaporated by high temperature and forms fuel vapour entering into cylinders. This phenomenon is called the film wet wall.

In the actual engine operation, the fuel system working principle is as followings: part of the fuel ejected by injector is mixed with the air and goes into the cylinder directly, while the remaining are adsorbed on the wall surface in the form of film. Fuel vapour is generated continually by film evaporation during engine operation, mixed with air and goes into the cylinder. When the engine works in steady state, the amount of film increasing and evaporated is equal in each working cycle, and the total mass of the film is unchanged, it can be considered that all of the fuel is injected into the cylinder and combusts [16]. When the engine works in transient operating conditions, the amount of film increasing and evaporated is different, which results in the fuel amount ejected by injectors is unequal to the fuel entering into the cylinder. Transient conditions contain two situations. One is that when opening degree of throttle becomes larger suddenly, the fuel injection and the film amount is increasing with the increase of the intake air amount, however, the acceleration of film evaporation requires some time, which will result in amount of evaporation is less than the deposited amount, and air-fuel ratio is too large. The other is that when throttle opening degree becomes smaller suddenly, which is contrary to the above situation [17]. Therefore, when modelling the fuel system, the dynamic characteristics of fuel transfer must be considered.

Based on Aquino model, Elbert Hendricks and Thomas Vesterholm made a further research and proposed that applying one-order inertial to describe the fuel transfer dynamic characteristics [15], which is presented as follows:

$$\begin{cases} \dot{m}_{ff} = \dfrac{dm_{ff}}{dt} = -\dfrac{m_{ff}}{\tau_{ff}} + x \cdot \dot{m}_{fi} \\ \dot{m}_{fv} = (1-x)\dot{m}_{fi} \\ \dot{m}_{f} = \dot{m}_{fv} + \dfrac{m_{ff}}{\tau_{ff}} \end{cases} \tag{10}$$

Where m_{ff} is the mass of the film; $\dot{m}_{fi}$, $\dot{m}_f$ and $\dot{m}_{ff}$ respectively are the fuel mass that is ejected by injector, actually entering into the cylinder and formed by film evaporation; $\dot{m}_{fv}$ is the fuel mass which is ejected by injector and goes into the cylinder directly; τ_{ff} is the time constant of the film evaporation, which is generally taken 0.25 ~ 0.4 s; x is the deposited fuel radio of the fuel injection, which is generally taken 0.5 ~ 0.8.

5. Establishment of dynamic system model

Engine power system operating state is modelled from perspective of flow of working fluid and energy transition. Engine output torque and power are computed according to the amount of air and fuel entering into the cylinders, and other important parameters (engine speed, the ignition advance angle, etc.). Mixture gas inside the cylinder combusts and swells quickly after ignited by sparks to promote the piston acting and torque is output through crankshaft. According to the conservation law of energy and Newton's second law, energy change rate pushing the piston movement is accelerating power of crankshaft rotation. Engine power model is built as follows [18-20]:

$$\dot{\omega} = \left[H_u \eta_i \dot{m}_f (t - \tau_d)/\omega - (T_f + T_p + T_b) \right] / J \tag{11}$$

Where ω is the angular velocity of crankshaft; J is rotational inertia of the engine; H_u is the fuel heat value; η_i is the engine indicated thermal efficiency; τ_d is the mean time delay of fuel feed; T_f, T_p and T_b respectively are resistance moment of frictional loss of the engine, resistance moment of pumping loss of the engine and load torque of the engine.

6. Validation of Engine Model

6.1. *Validation of Inlet Flow and Inlet Pressure*

Two groups of data are selected to validate inlet flow and inlet pressure. One is inlet flow and inlet pressure vary according to the throttle opening degree at the rotary speed is 2500r/min; the other is inlet flow and inlet pressure vary according to the throttle opening degree when the speed is 5000r/min.

For the rotary speed is at 2500r/min, as shown in figure 2, when the engine works in a state of heavy load and middle or low speed, the inlet pressure values measured in bench test is higher than simulation and both have large deviation; The inlet flow values measured in bench test is higher than simulation and both deviation is small. Maximum relative error of inlet pressure and inlet flow are respectively about 4% and 2% when the throttle opening degree is more than 60%. This is because the engine status is not stable and actual crankshaft speed fluctuation is large under the working condition, which results in intake noise and inlet pressure fluctuations are larger, so the values measured in bench test is higher than simulation. For intake flow, as the gas flow inertia in the air intake manifold and long response time of flowmeter measurement, which play a role as low-pass filter to inlet flow measurement, therefore, the deviation between bench test measurement values and simulation is small.

As shown in figure 3, the inlet flow values measured in bench test are lower than simulation in a state of small load and high speed. For example, maximum relative error of inlet flow is about 3% when the throttle opening is lower than 25%. That is because inlet flow rate is very high when the engine is in a state of high speed and small load. Actual intake flow loss increases, so the inlet flow value that measured by bench test is less than the simulation test. Figure 2 and figure 3 show that a deviation exists between bench test measurement values and simulation under the circumstances of low speed, heavy load and high speed, small load. For aspect of accuracy, the values of inlet flow and inlet pressure obtained by the engine model are consistent with values measured by bench test. The maximum error is less than 5% and is within the allowed error range. Air intake system model could accurately simulate the steady state parameters, such as inlet flow, inlet pressure and so on. Therefore, this model is utilized to compute fuel injection quantity when the engine is in steady state conditions.

6.2. *Validation of Engine Output Torque and Rotational Speed*

Performance data measured by the engine bench test is applied to further verify the engine model. Part of steady state conditions data measured by engine-dynamometer bench system is shown in table 1. The throttle opening range is 12%~16% and the crankshaft speed range is 2580r/min~5880r/min in the collected data. Engine simulation model is verified with data collected from a large scope of working condition.

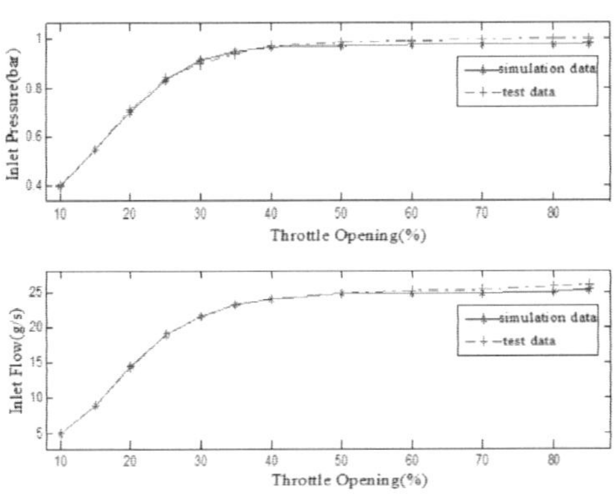

Figure 2. Engine model simulation results at 2500r/min

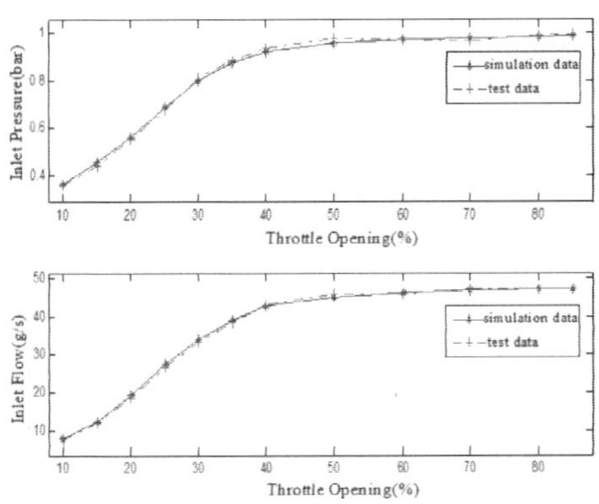

Figure 3. Engine model simulation results at 5000r/min

Table 1. Part of steady state conditions data measured by engine-dynamometer bench system

TPS (%)	Tb (Nm)	N (r/min)
12	32	2580
15	32	3260
20	39.8	4150
30	52.9	5280
40	68.6	5560
50	73.4	5750
60	76	5880

The values of throttle opening and load torque are put into the input of the engine model, the output torque and rotational speed of the engine could be computed through the model. From the simulation results of figure 4 and figure 5, the torque and rotational speed computed from different working conditions of engine model are very consistent with the data from bench test, and the error is within acceptable range. AFR is the throttle opening degree. Figure 6 shows the simulation result of crankshaft speed computed by the engine model when the throttle opening degree changes from 12° to 15°. From figure 6, after the sudden change of throttle opening degree, dynamic changes of the crankshaft speed are similar to actual situation of engine. The simulation results after the engine model is stable are consistent with data measured from bench test.

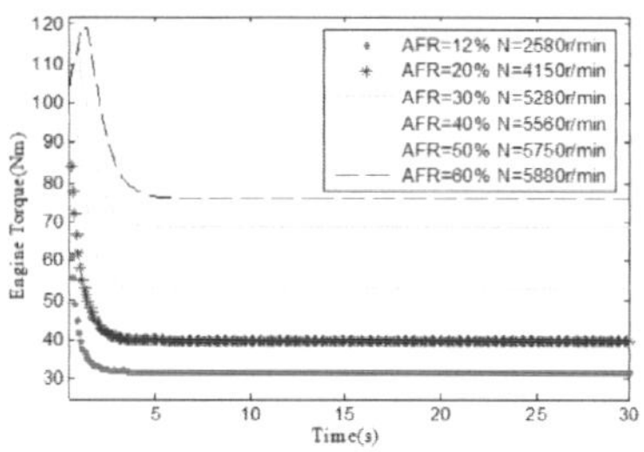

Figure 4. Engine model torque simulation curves

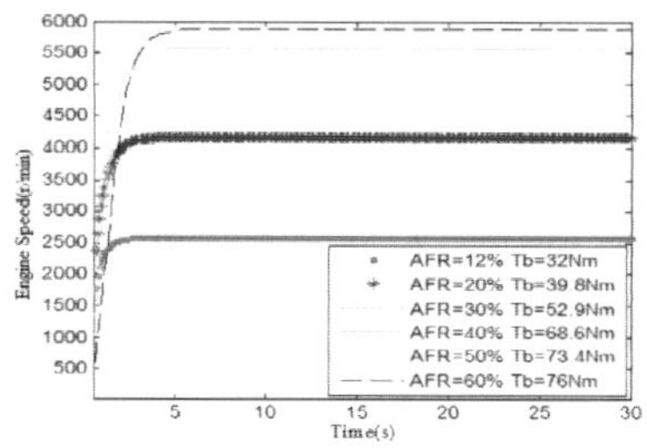

Figure 5. Engine model speed simulation curves

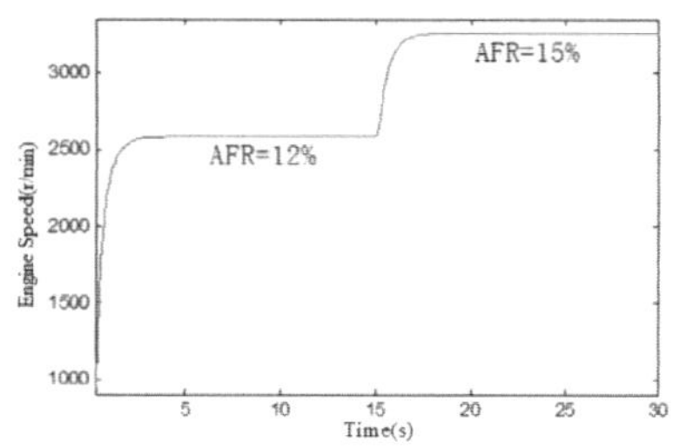

Figure 6. Engine model speed simulation curves when the throttle opening abrupt change

From analysis of the verification results of the inlet flow rate, inlet pressure, engine output torque and rotational speed. The engine model can truly reflect operating characteristics of the real engine. The errors between the simulation results and the bench test data are within the acceptable range. Therefore, this engine model could be applied to validate control strategy and logic performance of electronic controller, it can avoid the risk that the electronic controller is utilized to bench test directly and reduce the actual workload. For another aspect, errors existing in some certain conditions show that the actual engine working process is very complicated. Apart from the external environmental factors, the characteristics of the engine under different conditions are also diverse.

7. Conclusions

Mean valve model of aviation piston engine is built with MATLAB/SIMULINK platform from the energy conversion and working medium flow in the article. The engine model has advantages of simple structure, lower computing iteration amount, the overall accuracy can meet the requirements of simulation, it can describe the steady and dynamic characteristics of engine, and it is applicable to the hardware-in-the-loop simulation system. The engine model can truly reflect the operating characteristics of actual engine and the errors between the simulation results and bench test data are in the acceptable range by analysing the verification results of the inlet flow rate, inlet pressure, engine output torque and rotational speed. As the built model could be utilized as controlled plant of HIL system, it has great value to performance testing and logic validation of the electronic controller.

8. References

[1] Xu Jie. Research and Application on Gasoline Real-time Model for ECU HIL Testing Platform[D]. Hangzhou: Zhejiang University, 2011.

[2] Xu Zhixin, Zeng Guogui, Yan Feng, Zu Jiakui. Modelling Technique of Piston Engine System for Small Unmanned Helicopter. *Ordnance Industry Automation*. 2012,31(1):4~7.

[3] Zhou Wenxiang, Huang Jinquan, Huang Kaiming. Real-Time Simulation System for Aero Engine Based on Simplified Model. *Journal of Nanjing University of Aeronautics & Astronautics*. 2005, 37(2):251~255.

[4] Pai Song. Research on Control of Twin-Engine Power System[D]. Nanjing: Nanjing University of Aeronautics & Astronautics, 2008.

[5] Lin Haiying, Long Xiangyang. Simulation study of an aero-piston engine with single stage turbocharge[J]. *Journal of Aerospace Power*. 2009, 24(6):1332~1338.

[6] Hu Jinlong. Research of Air and Fuel Dynamic Model for EFI Gasoline Engine[D]. Hangzhou: Zhejiang University, 2011.

[7] Guzzella Lino, Christopher H Onder. Introduction to Modelling and Control of Internal Combustion Engine Systems[M]. Heidelberg: Scientific Publishing Services Pvt. Ltd., 2010.

[8] Qin Qifeng, Ye Xiang. Research of Aviation two-stroke piston engine modelling and rapid prototyping controller simulation. *Servo Control*. 2012(1):41~43.

[9] Guo Chunmei. Research on Modelling and Air/Fuel Ratio Control of Spark-Ignition Engine[D]. Tianjin: Tianjin University, 2003.

[10] Wang Xiaoqing. Research on Technology for Unmanned Helicopter Modelling and Control System Design[D]. Nanjing University of Aeronautics & Astronautics, 2009.

[11] Su Sanmai, Ma Rui, Leng WenBin. Semi-Physical Real Time Simulation of Digital Control System for Microturbo-Engine[J]. *Journal of Aerospace Power*. 2001, 16(1):93~96.

[12] C. Frangos. Control System analysis of A Hardware-in-loop Simulation. *IEEE Transactions on Aerospace and Electronic Systems*. 1990. 26(4):666~668.

[13] Shen Xiujuan. Research of Electronic control fuel injection system simulation[D]. Wuhan: Wuhan University of Technology, 2005.

[14] Lino Guzzella, Onder Christopher H. Introduction to Modelling and Control of Internal Combustion Engine Systems[M]. Heidelberg: Scientific Publishing Services Pvt. Ltd. 2010.

[15] Isermann R, Schaffnit J. Hardware-in-loop Simulation for the Design and Testing of Engine-control System. *Control Engineering Practice*. 1990. 7:643~653.

[16] Yan Yixin. Simulation for Air-Fuel Ratio Control of Electronic Controlled Gasoline Engine[D]. Xi'an: Chang'an University, 2012.

[17] Guo Qiang. Research of Engine Start Simulation for Hardware in the loop Test of ECU[D]. Hangzhou: Zhejiang University, 2011.

[18] CF Aquino. Transient A/F Control Characteristics of the 5 Litre Central Fuel Injection Engine. *SAE*. 1981. 810494:24~29.

[19] Du Changqing, Yan Fuwu. Methods of Engine Torque Estimation for Control Algorithms. *Transactions of CSICE*. 2008. 42(5):172~178.

[20] Wu Zhihu, Liu Shang, Yuan Run. Torque-based Engine Modelling and Feedback Linearization Tracking Control of Engine Speed. J*ournal of Harbin University of Science and Technology*. 2009, 19(2): 89~94.

Information model construction of MES oriented to mechanical blanking workshop

Jin-bo Wang[1,2], Jin-ye Wang[1], Yan-fang Yue[1] and Xue-min Yao[3]

[1]School of Mechanical Engineering, Hebei University of Science & Technology, Shijiazhuang, Hebei, 050018, China
[2]College of Mechanical Engineering, Inner Mongolia University of Technology, Huhhot, Inner Mongolia, 010051, China
[3]Hebei Supervision and Inspection Institute of Electronic Information Products, Shijiazhuang, Hebei, 050071, China

Abstract. Manufacturing Execution System (MES) is one of the crucial technologies to implement informatization management in manufacturing enterprises, and the construction of its information model is the base of MES database development. Basis on the analysis of the manufacturing process information in mechanical blanking workshop and the information requirement of MES every function module, the IDEF1X method was adopted to construct the information model of MES oriented to mechanical blanking workshop, and a detailed description of the data structure feature included in MES every function module and their logical relationship was given from the point of view of information relationship, which laid the foundation for the design of MES database.

1. Introduction

The blanking is the first process in product or parts manufacture, which is also the critical process influencing the production and products' quality. The mechanical blanking workshop, which is in the upstream of the part flow, is not only the initial link of the production chain, but also a important link of influencing the production and products' quality [1]. As the production task execution layer in manufacturing system, the mechanical blanking workshop plays a connecting role in material flow in the manufacturing system. On the one hand, it receives the production task from the upper management system in the enterprise, executes the task according to its own balance and capacity requirement and prepares for the following processing. On the other hand, it makes standardized management to the workshop resource and production status information according to the requirement of departments in the enterprise, which can ensure the smooth blanking production.

Building manufacturing execution systems (MES) oriented to mechanical blanking workshop is a key link to implement informatization in manufacturing enterprises, and it can not only improve the efficiency for workshop production and create economic benefit for the enterprise, but also play a driving role to the entire business management level [2, 3]. Therefore, construct the information model of MES oriented to mechanical blanking workshop on the base of analyzing its manufacturing process information and information demand is one of the key works in MES development and realization.

2. Manufacturing process information of the MES

As a management system oriented to mechanical blanking workshop, the core of MES is to implement the overall management of the production job, material, quality, workshop equipment, personnel and other key elements in the manufacturing process, support and realize visualization, real-time and transparency of the blanking process, which can realize the lean management in the entire

IOP Conf. Series: Materials Science and Engineering **157** (2016) 012005 doi:10.1088/1757-899X/157/1/012005

manufacturing process for the enterprise [4]. Therefore, MES must closely connect the top layer plan in enterprise and the bottom layer production to ensure the smooth workshop production. The complete activity chain of the manufacturing process information between different function modules in the MES is shown in figure 1.

The production planning department in the enterprise firstly assigns production task to the blanking workshop according to the master production schedule in the upper-layer ERP, and the job planning and scheduling module searches the relevant information from the drawings/documents management module after accepting the task, makes process planning and sends the working procedure of preparing the blanking procedure to the NC procedure management module.

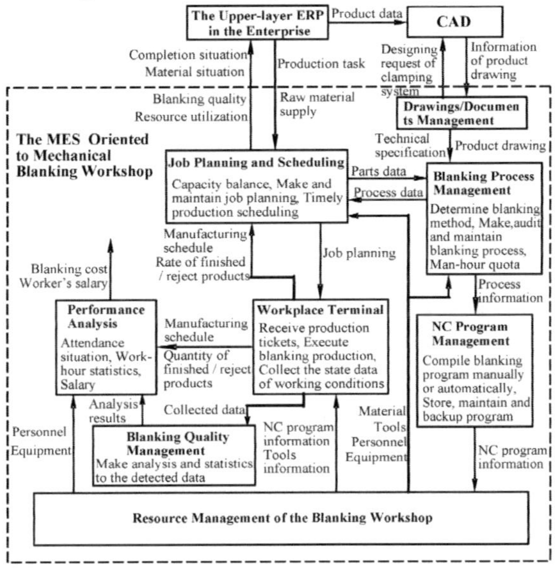

Figure 1. Manufacturing process information of MES

After obtaining the whole information needed for the job planning, the job planning and scheduling module makes a group of job planning and resource allocation instructions oriented to the workplace. Under constraint of resource condition of material, equipment, personnel, procedure and production condition of process, delivery date, batch quantity, etc., the planning and instructions need to optimization sequencing according to the process turnover time and clamping auxiliary time [5]. The sequenced job planning is confirmed by the workshop scheduler, and then converted to the production ticket and sent to the worker to execute the blanking production. When the specified job in the current production ticket is finished, the worker needs to feed back the production ticket recording the relevant manufacturing information to the job planning and scheduling module. Meanwhile, the feedback production tickets are confirmed by the workshop inspector and sent to the performance analysis module to make statistical analysis, generate the report data including the feedback information to ERP, the statistical report of every worker's attendance and work hour, the part's manufacturing process report, and so on. The quality management module detects and analyses the correlation data, and gives the product error distribution, the variation trend of the system error, the cause of the random error, etc., which can provide the necessary basis for eliminating error timely and improving the product's quality. During the blanking process, the bottom control system real time collects the running state data of equipments, and feeds the data back to the performance analysis module to obtain the operating rate, the actual utilization ratio and history operating data of each equipment. It not only provides necessary basis for making the following job planning, but also provides important reference information for maintenance of the equipment.

3. Information model construction of the MES

Information model is the foundation of database design in MES development, and the purpose of construction system information model is to provide a consistent definition of the data meaning and their mutual relation, and detail describe the structure feature of the data in the function modules from the point of view of information relationship, which can ensure the integrity of the integrated data, shared data and management data [6, 7].

3.1. IDEF1X modeling method

IDEF1X, a modelling technique for semantic data, is to apply the E-R (Entity-Relationship) method to the semantic data model to construct system information model [8]. The elements of IDEF1X mainly include: (1) Entity, including independent marking entity and subordinate marking entity. (2) Relationship, including calibration connection relation, non-calibration connection relation, classification relation and non deterministic relation. (3) Property / Keyword, including main keyword, creating keyword and exotic keyword. In the IDEF1X model, the rectangular box is used to represent the entity, the connecting lines between rectangular boxes are used to represent the links between entities, and the property names in the rectangular box are used to represent the entity attributes, as shown in figure 2.

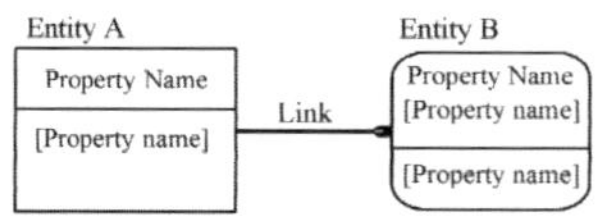

Figure 2. Basic structure of IDEF1X model

Using IDEF1X method to construct MES information model has many advantages compared with the traditional E-R method, which are mainly shown in the following aspects [9]: (1) IDEF1X semantic model is more abundant and precise and it can fully and clearly express the data information of the system and its connection; (2) IDEF1X model has the better consistency and higher degree of standardization; (3) The logic model defined by IDEF1X is more beneficial to convert into the physical model.

3.2. Information model of the MES

According to the established manufacturing process information in the MES oriented to mechanical blanking workshop, the entity analysis of the data information and its logical relationship was provided based on the characteristics of computer information processing, and IDEF1X modeling method was utilized to construct the information model of the MES every function module.

3.2.1. Information model of the fundamental data management module

The fundamental data is mainly static data, and its characteristics are that the data is stable in quite a long time and it is often queried and used. The information of fundamental data is usually the attribute information of the independent entity in the database design, and it is generally set as read-only in the application program. The fundamental data mainly includes product data information and workshop resource information. The product data information also mainly includes product BOM (Bill of materials) information and blanking process information, and its information model is shown in figure 3. The entity Product Assembly Drawing and Part Drawing generate the part detailed information (including the detailed information of homemade parts, standard parts, cooperation parts and purchased parts) according to the product structure configuration rule. The entity Part Detail, which is the basis for calculating the part work-hour, describes the structure size information of the blanking part. And the entity Part Detail and Calculate Work-hour are used to make the blanking process. When the part blanking mode is determined as NC machine processing, the technologist in the blanking workshop makes the relevant NC program according to the blanking process. Therefore, the uncalibrated relationship between the entity Blanking Process and NC Program is formed.

The workshop resource information, whose information model is shown in figure 4, mainly includes material information, personnel information and equipment information. The entity Material describes the material name, specification, number and other information, which including the relevant entities Material Batch and Material Inspection. The entity Personnel, which is the decision-maker and executor of the blanking production system, describes the basic information and functional role of the workshop personnel. And the entity Equipment describes the equipment's basic parameters, status data, maintenance record and other information.

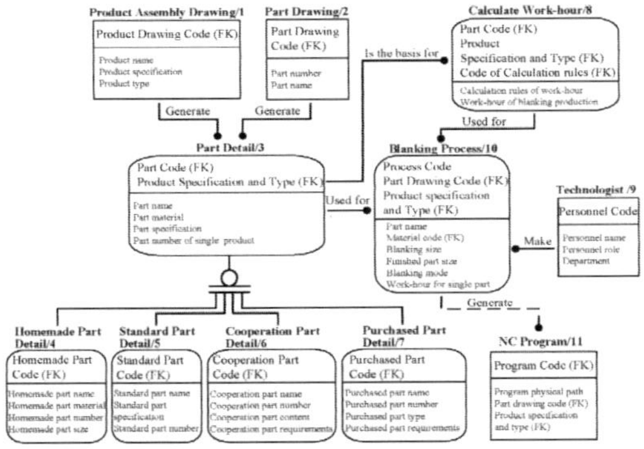

Figure 3. Information model of the product data

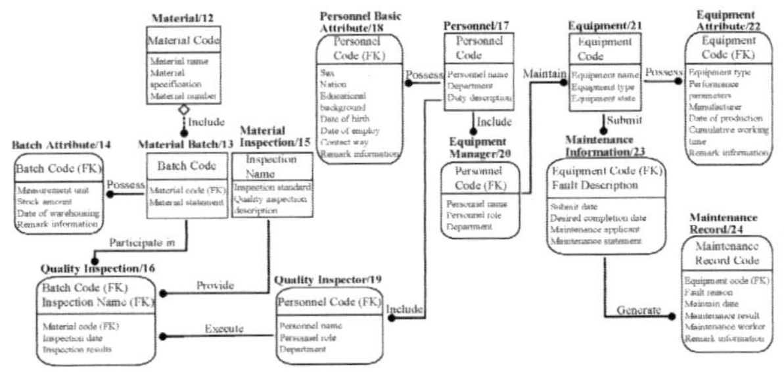

Figure 4. Information model of workshop resources

3.2.2. Information model of the job planning and scheduling module

The information of the job planning and scheduling is mainly the dynamic data in the blanking production, and its characteristics are that the data is changed constantly along with the time, which reflecting the status of the object at a certain time. The information of the job planning and scheduling is usually the attribute information of the dependent entity in the database design, and it is generally maintained dynamically in the application program, which supporting the transaction processing with user interaction. It includes production execution data and production monitoring data, and the former, whose information model is shown in figure 5, describes the executing information of the production order in the process from assigning production task, making job planning to completing the blanking production. The entity Production Planning, generated by the contract orders, decomposes the production task, makes the job planning and generates the entity Production Ticket to guide workers in blanking operation. The entity Blanking Quality Inspection feeds back the quality inspection information of the parts to the entity Production Ticket, and generates the entities Reject Record and Warehousing Record. The entity Blanking Production Preparation Report records the production preparing information of the materials, blanking process and equipments in the workshop, and it along with the entity Production Unexpected Situation Report describe the influence on the job planning.

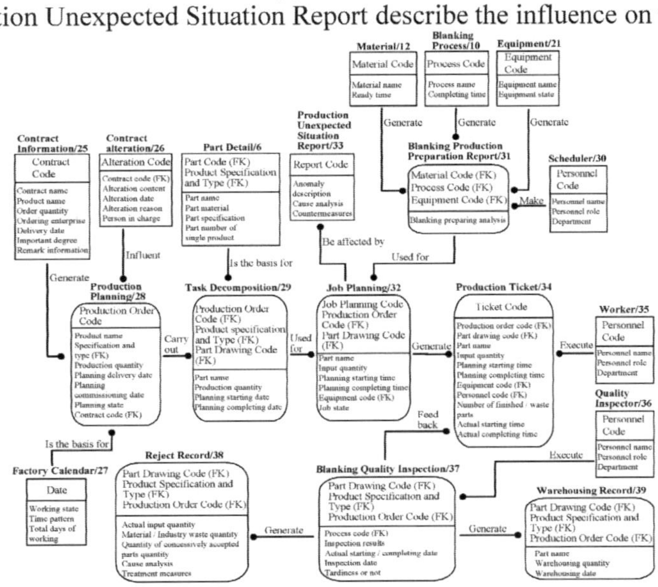

Figure 5. Information model of product executing

Production monitoring information records the relevant information in the blanking production process, including the job progress information, the collecting information of production resource state, work-in-process information, production unexpected situation, and so on. And its information model is shown in figure 6. The entity Blanking Process Data describes the information of work-hour, job process, inspection records of the blanking quality, material consumption, personnel status and equipment operation in the production process [10]. The entity Work-in-process Record describes the basic information of the work-in-process, including its process state, processing equipment, operator, quantity, actual starting/completing time, and so on. All of these information reflect the actual blanking processing course.

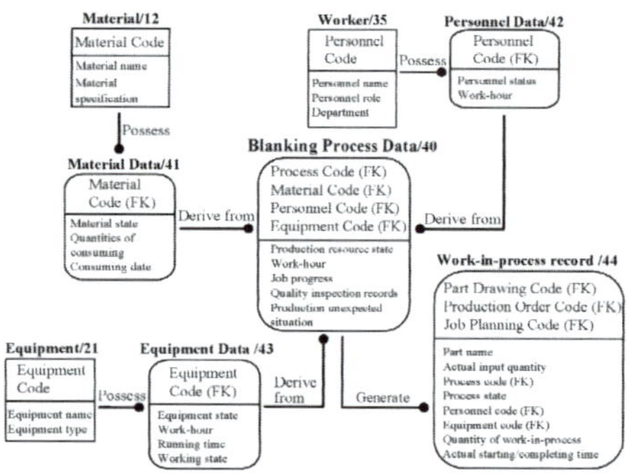

Figure 6. Information model of product monitoring

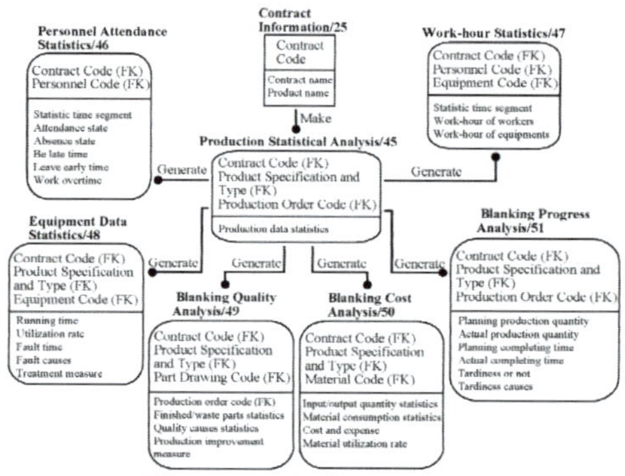

Figure 7. Information model of performance analysis

3.2.3 Information model of performance analysis module

Performance analysis information is mainly the middle data formed from statistical computing and analyzing the raw data source of the above static data and dynamic data. It is obtained and printed by the system program, and its output form is usually the statistic report in the database design. It contains the information of the blanking quality management module, and mainly performs a statistical analysis on the blanking production data of the order product to optimize the workshop blanking production. Its information model is shown in figure 7. The entity Personnel Attendance Statistics records workers' information of attendance/absence, be late/leave early, work overtime, and so on. The entity Work-hour Statistics records the actual work-hour of the workshop worker and processing equipment in the blanking production. The entity Equipment Data Statistics records the equipment running time, rate of the equipment utilization, equipment fault time, fault causes and treatment measure. The entity Blanking Quality Analysis records the information of finished/waste parts quantity, the existing quality problems and improvement measure. The entity Blanking Cost Analysis records the input/output quantity of the parts and material consumption, and analyzes the material utilization rate based on the cost and expense. The entity Blanking Progress Analysis records the actual blanking processing schedule, which is used to improve the blanking job planning.

4. Conclusion

The research and development of MES oriented to mechanical blanking workshop is the first step for the enterprise information management towards the workshop information management, and it is the key link for realizing enterprise production data sharing and optimization of the workshop management [11]. Meanwhile, it constructs an effective platform for other workshop to implement MES system. Adopt the IDEF1X semantic modeling method to carry on the data modeling of the MES oriented to the mechanical blanking workshop, and perform its requirement, logical and physical design with the modeling tool, which can fully ensure the consistency and integrality of the data, realize the organic combination of the database analysis, conceptual model design and physical database structure design, and greatly improve the development efficiency of the MES.

References

[1] Li H 2006 *Machinist Metal Forming.* **2** 35-36

[2] Yue Y F and Wang J Y 2011 *Automation & Instrumentation.* **26** 5-8
[3] Zhang Y D,Yang G and Yue Y F 2009 *Manufacturing Automation.* **31** 53-55
[4] Wang Q F and Shen D H 2009 *Industrial Engineering Journal.* **12** 70-74
[5] Kenneth N M and Vincent C S 2003 *Cogn Tech Work.* **5** 82-93
[6] Li T T 2007 *Study and development of E-kanban system in MES oriented to motor mould production.* (Jinan:Shangdong university) chapter 4 pp 31–45
[7] Wang J Y, Yue Y F, Wang J L, *et al* 2013 *Advanced Materials Research.* **622-623** 1479-83
[8] Chen Y L 1999 *IDEF modeling analysis and design method.* (Beijing: Tsinghua University Press) p 26
[9] Li J P 2005 *Microcomputer applications.* **26** 455-458
[10] Guang Y, Yue Y F, Wang J Y, *et al* 2010 *Applied Mechanics and Materials.* **26-28** 870-874
[11] Li Q 2007 *Manufacturing Execution System.* (Beijing: China Electric Power Press) p 77-107

IOP Conf. Series: Materials Science and Engineering **157** (2016) 012006 doi:10.1088/1757-899X/157/1/012006

Dual Operating Parameter Relationship with Engine Performance

Aparna Devi . I[1, a], R. Ramesh[2], N. Ravi Kumar[3] and G.Shiva Charan[4]

[1, 2, 3] MVGR College of Engineering and [4] Andhra University
MVGR College of Engineering, Vijayram Nagar Campus, Chintalavalasa, Vizianagaram, Andhra Pradesh- 535005, India. Phone: +91-9177501725

[a]aparna_devi@mvgrce.edu.in

Abstract. In order to reduce emissions and maximize performance of internal combustion engines, it is important to understand the relationship between operating parameters and performance. Estimation of engine performance under simultaneous variation of operating parameters is not well addressed in literature. In the present work an attempt has been made to develop a relationship between engine performance (brake thermal efficiency) and operating parameters such as load and compression ratio. Experiments were conducted varying load and compression ratio and Brake Thermal Effiency obtained was recorded. The results of the model were validated with the experimental results.

1. Introduction

The variation of performance, emission and combustion parameters is dependent upon the operating conditions of the engine. In a theoretical approach, it is generally taken from granted that the compression ratio is kept constant. However, in an actual engine, other processes which influence engine performance and efficiency vary with changes in compression ratio namely combustion rate and stability, heat transfer and friction. Empirical formulae derived earlier are in terms of a single operating parameter whereas a combined effect is produced in the output due to the simultaneous operation of the engine parameters like load, compression ratio etc. Although extensive research has been carried out to study the effects of such variation, not much work appears to have been done in the area of developing such generalized models. Especially so is the case with alternate fuels. An initiative towards this is made by creating a generalized formula for a performance parameter in terms of both load and compression ratio along with the validation of results.

2. Literature Survey

Ansari et al. [1] performed a study on the performance and emission characteristics of diesel and soybean blends and calculated the NOx formation based on temperature and concluded that higher NOx emissions are due to higher fuel consumption due to higher densities. Qi et al. [2] concluded that at higher engine loads, the peak cylinder pressure of biodiesel was almost similar to that of diesel, but the peak rate of pressure rise and the peak of heat release rate were lower for biodiesel. Sundarapandian [3] carried out experimental investigations and presented that the thermal efficiency of Jatropha was 26.48% and there was 44% of smoke reduction. NOx remained same as diesel fuel. It was concluded that Jatropha is the best alternative fuel than Mahua and Neem oils.

Moser et al. [4] concluded that the test engine consumed a greater amount of fuel operating on the SME and PHSME blends than on neat ULSD, but the increase was smaller for the PHSME blend. Shahabuddin et al. [5] reported that the heat release rate of biodiesel is slightly lower than diesel owing to the lower calorific value, lower volatility, shorter ID and higher viscosity. Liu et al's [6] paper presented emissions modelling and testing of a four-stroke single cylinder diesel engine using biodiesel fuels. A system level engine simulation tool developed by Gamma Technologies, GTPower, was used to perform predictive engine combustion simulations using direct-injection jet modelling technique. Rampure et al. [7] found out the effect of NOx emission and brake thermal efficiency upon varying the engine loads. Performance and emission analysis by Selvan and Maniysundar [8] indicates the reduction of NOx and brake thermal efficiency decreased with the application of EGR and Jatropha blends bio-diesel.

Balajee et al. [9] found out that the blends of biodiesel like jatropha and pongamia with diesel could substitute in the place of pure diesel and be used as an alternate source of fuel in the near future, thus saving the natural resources for the future generation. Chotai[10]conducted a review on the effect of varying compression ratio on performance and emissions of diesel engine fuelled with bio-diesel. The paper deals with effects on engine fuelled with diesel, blend of diesel with biodiesel and purely on biodiesel with a view to provide a platform for comparison of the parameter on various fuels.

Although extensive research has been carried out to study the individual effects of variation of load or compression ratio or speed on engine efficiency, not much work appears to have been done in the area of developing generalized models where a combined effect on the engine performance. Hence the present selection focuses on developing a relationship between brake thermal efficiency and dual operating parameters: load and compression ratio.

3. Results and Discussion

Experiments were carried out with soyabean methyl ester (SME) under compression ratios 15 to 18 and load varying from 0 to 12kg on a variable combustion engine. The engine emissions were measured during running. As the engine is computerised, the operating parameters were automatically saved in the system enabling computation and evaluation of performance characteristics as well as combustion characteristicsThe combined effect of variation of load and compression ratio on some of the performance parameters (namely specific fuel consumption and volumetric efficiency) and some emission parameters (namely Carbon Monoxide and smoke number) are plotted in figures 1 to 4 respectively.

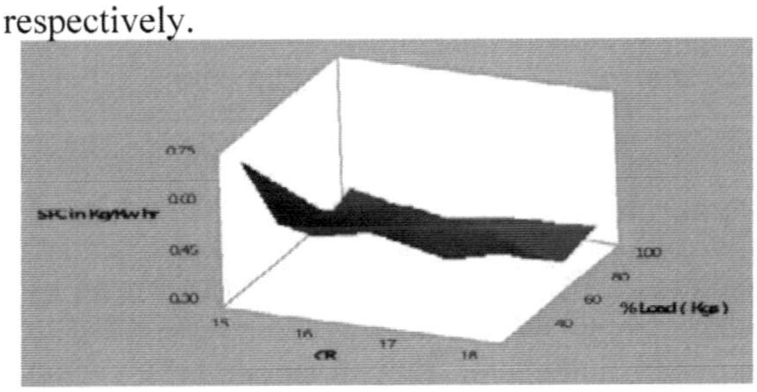

Figure 1 Variation of Specific Fuel Consumption with Compression Ratio and Load

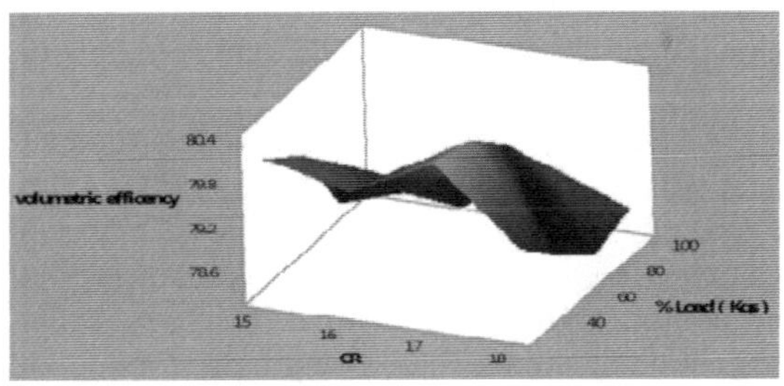

Figure 2 Variation of Volumetric Efficiency with with Compression Ratio and Load

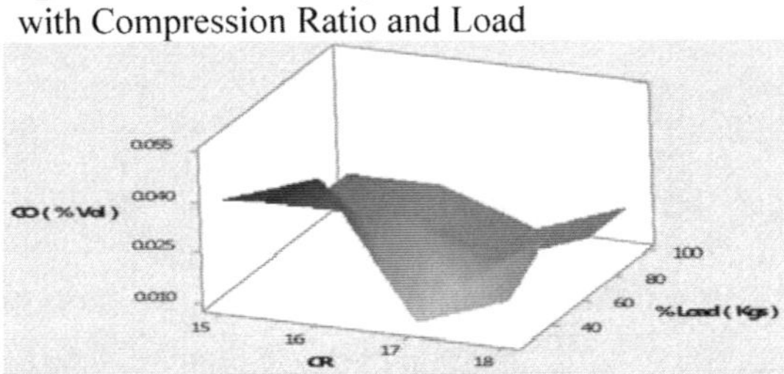

Figure 3 Variation of Carbon Monoxide Emissions with Compression Ratio and Load

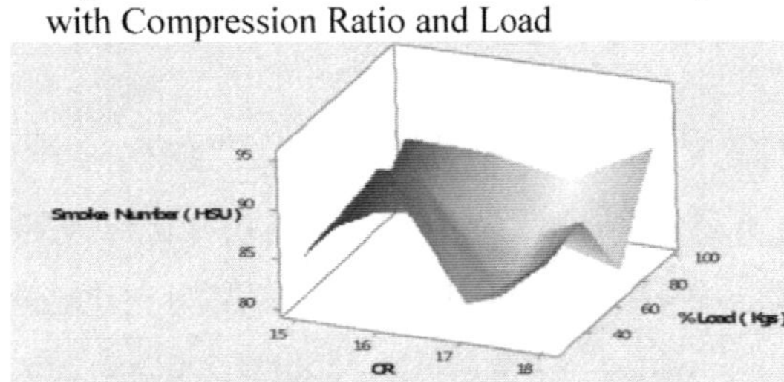

Figure 4 Variation of Smoke Number with with Compression Ratio and Load

The following figures 5 and 6 depict the variation of four performance parameters (brake mean effective pressure, brake thermal efficiency, specific fuel consumption and volumetric efficiency) for SME and diesel respectively. These are plotted together in order to get a quick overview of the comparative performance of the two fuels.

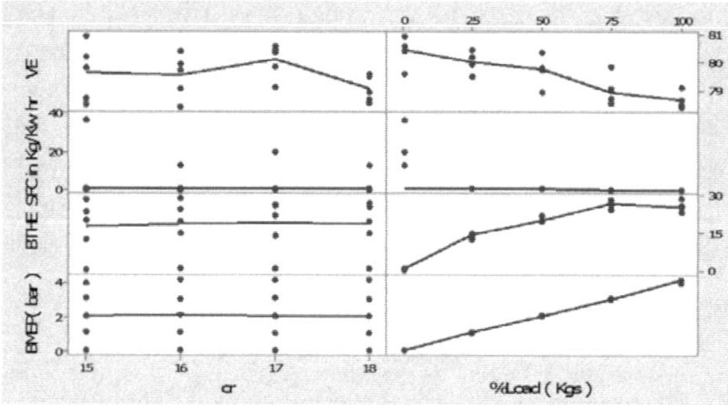

Figure 5 Variation of Performance Parameters for SME

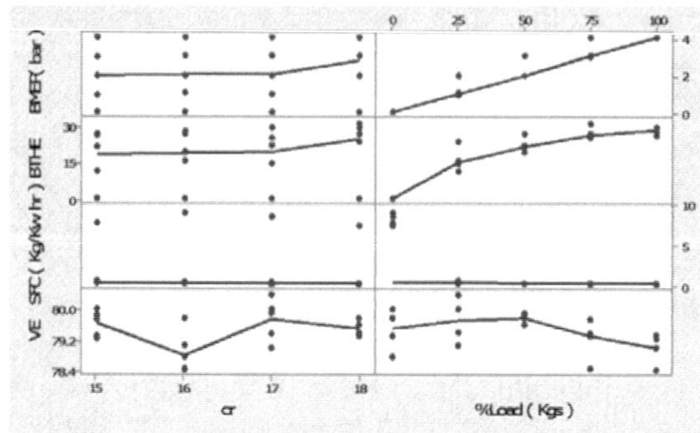

Figure 6 Variation of Performance Parameters for Diesel

3.1 Model Validation

Experiments were conducted for varying loads (25%, 50%, 75% and 100%) and compression ratios (15, 16, 17 and 18). The variation of BTE with load for all the compression ratios is first plotted and then the relationship modelled in this section.

3.1.1 Comparison of BTE at different compression ratios. The variation of BTE of the engine with load for different compression ratios is plotted in figure 7.

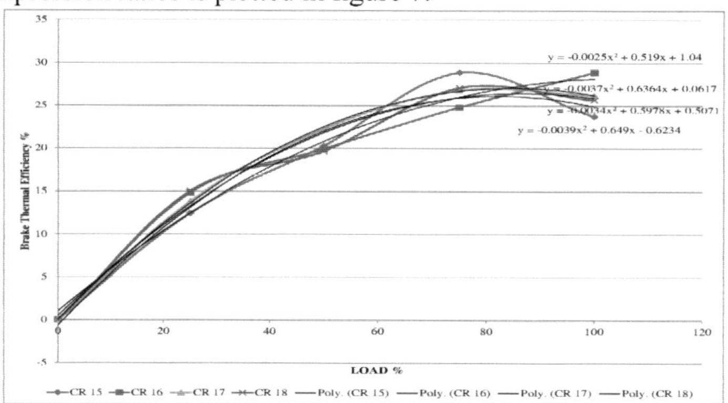

Figure 7 Variation of BTE with Load for Different Compression Ratios

It can be seen from the above graph that the variation in BTE is more or less the same for all the four compression ratios. In order to arrive at a generalized model, mathematical models were fitted for each of the four individual compression ratio variations. The equations of the best fit curve for each of the compression ratios are given below:

CR = 15 $BTE = -0.0039x^2 + 0.649x - 0.6234$

CR = 16 $BTE = -0.0025x^2 + 0.519x + 1.04$

CR = 17 $BTE = -0.0037x^2 + 0.6364x + 0.0617$

CR = 18 $BTE = -0.0034x^2 + 0.5978 + 0.5071$

The above relationships capture only the variation of BTE with load for each compression ratio. The effect of compression ratio needs to be included into this equation as one of the parameters so as to be able to compute the BTE for simultaneously varying load and compression ratio values.

3.1.2 Generalised equation for BTE. In order to arrive at a generalized equation for the above set of simultaneous equations, the technique of interpolation is adopted. The algorithm was coded in *Matlab*

using the coefficients obtained from the above equations (Appendix II). The final equation obtained is of the form:

$$BrTE = Az*Load^2 + Bz*Load + Cz$$

where BrTE is the Brake Thermal Efficiency and Az, Bz and Cz are derived coefficients that are functions of compression ratio (CR) and given as:

$Az = A(1,1) + (CR-15)*A0(1,1) + ((CR-15)*(CR-16)/factorial(2))*A1(1,1) + ((CR-15)*(CR-16)*(CR-17)/factorial(3))*A2;$

$Bz = B(1,1) + (CR-15)*B0(1,1) + ((CR-15)*(CR-16)/factorial(2))*B1(1,1) + ((CR-15)*(CR-16)*(CR-17)/factorial(3))*B2;$

$Cz = C(1,1) + (CR-15)*C0(1,1) + ((CR-15)*(CR-16)/factorial(2))*C1(1,1) + ((CR-15)*(CR-16)*(CR-17)/factorial(3))*C2;$

where A, B, C, A1, B1, C1 and A2, B2, C2 are forward difference values obtained from the original coefficients.

3.2 BTE model validation

Using the above equation, the BTE was computed for all the experimental conditions tested and compared with the actual experimental values.

It can be seen from figures 8 and 9 that for compression ratio 17, the error in prediction using the Newton Forward Difference Interpolation model is less than 5% for all load conditions. It is also seen that the prediction error is least (6.61%) at full load. Thus performance at compression ratio 17 and full load is best predictable using this model.

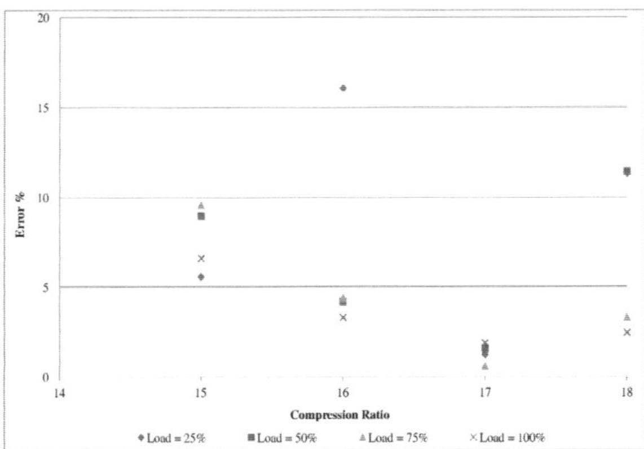

Figure 8 Prediction Error v/s CR at Various Loads

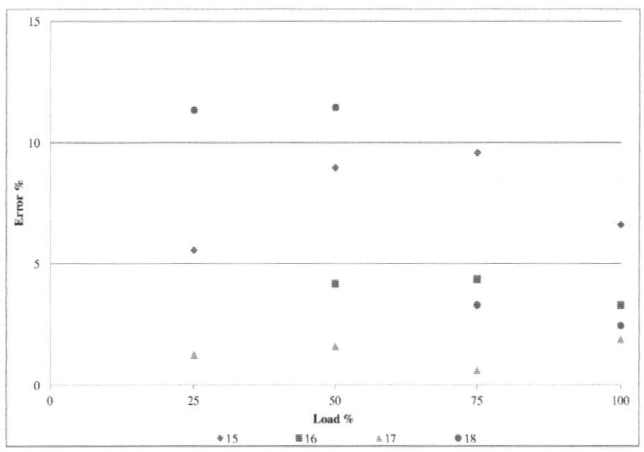

Figure 9 Prediction Error v/s Load at Various CRs

4. Conclusion

An attempt was made to model the combined effect of the engine operating parameters on the performance characteristics. This is very much important to be able to accurately predict the performance of the engine at given conditions. Based on the experimental results obtained for soyabean methyl ester, a generalised polynomial model was developed for brake thermal efficiency (BTE) using the Newton Forward Difference Interpolation approach. This model was able to fairly accurately predict the BTE given the load and compression ratio especially for compression ratio of 17 and full load condition. This model gives an empirical relationship for estimating BTE of SME under varying operating conditions.

References

[1] N. A. Ansari, J. Kumar, A. Kumar and D. Trivedi, 2013, *Emission Characteristics of A Diesel engine using Soyabean Oil and Diesel Blends*, International Journal of Research in Engineering and Technology, Vol. 2, Issue 5, pp. 793-798.

[2] D. H. Qi, L. M. Geng, H. Chen, Y. Zh. Bian, J. Liu and X. Ch. Ren, 2009, *Combustion and Performance Evaluation of a Diesel Engine Fueled With Biodiesel Produced from Soybean Crude Oil*, Renewable Energy, Vol. 34, Issue 12, pp. 2706-2713

[3] S. Sundarapandian, 2013, *Experimental and Theoretical Evaluation of the Performance Parameters and Emission Characteristics of Bio Diesel Using C.I Engine for Various Injection Pressure*, Journal of Petroleum Technology and Alternative Fuels, Vol. 4(7), pp. 115-120.

[4] B. R. Moser, A. Williams, M. J. Haas and R. L. Mccormick, 2009, *Exhaust Emissions and Fuel Properties of Partially Hydrogenated Soybean Oil Methyl Esters Blended with Ultra Low Sulfur Diesel Fuel*, Fuel Processing Technology, Vol. 90, pp. 1122-1128.

[5] M. Shahabuddin, A. M. Liaquat, H. H. Masjuki, M. A. Kalam and M. Mofijur, 2013, *Ignition Delay, Combustion and Emission Characteristics of Diesel Engine Fueled With Biodiesel*, Renewable and Sustainable Energy Reviews, Vol. 21, pp. 623-632.

[6] H-P. Liu, S. Strank, M. Werst, R. Hebner and J. Osara, 2010, *Combustion Emissions Modeling and Testing of Neat Biodiesel Fuels*, Proc. of the ASME 2010 4th International Conference on Energy Sustainability, May 17-22, Pheonix, AZ, USA, pp. 1-10.

[7] P. B. Rampure, C. V. Reddy and K. V. Reddy, 2012, *Effect of Injection Pressure on the Emission and Performance of Diesel Engine Using Non Edible Rice Bran Oil [NERBO]*, International Journal of Engineering Research and Development, Vol. 4, Issue 3, pp. 28-34

[8] R. Selvan and K. Maniysundar, 2014, *Performance and Emission Analysis of Single Cylinder Diesel Engine Using Jatropha Oil With EGR*, International Journal of Science, Engineering and Technology Research, Vol. 3, Issue 1, pp. 14-18.

[9] D. Balajee, G. Shankarnarayanan, P. Harish and N. Jeevarathinam, R. Jalili Saffar, M. R. Razfar, A. H. Salimi, and M. M. Khani, 2009, *Optimization of Machining Parameters to Minimize Tool Deflection in the End Milling Operation Using Genetic Algorithm*, World Applied Sciences Journal, Vol. 6, No. 1, pp. 64-69.

[10] H. Chotai, 2013, *Review on Effect of Varying Compression Ratio on Performance & Emissions of Diesel Engine Fueled With Bio-diesel*, International Journal of Engineering Science and Innovative Technology, Vol. 2, Issue 4, pp. 561-567.

Experimental Investigation of process parameters influence on machining Inconel 800 in the Electrical Spark Eroding Machine

K.Karunakaran[1,a] and M.Chandrasekaran[2,b]

[1]Research Scholar & Assistant Professor, Vels University, Chennai,TN, India
[2]Professor& Director, Dept. of Mechanical Engineering, Vels University, TN, India.

[a]ksivasakthi1966@gmail.com; [b]dr.chandrasekar@yahoo.in

Abstract. The Electrical Spark Eroding Machining is an entrenched sophisticated machining process for producing complex geometry with close tolerances in hard materials like super alloy which are extremely difficult-to-machine by using conventional machining processes. It is sometimes offered as a better alternative or sometimes as an only alternative for generating accurate 3D complex shapes of macro, micro and nano-features in such difficult–to-machine materials among other advanced machining processes. The accomplishment of such challenging task by use of Electrical Spark Eroding Machining or Electrical Discharge Machining (EDM) is depending upon selection of apt process parameters. This paper is about analyzing the influencing of parameter in electrical eroding machining for Inconel 800 with electrolytic copper as a tool. The experimental runs were performed with various input conditions to process Inconel 800 nickel based super alloy for analyzing the response of material removal rate, surface roughness and tool wear rate. These are the measures of performance of individual experimental value of parameters such as pulse on time, Pulse off time, peak current. Taguchi full factorial Design by using Minitab release 14 software was employed to meet the manufacture requirements of preparing process parameter selection card for Inconel 800 jobs. The individual parameter's contribution towards surface roughness was observed from 13.68% to 64.66%.

1. Introduction

Electrical discharge machining is an unconventional Spark eroding machine, which removes metal from a specimen by means of Electrical sparks (heat energy). Electrically conductive materials are only machined in electrical discharge machining. Comparing to conventional method of machining, it can machine the high hardened material. But the material removal rate (MRR) is very low and good surface finish (less surface roughness). Tzeng and Lee[1], highlighted that the handling of process parameters in EDM is helping to get optimum value of MRR, TWR and SR. It is clearly stated that the effect of increasing the current will increase MRR. And MRR depends on pulse on & pulse off time also. Initially, increasing pulse on time reason for decreasing MRR. TWR is based on current. The peak current increases in directly proportional to the TWR. The surface roughness is increasing with increasing peak current and pulse on time. Jothimurugan et al.[2] studied that by varying the pulse off time, discharge current, voltage had significant of MRR. By giving the pulse on time of 400μs, pulse of time of 1600μs, current of 15A and voltage 40v was given to get a maximum MRR using response surface methodology (RSM) in EDM process. Ay et al.[3] varied tools configuration like hexagonal and square profile copper electrode for machining of Inconel 718 to investigate TWR, MRR and SR in micro EDM. Assarzadeh and Ghoreishi[4] investigated based on desirability value is 1.0 as the value is completely dependent on how closely the lower and upper limits are set relative to the actual optimum in EDM with alumina power mix in dielectric fluid. Gopalakannan and Senthilvelan[5] investigations were on machining of Al–SiC nano-composites, Kumar et al.[6] concentrated Electrical Discharge Machining of Al (6351)-5% SiC-10% B4CnHybrid Composite in their studies and Goyal and Choudhary[7] considered inconel 800 machining in WEDM Machining. But here inconel 800 material is preferred for investigation due to its wide application in high temperature and high pressure applications. Dhanabalan et al.[8] used dielectric fluid as clean kerosene in EDM of Al Alloy (Grade HE9). In this research the kerosene dielectric is enriched with servotherm EDM oil. In this research Taguchi full factorial design is used. The experimentations were completely done to design the parameter setting card to the manufacture.

2. Method and Experimental Setup

These experimental runs were conducted with three factors of three levels to investigate process parameters' influence in machining in die sinking EDM machine. The tool was electrolytic copper. The work piece was Inconel 800 with servotherm EDM oil enriched kerosene as dielectric medium. Herein, the influences of machining parameters (Current, pulse on time and pulse off time) on machining performances are discussed. Material removal rate (MRR) and Tool Wear Rate were observed through the weight loss method for both work piece and tool for finding MRR and TWR. In this experiment, the work piece with size of ϕ22x20mm Inconel 800. Additionally, the tool material

Table 1 Parameter Description

Working parameter	Description
Work piece	Inconel 800
Tool	Electrolytic copper
Dielectric medium	kerosene enriched with Servotherm EDM oil
Peak Current	5,10 &15 (A)
Pulse on time	6,7& 8 (μs)
Pulse off time	3,4 & 5 (μs)
Pulse off time	5 (μs)
Supply voltage	240 (V)
Polarity	Straight

(ϕ25x23mm) was electrolytic copper The experiments have been performed on Electronica Machine Tools make Xpert1 model die sinking type CNC EDM machine. The Taylor Hobson makes, Surtronic3+ branded contact type profile-meter was employed in surface roughness measurement with 0.8mm cut length and sampling number is three. The laboratory balance of semi micro with an accuracy of 0.00001g, was employed in measuring weight loss tool and work material in before and after every run to compute MRR and TWR. Here pulse on time, Pulse off time and peak current value has been varied, to analyze the responses of MRR, SR and TWR with electrolytic copper as tool for machining 5 minutes.

3. Experimentation

At this experimental setup, Pulse on value has been maintained at 6 μs, Pulse off time 3μs and Peak current value is varied 5A, 10A and 15A. The MRR, SR and TWR have been observed after machining 5 minutes. Then similarly the experimentations were carried out for Pulse off time 4μs and for 5μs for peak current 5A, 10A and 15A at 6 μs. The observations recorded for pulse on time 6 μs in Table 2, for 7 μs in Table 3 and for 8 μs in Table 4.

Table 2 Observation for the Experimental Set 1

RUN	Current (A)	Pulse On (μs)	Pulse Off (μs)	MRR (g/min)	TWR (g/min)	SR (μm)
1	5	6	3	0.07219	0.00037	1.21
2	10	6	3	0.12975	0.00057	1.30
3	15	6	3	0.18745	0.00117	1.43
4	5	6	4	0.08240	0.00042	1.39
5	10	6	4	0.15471	0.00066	1.49
6	15	6	4	0.23205	0.00130	1.57
7	5	6	5	0.08968	0.00055	1.63
8	10	6	5	0.16180	0.00076	1.76
9	15	6	5	0.25684	0.00147	1.86

Table 3 Observation for the Experimental Set 2

RUN	Current (A)	Pulse On (μs)	Pulse Off (μs)	MRR (g/min)	TWR (g/min)	SR (μm)
10	5	7	3	0.30573	0.00138	1.34
11	10	7	3	0.34230	0.00228	1.39
12	15	7	3	0.37632	0.00310	1.67
13	5	7	4	0.32676	0.00174	1.65
14	10	7	4	0.36833	0.00269	1.78
15	15	7	4	0.43249	0.00342	1.98
16	5	7	5	0.34753	0.00225	1.93
17	10	7	5	0.38756	0.00300	2.09
18	15	7	5	0.44281	0.00388	2.19

Table 4 Observation for the Experimental Set 3

RUN	Current (A)	Pulse On (μs)	Pulse Off (μs)	MRR (g/min)	TWR (g/min)	SR (μm)
19	5	8	3	0.50210	0.00352	1.35
20	10	8	3	0.61234	0.00467	1.51
21	15	8	3	0.64130	0.00635	1.68
22	5	8	4	0.53725	0.00407	1.72
23	10	8	4	0.63530	0.00557	1.89
24	15	8	4	0.67268	0.00705	2.05
25	5	8	5	0.56255	0.00486	2.01
26	10	8	5	0.67892	0.00641	2.25
27	15	8	5	0.74297	0.00859	2.54

4. Result and Discussion

The parameters influence investigation was carried out in three aspects such as the influence of input parameters with respect to the performance of MRR, TWR and SR. The first row reveals the MRR with respect to Pulse on Time in the order of 6μs, 7μs and 8μs. It is observed that if the peak current increases the MRR increase linearly as well as the increase of pulse on time. But the MRR response with respect to pulse off time minimal compared to peak current. If the pulse on time increases the MRR increase steeply. Similarly second row the MRR for Pulse on Time 6μs, 7μs and

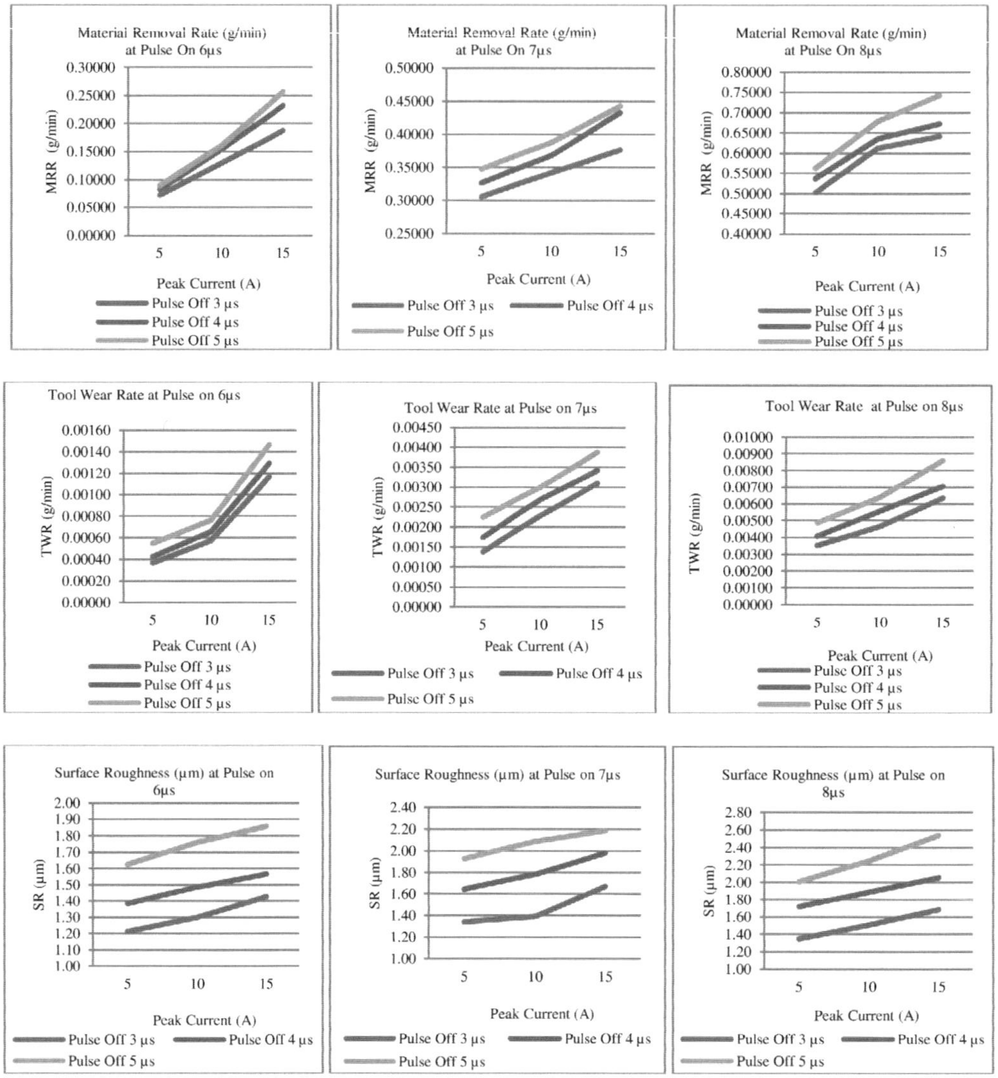

Figure 1 Influence of Input Parameters

8μs. The TWR increases with increase of peak current, as well as pulse on time. The increase, if pulse off time. the increases of TWR less than MRR. The TWR is also increases with increase of pulse on time as well as peak current and some extent to increase of pulse off time. The third row is for SR, in which the pulse off time also influence in the SR and it is directly proportional to SR. The increase of pulse on time steeply increases the surface roughness. The significant differences were observed with increases of peak current. The influence of pulse off time is increase of pulse off time increases the SR.

5. Conclusion

The experimental study of influence of input parameters like pulse on time, pulse off time and peak current in EDM on inconel 800 was discussed. The pulse on time and peak current contribution is more than pulse off time. The influences of various parameters were analyzed in detail in results and discussion. The observation table can be used as a parameter selection card to obtain the suitable roughness. The contribution is the ratio of individual experimentation set rage of surface roughness variation to total surface roughness range of variation. Hence the lowest variation was obtained at pulse on time 6μs, Pulse Off time 4 μs at 13.68%.

References

[1] Y.-F. Tzeng and C.-Y. Lee, 2001. Effects of Powder Characteristics on Electro discharge Machining Efficiency, International Journal Advanced Manufacturing Technology (2001) 17:586–592

[2] R. Jothimurugan, K.S. Amirthagadeswaran and Joel Daniel, 2012. Performance of Silver Coated Copper Tool with Kerosene-Servotherm Dielectric in EDM of Monel 400 TM. Journal of Applied Sciences 12 (10): 999-1005.

[3] M. Ay, U. Çaydaş and A. Hasçalık, "Optimization of micro-EDM drilling of inconel 718 Superalloy," Int J Adv Manuf Technol 66:1015–1023, (2013).

[4] R. S. Bharathi, and N. Baskar, "Optimization techniques for machining operations: A retrospective research based on various mathematical models," International Journal of Advanced Manufacturing Technology, 48(9–12),1075–1090 (2010).

[5] R. V. Rao, and P. J. Pawar, "Parameter optimization of a multi-pass milling process using non-traditional optimization algorithms," Applied Soft Computing, 10(2), 445–456. ISSN 1568-4946, DOI: 10.1016/j.asoc.2009.08.007, (2010).

[6] S. Yang, J. Srinivas, S. Mohan, D.M. Lee and S. Balaji, "*Optimization of electric discharge machining using simulated annealing*," Journal of Materials Processing Technology, Volume 209(9) (pp. 4471–4475). ISSN 0924-0136, DOI: 10.1016/j.jmatprotec.2008.10.053, (2009)

[7] H.Chen, J.Lin, Y.Yang, and C.Tsai, "Optimization of wire electrical discharge machining for pure tungsten using a neural network integrated simulated annealing approach," Expert Syst. Appl., 37(10), 7147–7153 (2010).

[8] M. Zain, H. Haron and S. Sharif, "Simulated Annealing To Estimate The Optimal Cutting Conditions For Minimizing Surface Roughness In End Milling Ti-6Al-4V," Machining Science and Technology, 14, 43–62, (2010a).

Engineering Change Management Method Framework in Mechanical Engineering

Alexander Stekolschik

Engineering Design and CAD/CAE, Cologne University of Applied Sciences, Betzdorfer Str. 2, 50679 Cologne, Germany

alexander.stekolschik@th-koeln.de

Abstract. Engineering changes make an impact on different process chains in and outside the company, and lead to most error costs and time shifts. In fact, 30 to 50 per cent of development costs result from technical changes. Controlling engineering change processes can help us to avoid errors and risks, and contribute to cost optimization and a shorter time to market. This paper presents a method framework for controlling engineering changes at mechanical engineering companies. The developed classification of engineering changes and accordingly process requirements build the basis for the method framework. The developed method framework comprises two main areas: special data objects managed in different engineering IT tools and process framework. Objects from both areas are building blocks that can be selected to the overall business process based on the engineering process type and change classification. The process framework contains steps for the creation of change objects (both for overall change and for parts), change implementation, and release. Companies can select single-process building blocks from the framework, depending on the product development process and change impact. The developed change framework has been implemented at a division (10,000 employees) of a big German mechanical engineering company.

1. Introduction into Engineering Change processes in Mechanical Engineering

There is a daily rise in product and process complexities relating to mechanical engineering, not the least because of stronger customer orientation and individualization of customer-specific solution packages [1]. Additionally, increasing quality standards and shortened delivery periods have necessitated the introduction of integrated processes in the product lifecycle, in combination with high process reliability and stability. Manufacturers must respond quickly with engineering changes to maintain and increase their market shares. Such changes might be necessary to respond to market demand, governmental requirements, safety issues, service requirements, or functional and competitive reasons. Changes can be initiated in the different operative structure levels and parts (Figure 1) of the product lifecycle [1].

Such changes impact different process chains in and outside the company, and result in most error costs and time shifts. Most of the operative product structures are dynamic during the product lifecycle and will be reworked due to changed customer requirements. In early stages of product development, correction loops are intentionally planned to optimize the product and to make it ready for series production. Therefore, an integrated change management process is important for all phases of the product lifecycle, thereby automatically involving all affected parties across product definition, development, and manufacturing.

IOP Conf. Series: Materials Science and Engineering **157** (2016) 012008 doi:10.1088/1757-899X/157/1/012008

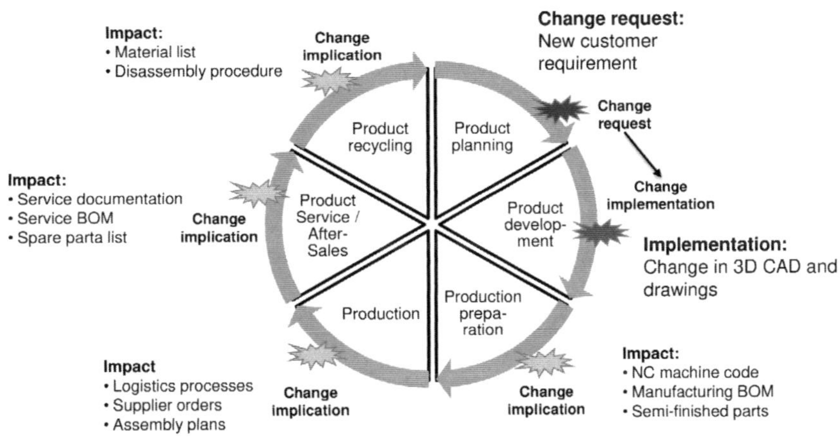

Figure 1. Engineering changes within the product lifecycle

Controlling change processes can help us to avoid errors and risks; it contributes to the cost optimization and a shorter time to market. According to research studies [1,2], up to 85 per cent of companies report that their change management procedures are either broken or could be improved. Change management, when done effectively, provides a major opportunity for improving development efficiency. Change management can lead to 'lessons learned' and process optimization. When done incorrectly, it can lead to delay as stakeholders lose time in waiting for decisions and updated information or work that must be redone because it was based on outdated design details. It can also lead to higher costs when errors lead to scrap and rework.

Engineering change management can be defined according to different standards [3] such as the coordinated management and uniform tracking of changes—i.e., collecting ideas or need for product and product-related changes, elaborating one or more possible solutions, evaluating them with respect to technical and cost aspects, and implementing them with respect to engineering and manufacturing [3]. The ECM process can be broken down into a number of key stages as follows [4, 5]:

- Identification of engineering change requirement, evaluation of benefits, implications, and affected products
- Preliminary authorization of change request
- Engineering of detailed solution
- Final authorization of the proposal
- Implementation of the solution and final review

2. Engineering change classification and requirements on engineering change management in product data management systems in mechanical engineering

The first step of the research has been the classification of different change types. In different projects, conducted at the institution, engineering changes were analysed and classified (Figure 2). The main criteria are change impact and change implementation effort. The impact of a change can be the adjustment or even complete stop of production, initialization, or change of purchase orders. Moreover, external parties can be affected. The change can lead to production stops at suppliers or new plant designs at customer site. The change implementation effort mainly depends on the number of affected parts and assemblies, the required change activities, and the impact on other products. The consequence for an enterprise is the risk related to high costs. Both change efforts and generated process chain costs (change impact) lead to the monetary project risk. The management demands a clear and transparent classification of the change risk, which can be provided by the ABC analysis (Figure 2).

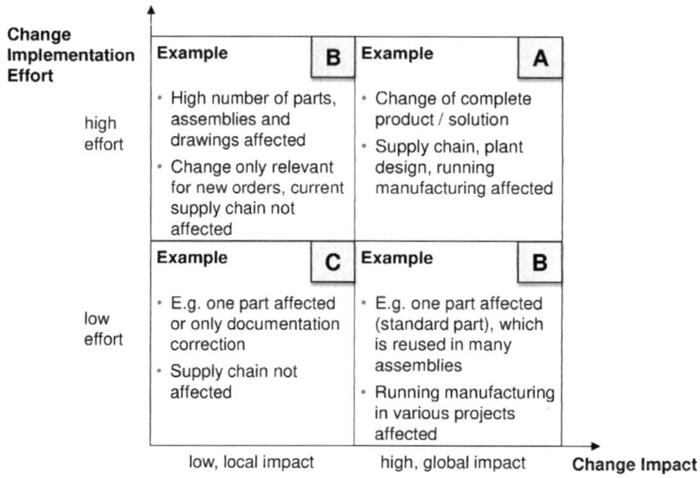

Figure 2. Engineering change classification

The change categories can be defined as follows:

- **'A' changes**: high and very high costs both within and outside the enterprise, internal, and external supply chain is affected
- **'B' changes**: medium change costs, caused by high change implementation efforts or the affected supply chain
- **'C' changes**: minor or no cost affects, local change implementation

Both classifications are the basis for the engineering change method framework. The next step is the definition of the requirements on the engineering change support in engineering IT tools. Today, the storage and management of product data and engineering changes is dependent on the used data management system like product data management (PDM) or enterprise resource planning (ERP) systems [6]. Additionally, specific product data can be found within product configuration and sales systems, manufacturing systems, or service systems. Each of these systems uses their own use case specific form of product representation and storage. In engineering systems, primarily, technical product structures are maintained, while logistical systems focus on process-related structures (e.g., order transactions) and cost structures. Especially business processes producing mechanical product enterprises are the basis for the main requirements. The following requirements of the engineering change management were acquired from engineering departments within the product creation chain:

- Management of problem reports, change requests, change orders, consequences and impacts
- Support of different change complexities (fast, standard, complex, complete)
- Workflow support, involvement of all stakeholders and decision-makers, access to all relevant product data, decisions and the current valid product configuration
- Legally compliant documentation and traceability of all changes and decisions
- Status-driven synchronization of revisions and validity between IT systems (PDM and ERP)

The complex engineering-to-order product development process [7] leads to additional requirements regarding change management, both internally and with suppliers and customers. Product and change data has to be available for decades due to legal requirements and service needs.

3. Development of a method framework

The developed method framework consists of two main areas: special data objects managed in different engineering IT tools and process framework. Objects from both areas are building blocks that can be selected to the overall business process based on the engineering process type and change classification.

3.1. Development of data objects for engineering IT systems

Data objects in engineering systems store part and process information related to engineering changes. Research projects have identified and proved the need for different types of necessary change data objects (Figure 3). The developed data objects are created and managed in data management systems within the engineering change process, independent of the change complexity.

- **Change-related data**: The change object is an 'information folder' that contains information necessary for the management and control of changes. The change object contains data of a descriptive character (such as the reason for the change) and control data (such as effectivity data, object type indicators). Besides this data that the user maintains, there is also data that is automatically updated by the system (management data). Change objects should be identified by a unique change number. Number Assignment is made according to the defined criteria. Such objects contain information to all parts, assemblies, and other product data related to the change.
- **Part-related data**: Some change information concerns only singular parts or assembly. Such data is known as part change description, which describes modifications of a concrete part in detail. Such data cannot be applied to overall product change because it does not concern other parts or assembly. In many cases, part-related change data is linked with the overall change-related data (overall change objects).
- **Run-time-objects**: Stakeholders within the product creation process often require different types and scopes of change's describing information. In many cases, additional information, such as project performance indicators or cost overviews, is needed to the change description. It can be implemented by using run-time-objects like business reports. Such reports contain information from different sources—e.g., change objects, part descriptions, and cost database. These reports can be defined with stakeholder requirements on the information scope.

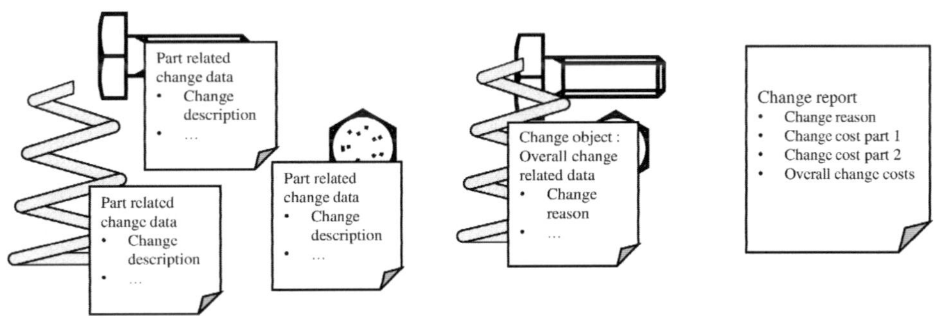

Figure 3. Types of Engineering Change data objects

3.2. Engineering change process framework

The developed engineering change process framework is the generic business process through engineering IT tools. It is the so-called end-to-end process, beginning at the change initiation and ending at the change impact implementation at production (Figure 4).

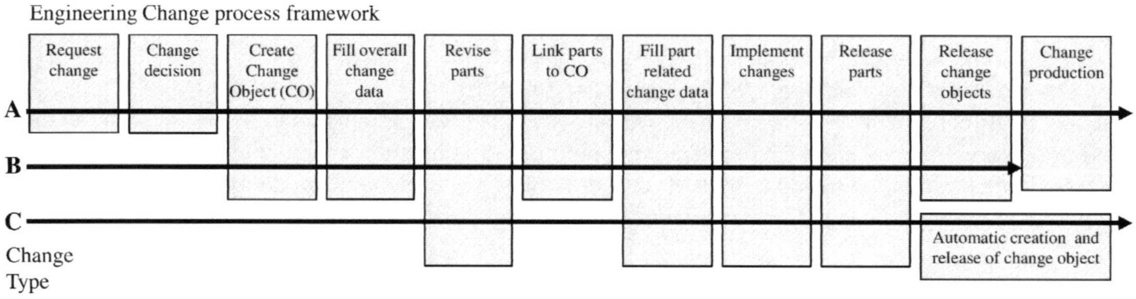

Figure 4. Use of the engineering change process framework

The process framework contains steps for the creation of change objects (both for overall change and for parts), change implementation, and release. Companies can select single process building blocks from the framework, depending on the product development process and change impact. Complex changes (Classification A) will be carried out by using all the steps for the effective change control, while changes pertaining to Classification B could be done without formal change request by the direct change start at the department. Changes pertaining to Classification C without impact can be done without creating the overall change object by only filling part change descriptions (the overall change object is created for each part automatically for tracking purposes in this case). The flexibility of the developed framework enables its use at both small and big enterprises.

4. Implementation and outlook

The developed change framework was implemented at a division (10,000 employees) of a big German mechanical engineering company. Engineering Systems NX (for CAD modelling), Teamcenter (PDM), and SAP (ERP) were used. Special system-based process workflows were implemented for different change types (A-B-C). Guidelines and access right system supported the selection of the proper workflows. The list of affected parts and departments were managed in the workflow so that automatic processes at production and the supply chain could be started on demand. Various implemented reports were created for project managers and the management. The implementation proved the practical relevance and usability of the framework.

Next steps in its research and implementation will be developing algorithms for automatic selection of change types, depending on different criteria (e.g., part referencing in assembles or production processes). Moreover, the process framework will be extended for being used outside the enterprise—e.g., for engineering change processes involving engineering and production suppliers and customers.

References

[1] Stekolschik A and Schulte 2014 Approach for an integrated reference structure in the product lifecycle 2014 *Tools and methods of competitive engineering* 1053
[2] Burneika L 2009 Managing Changes in Product Configurations *Solid State Phenomena* **144** 208
[3] VDA 2010 *Engineering Change Management* Verband der Automobilindustie
[4] Lashin G 2006 *Engineering Change Management für globale Produktentwicklung* Product Life live
[5] Malatesta M Raffaelo R Mengoni M and Germani M 2013 Supporting the modification process of products through a change management tool *International Conference on Engineering Design - ICED* p 444
[6] Vajna S 2014 *Integrated Design Engineering* (Berlin: Springer) p 433
[7] Feldhusen J and Grote K-H 2013 *Konstruktionslehre: Methoden und Anwendung erfolgreicher Produktentwicklung* (Berlin: Springer) p 110

Gait Generation for a Small Biped Robot using Approximated Optimization Method

Tinh Nguyen, Linh Tao, and Hiroshi Hasegawa

Graduate School of Engineering and Science, Shibaura Institute of Technology, 307 Fukasaku, Minuma-ku, Saitama-shi, Saitama, 337-8570, Japan

E-mail: mf15422@shibaura-it.ac.jp

Abstract. This paper proposes a novel approach for gait pattern generation of a small biped robot to enhance its walking behavior. This is to aim to make the robot gait more natural and more stable in the walking process. In this study, we mention the approximated optimization method which applied the Differential Evolution algorithm (DE) to objective function approximated by Artificial Neural Network (ANN). In addition, we also present a new human-like foot structure with toes for the biped robot in this paper. To evaluate this method achievement, the robot was simulated by multi-body dynamics simulation software, Adams (MSC software, USA). As a result, we confirmed that the biped robot with the proposed foot structure can walk naturally. The approximated optimization method based on DE algorithm and ANN is an effective approach to generate a gait pattern for the locomotion of the biped robot. This method is simpler than the conventional methods using Zero Moment Point (ZMP) criterion.

1. Introduction

Humans have a complicated physical structure and can implement difficult movements. Inspiration from human, during the past several decades, a number of researchers in the world have concentrated on the field of the biped robot (also known as humanoid robot) such as Asimo [1] of Honda, Johnnie [2] of Technical University of Munich, KHR robot series [3] of Kondo Kagaku Co., Qrio [4] of Sony Intelligence Dynamics Laboratory, Inc. The first aim of research carried out in the field of the biped robot attempts to solve the following problem: How can we build a biped robot prototype able to walk as the humans are doing.

To reach this target, there have been some studies to focus on generating gait patterns of the robot based on machine learning methods. For instance, Tang Z. et al. have proposed genetic algorithm based optimization for human walking [5]. Krissana K. and Hasegawa H. have developed the gait generation method based adaptive plan system with genetic algorithm [6]. Kuo P. H. et al. have presented the gait learning by using particle swarm optimization for humanoid robot [7]. Lee B. H. et al. have proposed optimal trajectory generation for a humanoid robot based on fuzzy and genetic algorithm [8]. Such the goal requires a biped robot that is considerably similar to humans in term of its mechanism and motion pattern. However, it cannot be said with certainty that the biped robots developed in recent studies realize a human-like walk. Therefore, the study on robot walking is still continuing.

In this study, to get flexible and natural gait pattern for a small biped robot when it walks on flat ground, we propose a novel approach for gait generation problem based on approximated optimization method. We confirmed the success of this approach through the dynamic simulation of the robot walking process by Adam software.

IOP Conf. Series: Materials Science and Engineering **157** (2016) 012009 doi:10.1088/1757-899X/157/1/012009

2. Simulation model

2.1. Proposed model
In this paper, the KHR-3HV robot of Kondo Kagaku Company was used to build an experiment model. Figure 1 shows the picture of the real robot and the simulation model.

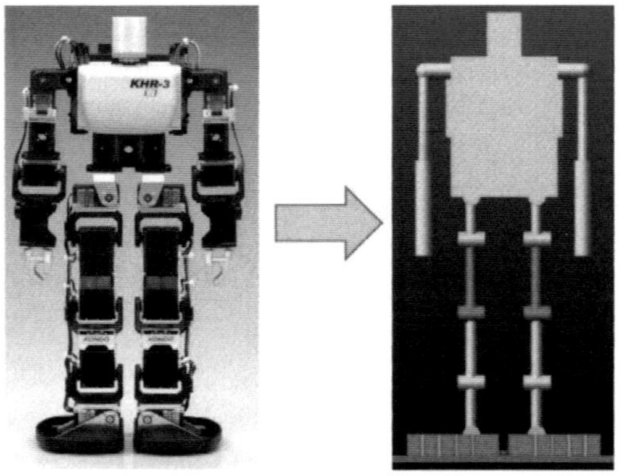

Figure 1. Real robot and simulation model

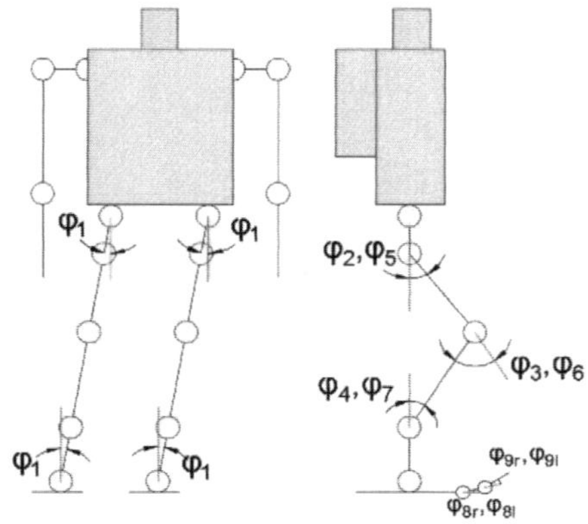

Figure 2. Robot linkage model

A new foot structure composes of heel and many toes is proposed in this study. Each foot has 10 degree of freedom as described in Figure 2. As proven in [9], the biped robot whose big toe width ratio per foot equals 0.28, has the longest walking distance when big toe length is fixed and this ratio is similar to the ratio of human's foot. Thus, the width of big toes and foot were designed of 22mm and 78mm, respectively. The other smaller toes have the same width of 12mm.

In [10], Chockalingam et al. have proved that an average ratio between the foot length and heel varies from 1.196 to 1.426. Thus, in this paper, the length of heel and foot are 95mm and 123mm, respectively. To reduce energy consumption, toe joints were applied by torsion spring that based on principles of torsion and twisting. Spring stiffness coefficient is respectively set of 0.26N.mm/deg, 0.52N.mm/deg to lesser toe joint and big toe joint.

2.2. Motion pattern
Basing on the human walking pattern as depicted in [11], we assumed the robot control data was generated by the gait function as trigonometric function shown in Equation (1). By changing a, b, c, d coefficients, the gait functions will be created to allocate to each joint of the biped robot.

$$\varphi_i(t) = a_i + b_i\cos(\omega t) + c_i\sin(\omega t) + d_i\cos(2\omega t) \qquad (1)$$

Where a, b, c, d are coefficients, t is time, ω is angular velocity and i is index of joint.

3. Optimization procedure

3.1. Gait generation
In this study, one cycle was set up to 1.2 seconds. Therefore, angular velocity was calculated as described in Equation (2). We simulated the biped robot in 3 cycles which spent on 3.6 seconds and 1.2 seconds was used for checking robot stability. In simulation, one step took 0.02 second, so the total number of step were 240.

$$\omega = \frac{2\Pi}{1.2} = 5.236\ (\text{rad/s}) \qquad (2)$$

Gait functions are assigned to each joint as known in Equation (3-9).

$$\varphi_1 = \begin{cases} 0, \ t = 0 \text{ or } t \geq 3.6 \\ \pm 1.5, \ t = 0.3 \text{ and } t = 3.3 \\ \varphi_1(t), \ 0.3 < t < 3.3 \end{cases} \quad (3)$$

$$\varphi_5 = \begin{cases} 0, \ t = 0 \text{ or } t \geq 3.3 \\ 15, \ t = 0.3 \\ \varphi_2(t), \ 0.3 < t < 3.3 \end{cases} \quad (7)$$

$$\varphi_2 = \begin{cases} 0, \ t \leq 0.3 \text{ or } t \geq 3.6 \\ \varphi_2(t+0.6), \ 0.3 < t < 3.3 \\ 15, \ t = 3.3 \end{cases} \quad (4)$$

$$\varphi_6 = \begin{cases} 0, \ t = 0 \text{ or } t \geq 3.3 \\ 30, \ t = 0.3 \\ \varphi_3(t), \ 0.3 < t < 3.3 \end{cases} \quad (8)$$

$$\varphi_3 = \begin{cases} 0, \ t \leq 0.3 \text{ or } t \geq 3.6 \\ \varphi_3(t+0.6), \ 0.3 < t < 3.3 \\ 30, \ t = 3.3 \end{cases} \quad (5)$$

$$\varphi_7 = \begin{cases} 0, \ t = 0 \text{ or } t \geq 3.3 \\ 15, \ t = 0.3 \\ \varphi_4(t), \ 0.3 < t < 3.3 \end{cases} \quad (9)$$

$$\varphi_4 = \begin{cases} 0, \ t \leq 0.3 \text{ or } t \geq 3.6 \\ \varphi_4(t+0.6), \ 0.3 < t < 3.3 \\ 15, \ t = 3.3 \end{cases} \quad (6)$$

3.2. Problem formulation

The concept of the optimization process is shown as in Figure 3. This figure describes the parameters of the optimization process. Where Z_f and X_f denote the distances of the biped robot from initial position to final position along z-axis and x-axis. R_f is angle of rotation. Design variable vector, objective function, constraint functions and penalty function are defined as described in Equation (10-16).

Design variables (DVs):

$$x = [a_1, b_1, c_1, d_1, a_2, b_2, c_2, d_2, a_3, b_3, c_3, d_3, a_4, b_4, c_4, d_4] \quad (10)$$

Constraint functions:

$$g_1(x) = 20 - |X_f| \geq 0 \quad (11))$$

$$g_3(x) = 243 - Y_f \leq 0 \quad (13)$$

$$g_2(x) = 5 - |R_f| \geq 0 \quad (12))$$

$$h(x) = N - 240 = 0 \quad (14)$$

Penalty function:

$$P(x) = \sum_{i=1}^{3} max(g_x(x), 0) + h(x) \quad (15)$$

Modified objective function:

$$F(x) = - Z_f + \gamma.P \rightarrow \min \quad (16)$$

The objective function is minimized. Where a_i, b_i, c_i, d_i (i= 1,2,3,4) variables are coefficients of gait function. There are four constraint functions. In Equation (11-12), X_f distance and R_f angle are constrained under ±20mm and ±5° to ensure that the biped robot can walk straight. In Equation 13, Y_f which is the distance from Center of Mass (CoM) to the ground, must be more than 243mm to ensure the robot not to slip and fall down at the final framework. In Equation 14, N is a total simulation step which is equal 240 to check the success of the simulation. In Equation (16), γ is the penalty coefficient set to 1000. Equation (11-14) will be also checked again when the simulation finishes.

3.3. Optimization solution

The optimization process was implemented by the steps as shown in Figure 4.

4. Simulation result

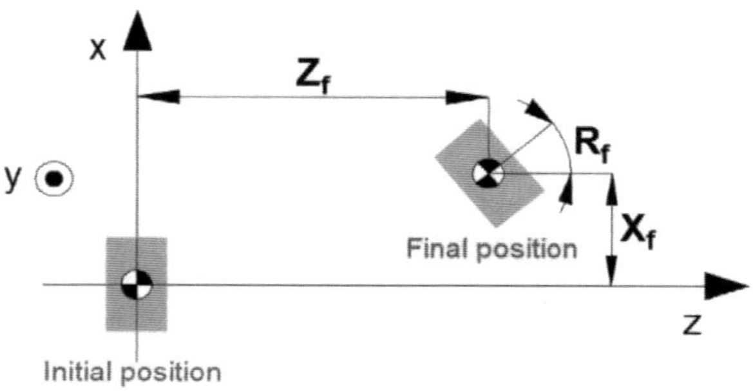

Figure 3. Overview of optimization

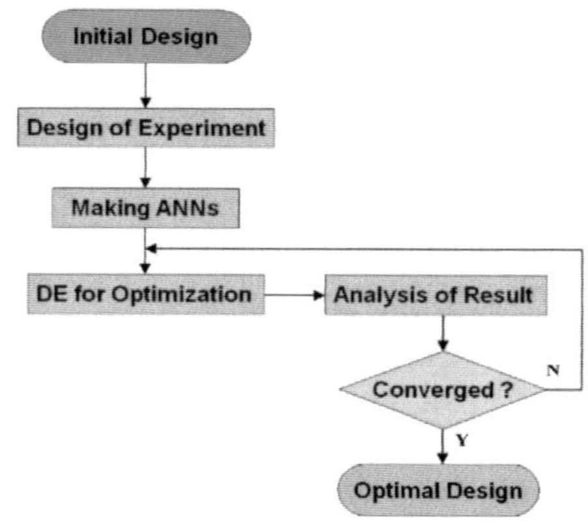

Figure 4. Approximated optimization process

The optimal result is shown as in Figure 5 with X_f, Z_f, R_f of 0.044mm, 177.96mm and 1.203°, respectively. The trajectory of the robot's CoM is depicted as in Figure 6. As a result, the trajectory of the CoM is close to a waveform of a circular function which is similar to the trajectory of the human's CoM.

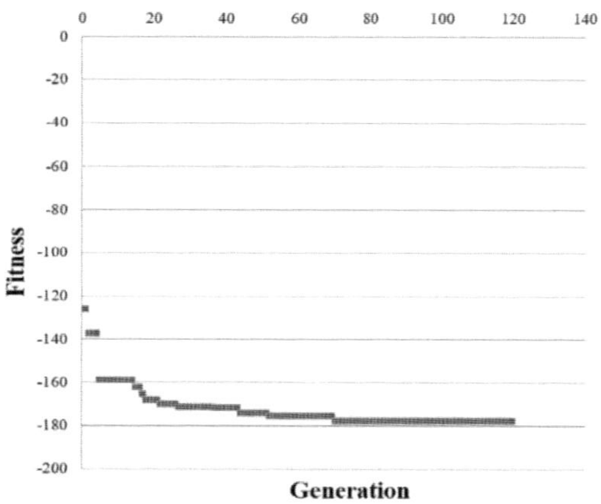

Figure 5. Optimization process convergence

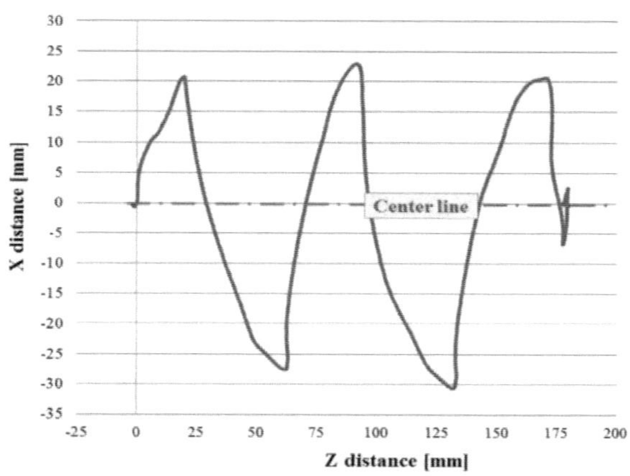

Figure 6. Robot's CoM trajectory

The waveform of the gait function allocated to all joints of the biped robot are shown in Figure 7. We can see that the waveforms of hip and knee joint gait functions are comparable to that of human.

5. Conclusion

In this paper, to enhance walking gait of a small biped robot in the locomotion on flat ground, a novel gait pattern generation method is proposed. The gait pattern is generated by the approximated optimization method based on DE algorithm and ANN. We verified the effectiveness of this approach by dynamic simulation on Adams software. Moreover, the new gait pattern generation method reduced the complexity of the gait pattern generation problem for the biped robot in comparison to the conventional methods using ZMP criterion.

Finally, this study plans to do improvements with the optimization for ankle joint position and length of robot toe.

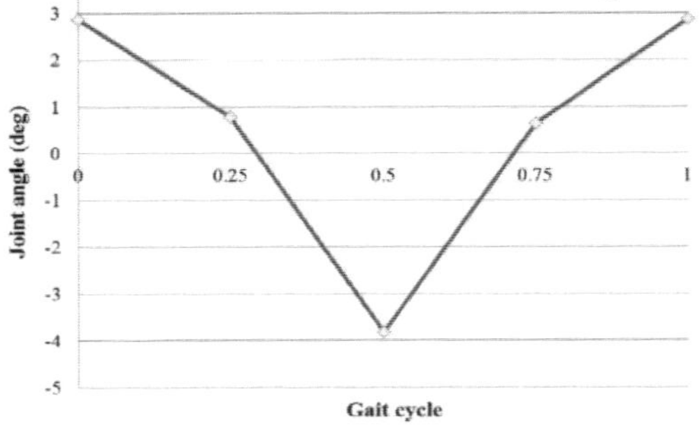

(a) A cycle of gait function (hip and ankle roll joint angle)

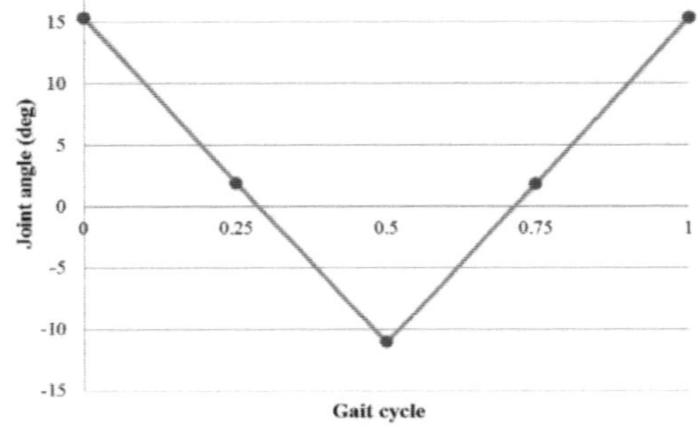

(b) A cycle of gait function (hip pitch joint angle)

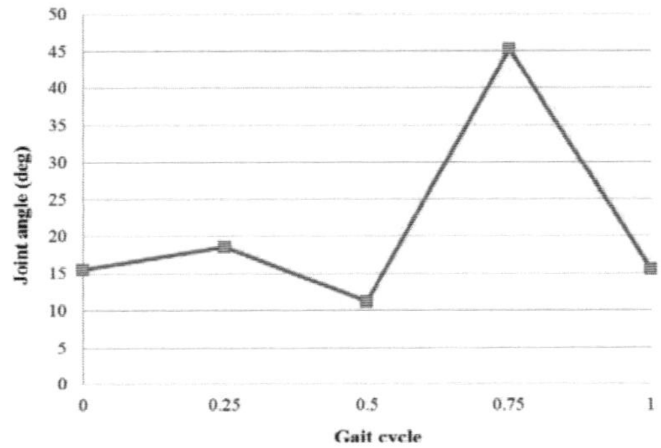

(c) A cycle of gait function (knee pitch joint angle)

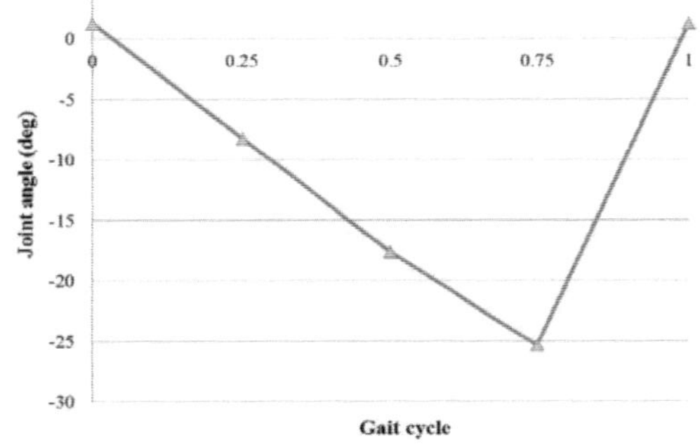

(d) A cycle of gait function (ankle pitch joint angle)

Figure 7. Waveform of the gait functions

References

[1] Sakagami Y, Watanabe R, Aoyama C, Matsunaga S, Higaki N, Fujimura K 2002 The intelligent ASIMO: System overview and integration. *Proc. of the IEEE/RSJ Int. Conf. on Intelligent Robots and Systems,* Switzerland, pp 2478-83

[2] Lohmeier S, Loffler K, Gienger M, Ulbrich H, Pfeiffer F 2004 Computer system and control of biped "Johnnie". *Proc. of IEEE Int. Conf. on Robotics and Automation,* New Orleans, LA, vol 4 pp 4222-27

[3] KHR-3HV biped robot. Website: http://kondo-robot.com/product/khr-3hv-ver-2-life

[4] Ishida T 2004 Development of a Small Biped Entertainment Robot QRIO. *Proc. of The 4th Symp. Micro-Nanomechatronics for Information – Based Society,* pp 23-28

[5] Tang Z, Sun Z, Zhou C 2006 GA Based Optimization for Humanoid Walking. *ICGST-ARAS Journal,* **5** 1-10

[6] Nerakae K, Hasegawa H 2011 Simulation Based Design Optimization Framework for a gait pattern generation of a small biped robot with tiptoe mechanism. *The 10th Int. Conf. on Modeling and Applied Simulation,* Rome, Italy, pp 295-302

[7] Kuo P H, Ho Y F, Lee K F, Tai L H, Li T H S 2013 Development of Humanoid Robot Simulator for Gait Learning by Using Particle Swarm Optimization. *Proc. of IEEE Int. Conf. on Systems, Man and Cybernetics (SMC),* pp 2684-88

[8] Lee B H, Kong J S, Kim J G 2006 Optimal Trajectory Generation for a Humanoid Robot Based on Fuzzy and Genetic Algorithm. *IEEE Congress on evolutionary computation,* Canada, pp 1968-74

[9] Nerakae K, Hasegawa H 2014 Big toe sizing design of small biped robot by using gait generation method. *Applied Mechanics and Materials,* **541-542** 1079-86

[10] Chockalingam N, Ashford R 2007 Selected foot length ratios in a non-pathological sample. *Revista Internacional de Ciencias Podológicas,* **1** 25-30

[11] Whittle M W 2007 *An Introduction to Gait Analysis.* 4th ed., Oxford, p 59

Design of wireless temperature measurement system for piston based on CC2530

Ge Rongyu[1,*], Wang Jin[1], Wu Yimin[2], Wu Guodong[2],

[1]School of Mechanical Engineering, University of Jinan, Jinan 250022, China
[2]Shandong Binzhou Bohai Piston Co., Ltd, Binzhou 256602, China

* me_gery@163.com

Abstract. Temperature measurement of piston is important to analysis its temperature field. Since it is difficult to extract the temperature signal in the piston by hard line, a multi-channel wireless measurement system of piston temperature is developed. The wireless SOC CC2530 is used. NTC thermistors with glass package are selected as temperature sensors. The realization of hardware and the design of software flowchart are explained in detail. The actual test demonstrates that the system has high measurement accuracy and strong practical value.

1. Introduction

The piston, as the key part of the internal combustion engine, is always running with the largest thermal load. As piston durability has an important influence on the lifetime, reliability, emission performance and economy of the internal combustion engine, piston temperatures are useful information for engine development. However, the current limitations of piston-temperature measurement systems make accurate measurements a difficult task. The difficulties are mainly due to piston revolution speeds over 6,000 r/m, accelerations over 2000g and temperatures in excess of 300 degrees Celsius[1].

The piston temperature can be measured in five ways: templugs, contact method, electromagnetic induction ,telemetry and optical method. Templug[2], a traditional measurement tool, has a short and simple preparation but the disadvantages are that it can only measure a maximum temperature and its error range is quite large. Contact method[3] can get most accurate temperature measurement but a disadvantage is that many engine parts need to be redesigned and a rather long preparation time is required. By the electromagnetic induction method[4], the signal transmission is not continuous and limited at the BDC. In addition, the number of measurement channels is limited. By the telemetry method[5], the electronic circuits connected to the temperature sensor are installed in the piston. In this method, signals are transferred via various media, such as infrared rays, radio frequency (RF), Bluetooth and so on. The method for optical measurement[6] is an innovative method, which works by detecting the thermal radiation emitted from the hot piston surface. It does not need any contact with the measured surface and can derive the surface temperature without temperature gradient. But it needs complicated preparation and special instruments. The emergence of wireless system-on-chip (SOC) provides a new solution to measure the piston temperature making the system simple and easy to develop. A wireless temperature measurement system for piston based on CC2530 is developed. This measurement system can be installed in the in real commercial engines and thus is small and compact.

2. General design of measurement system

The wireless temperature measurement system for piston, as shown in Figure 1, is mainly composed of power unit, temperature signal acquisition unit, wireless transmitting unit, wireless receiving unit and host computer.

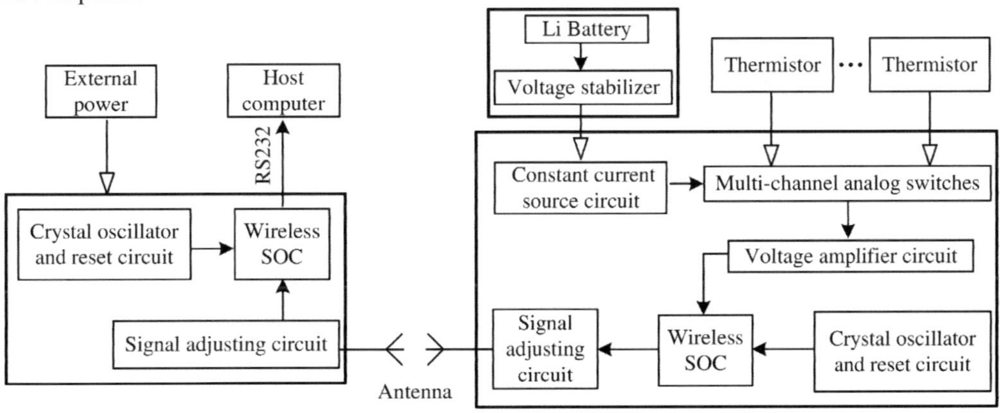

Figure 1. General design diagram of system

The power unit provides stable power for the whole system including Li battery and voltage stabilizer. The temperature signal acquisition unit is used to transfer the temperature signal in piston into voltage signal including temperature sensors, multi-channel analog switches, constant current source circuit and voltage amplifier circuit. Both the wireless transmitting and receiving unit use the wireless SOC CC2530. The voltage signal of piston temperature is adjusted with suitable level and connected to A/D pin of the master CC2530. After A/D transformation, the master CC2530 will transmit the numerical values to slave CC2530 with the point-to-point protocol. The slave CC2530 will sent the signals to host computer by the serial port. Eventually, the temperature valued will be arrived when the host computer gets and then processes the data with communication software.

3. Hardware design

3.1. Power unit design

The operating voltage is 3.3V for the whole system, so in this way it is avoided to convert different voltages and it have also some advantages such as low power consumption and easy development. As is shown in Fig.2, The system power is provided by Li battery with 3.6Voutput voltage and high temperature resistance. The stable 3.3V voltage can be output throughout the Power Management IC TPS63020.The TPS63020 is the industry's smallest and highest performance buck-boost converter with a 4-A switch and up to 96 percent efficiency, whose input voltage range is 1.8V to 5.5V, fit to manage the Li battery.

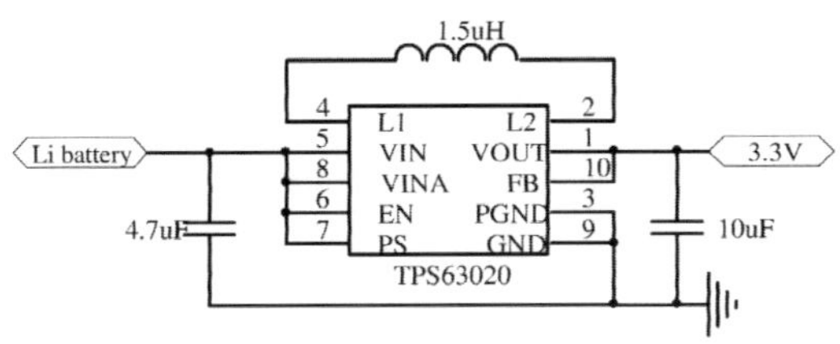

Figure 2. Schematic of power unit

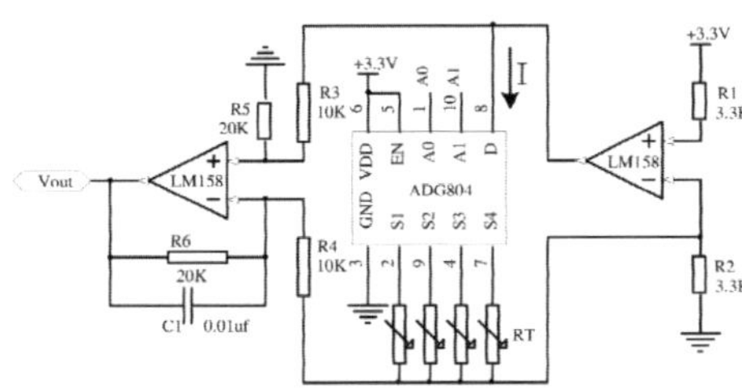

Figure 3. Schematic of temperature signal input unit

3.2. Temperature signal acquisition

3.2.1. Temperature sensor. There are three most widely-used temperature transducers: thermocouples, resistance temperature detectors (RTDs), and thermistors. A thermocouple produces a temperature-dependent voltage as a result of the thermoelectric effect, and this voltage can be interpreted to measure temperature. Thermocouples can measure a wide range of temperatures up to 1700°C, but the main limitation with thermocouples is bad accuracy and big volume. RTDs are made from a pure material, typically platinum. The material has a predictable change in resistance as the temperature changes and it is this predictable change that is used to determine temperature. RTDs have also a wide measurement range up to 800°C, but because of their big volume, they are not suitable to be assembled in pistons. Thermistors are similar to RTDs because they change resistance between their terminals with a change in temperature. However, they can be made with either a positive or negative temperature coefficient. The negative temperature coefficient (NTC) thermistor is widely used , which means that the resistance decreases with increasing temperature. Thermistors have small volume and are more sensitive and accuracy than RTDs but have a smaller temperature range up to 350°C. Since the assembling room is limited in piston and the maximum working temperature is less than 350°C, so NTC thermistors with glass encapsulated are used as the temperature sensors in this system. This kind of sensor with only 2mm diameter is easy to be assembled in piston and can also satisfy the temperature testing range of pistons.

The temperature inside pistons is close to the maximum test value of NTC and NTC has a much lower ratio of resistance change per °C in high temperature. Because the higher nominal resistance value, the higher ratio, so NTCs with 100K nominal resistance at 25°C, whose resistance is about 0.1KΩ with about 3Ω ratio of resistance change per °C at 300°C, are applied in this testing system.

3.2.2. Multi-channel analog switch Temperatures of many points inside piston need to be tested such as bottom of combustor, oil ring groove, skirt and pine hole. So method of time division multiple access is executed by applying the multi-channel analog switch IC ADG804, which decides voltages from all thermistors will be transited in turn to differential amplifier, as is shown in Fig.3. The ADG804 switches one of four inputs to a common output, as determined by the 2-bit binary address lines, A0 and A1.In this way, PCB of whole system has small volume and simple interface circuit. The inherent resistance of ADG804 is only 0.5Ω, almost no influence to the test results.

3.2.3. Make a Thermistor Measurement Because thermistors are resistive devices, you must supply them with an excitation source and then read the voltage across their terminals. This source must be constant and precise. Thermistors are usually made a measurement with either three-wire, four-wire or constant current source configurations. In order to improve the test accuracy and simply the design, constant current source supplied by a operational amplifier is adopted as depicted in Figure 3. The constant current can be derived as I=3.3/R2=1mA and R1 component is balance resistance. In this way, the resistance of thermistors can be transferred into voltage, and output voltage Vout derived after differential amplification is connected to A/D pin of CC2530.

The rang of piston temperature is 150°C through 300°C and during this range the change of thermistor resistance is from 1K to 1.5K and that of voltage is from 1V to 1.5V. So it works for A/D pin of CC2530 with twice of differential amplification. Because thermistors have a high nominal resistance, lead-wire resistance does not affect the accuracy of their measurements.

3.3. Wireless transmitting and receiving unit

Both the wireless transmitting and receiving unit have SOC CC2530 and its peripheral circuit and RF filter as shown in Fig.4. In addition, RS-232 communication interface between the wireless receive unit and host computer is needed to be designed so that the received data can be sent to the host computer.

3.3.1. SOC CC2530 CC2530 is a cost-effective, low-power and true system-on-chip (SOC) solution developed by TI. Its key features are described as extended operating temperature range from –40 to 125°C, less than 1μA current consumption in power down with sleep timer running, wide supply voltage range from 2 to 3.6V, which agree with the high working temperature and low consumption request of piston measurement system. There is a ADC in the CC2530 supporting analog-to-digital conversion with up to 12 effective number of bits. The effective resolution is 1.6mV for the reference voltage 3.3V, reaching requirement of test system of piston temperature. A 2.4-GHz IEEE 802.15.4

compliant RF transceiver is integrated in CC2530 to build Zigbee wireless net as well as communicate with point-to-point agreement.

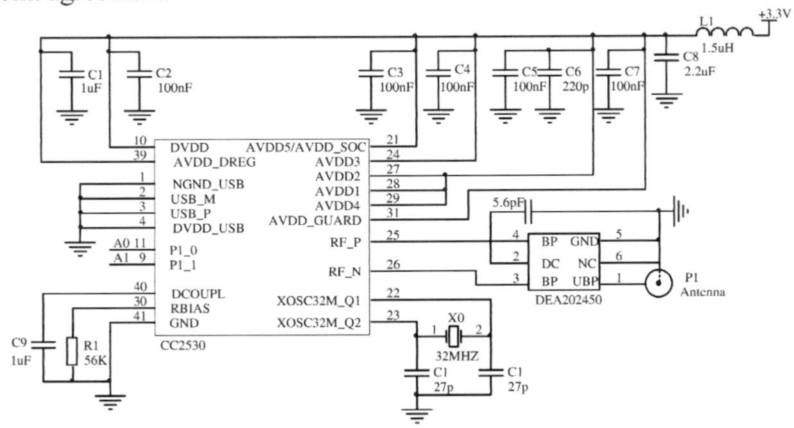

Figure 4. Schematic of wireless transmit unit

3.3.2. RF filter. The RF output of CC2530 is high impedance differential as seen from the RF ports (RF_P and RF_N), but the antenna is Single-Ended, so a balun convert between unbalanced and balanced signals is necessary. Balun circuit is usually composed of two capacitors and two resistances, but a integrating balance filter chip DEA202450 is used in the test system to reduce the peripheral design of CC2530, save the room of PCB and improve the RF performance.

4. Software design

The software design of the piston temperature measurement system accomplishes the program design of lower computer CC2530 on wireless transmitting unit and wireless receiving unit and the program design of the upper host computer as well.

4.1 Software design of wireless transmitting and receiving unit.

Wireless transmission unit mainly completes the wireless transmission of temperature data from all channels and the collection. The system collects and sends the temperature data every five minutes in order to reduce the power consumption and prolong the service life of the system and the rest of the time for CC2530 will work in sleeping mode. The wireless receiving unit mainly completes the data receiving and transmits them to the host computer through the serial port .

We will use the point-to-point communication mode when it receives and sends data, because the piston temperature testing system only comprises a sending end and a receiving end. And the program code is simple and clear by this mode. Besides, TI company provides ZigBee standard based on the protocol stack, and provides a series of API interface for the developer, so we need only to call the corresponding API function when we develop the wireless transceiver procedures. It is easy to editor program, because we should not set up ZigBee network.

4.2 Software design of host computer

The upper host computer mainly fulfills data processing, recording and displaying and so on, among which the data processing is most complex. On the one hand, we should calculate the resistance of the thermistor by the received voltage value Vout. On the other hand, we should determine the measured temperature by the R-T (resistance vs. temperature) tables thermistors.

5. Performance testing

It is put into the constant temperature furnace to carry out the simulation test in order to verify the testing precision of the system. Test results show that in the range of 150 degrees through 300 degrees, test accuracy of the piston wireless temperature measurement system is ± 1℃, and it can fully meet the test requirements. On the power consumption, when we send the data of testing, the instantaneous working current of the wireless transmitting unit is 28mA, and the current is only 0.4uA in the sleeping mode. Thermistors are respectively buried in four tested points, which are the bottom of the piston combustion chamber, the first ring groove, the piston pin hole and the skirt. Then, the Li battery and the PCB are fixed on the inner cavity of the piston with high temperature glue. Assembly of the entire measurement system is shown in Fig.5. Finally we can proceed bench test and the real data of testing can meet the requirements.

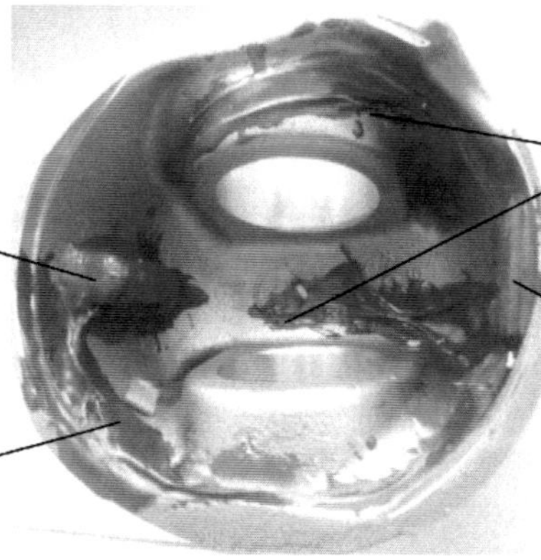

Figure 5. Assembly of entire measurement system

6. Conclusions

Temperatures inside the piston is difficult to measure because the temperature value is very high and installing room is limited as well as the reciprocating motion of piston with very high speed makes the signal get out hardly. A multi-channel wireless temperature measurement system for piston based on CC2530 is designed in this paper with the advantages such as low power consumption, compact PCB, small volume, high test accuracy and no lining out. This system can meet the dynamic test requirements under high temperature, high speed and small installing space, which has a good extensive application prospect in the engine measuring field.

References

[1] Wang Qingsheng, Liu Kun. Thermal analysis of piston in high-speed gasoline engines. *Journal of Hefei University of Technology(Natural Science)*, 2008, 31(6): 839-842.

[2] Lo K H, Shek C H, Lai K L. Metallurgical Temperature Sensors. *Recent Patents on Mechanical Engineering,* 2008, (1):225–232.

[3] TAKAMATSU H. Piston temperature measurement method for high-speed gasoline engines. *JSAE Review*, 1999, 53(20): 259-261.

[4] Lee H H, Bang W, Kauh S K, et al. Piston-temperature measuring system using electromagnetic induction with verification by a telemetry system. *International Journal of Automotive Technology*, 2011, 12(3):351-358

[5] LEE H W, OH C H, KAUH S K. Development of a multipiston temperature telemetry system using bluetooth networks. *Journal of Automobile Engineering*, 2006, 220(2):127- 138.

[6] Husberg T, Gjirja S, Denbratt I. Piston Temperature Measurement by Use of Thermographic Phosphors and Thermocouples in a Heavy-Duty Diesel Engine Run Under Partly Premixed Conditions, *SAE paper*, 2005-01-1646.

IOP Conf. Series: Materials Science and Engineering **157** (2016) 012011 doi:10.1088/1757-899X/157/1/012011

Combined Effect of Load Waviness and Auxeticity on the Shear Deformation in a Class of Rectangular Plates

Teik-Cheng Lim

School of Science and Technology, SIM University, Singapore.

tclim@unisim.edu.sg

Abstract. Auxetic materials are solids that exhibit negative Poisson's ratio. Due to its uniqueness, it is expected that auxetic solids and structures manifest certain mechanical responses that are distinct from conventional materials. So far classical plate theory (CPT) has been used to investigate the behavior of thin auxetic plates and, for thick auxetic plates, the first-order shear deformation theory (FSDT) has been recently employed. The third-order shear deformation theory (TSDT) is adopted herein as it employs a realistic transverse shear stress and strain distribution across the plate cross section, and therefore removes the need for a correction factor. The results obtained herein reveal that as a plate material's Poisson's ratio becomes more negative, the plate deflection characteristic mimics that of (a) reduced plate thickness, (b) reduced plate aspect ratio, and (c) reduced load waviness. The results of this investigation support (i) the use of auxetic materials in applications where there is a need to reduce the extent of transverse shear deformation, and (ii) the use of the simpler CPT instead of any shear deformation plate theory if the Poisson's ratio of the plate material is sufficiently negative.

1. Introduction

The use of the classical plate theory (CPT) [1, 2] for predicting the deflection of plates is valid when the plate is thin due to the insignificant amount of transverse shear deformation. Its deflection modeling neglects transverse shear strain and stress. However the transverse shear deformation increases with the plate thickness, thereby increasing the extent of deflection. To take into consideration the transverse shear deformation, the first-order shear deformation theory (FSDT) [3, 4] was developed. Since FSDT assumes constant transverse shear strain (and hence constant transverse shear stress) along the plate thickness, a shear correction factor is a required compensation because the actual transverse shear stress is parabolically distributed along the plate thickness. On the other hand, the third-order shear deformation theory (TSDT) does away with the shear correction factor, as it caters to a quadratic distribution of the transverse shear strains (and hence transverse shear stress) with diminishing transverse shear stresses at the top and bottom surfaces of the plate [5]. Although TSDT of plates have been increasingly investigated (e.g. [6,7]), it has not been used for analyzing auxetic (negative Poisson's ratio) plates. The reader is referred to analyses of auxetic plates using CPT [8-11] and FSDT [12-15], as well as a recent monograph for an introduction to auxetic materials [16]. In a departure of previous analysis that adopt FSDT for understanding the extent of shear deformation in auxetic plates, this paper employs the use of TSDT to understand the negativity effect of a rectangular plate's Poisson's ratio.

2. Analysis

Using the superscript notations $^{(K)}$ and $^{(R)}$ for indicating Kirchhoff and Reddy plates respectively, while $\mathrm{M}^{()}$ and M indicate the marcus moment and the bending moment respectively, Wang et al. [17] showed that the Kirchhoff and Reddy plate deformations for a simply supported rectangular plate are related as

$$\nabla^2 w^{(R)} - \lambda_0^{\ 2} w^{(R)} = -\lambda_0^{\ 2}\left(w^{(R)} + \frac{C}{Gh}\mathrm{M}^{(K)} \right) \tag{1a}$$

where

$$\lambda_0^{\ 2} = \frac{70Gh}{D} \tag{1b}$$

and $C = 17/14$, in which G and D denote the shear modulus and flexural rigidity of the plate, respectively. To re-cast the above relationship in a form that is more readily executed, we substitute

$$\frac{G}{D} = \frac{6(1-v)}{h^3} \tag{2}$$

so that

$$\lambda_0^{\ 2} = \frac{420(1-v)}{h^2}, \tag{3}$$

and

$$\mathrm{M}^{(K)} = -D\nabla^2 w^{(K)} \tag{4}$$

to give the general deflection relationship between Reddy and Kirchhoff simply supported rectangular plates as

$$w^{(R)} - \frac{h^2}{420(1-v)}\nabla^2 w^{(R)} = w^{(K)} - \frac{17h^2}{84(1-v)}\nabla^2 w^{(K)} \tag{5}$$

with the error from using the CPT, with reference to the TSDT, as

$$error = w^{(K)} - w^{(R)} = \frac{h^2}{420}\left(\frac{85\nabla^2 w^{(K)} - \nabla^2 w^{(R)}}{1-v} \right). \tag{6}$$

For a general sinusoidal transverse load of

$$q(x,y) = q_0 \sin\frac{m\pi x}{a}\sin\frac{n\pi y}{b} \tag{7}$$

acting on a simply supported rectangular plate of sides a and b measured along the x and y axes, respectively, where m and n are positive odd integers, the CPT and TSDT plate deflections of the form

$$\begin{Bmatrix} w^{(K)} \\ w^{(R)} \end{Bmatrix} = \begin{Bmatrix} A_{mn}^{(K)} \\ A_{mn}^{(R)} \end{Bmatrix} \sin\frac{m\pi x}{a}\sin\frac{n\pi y}{b} \tag{8}$$

satisfies all boundary conditions:

$$w_{x=0,a} = w_{y=0,b} = \left(\frac{\partial^2 w}{\partial x^2} \right)_{x=0,a} = \left(\frac{\partial^2 w}{\partial y^2} \right)_{y=0,b} = 0. \tag{9}$$

Substituting Eq.(8) into Eq.(5) leads to

$$A_{mn}^{(R)}\left[1 + \frac{\pi^2 h^2}{420(1-v)}\left(\frac{m^2}{a^2} + \frac{n^2}{b^2} \right) \right] = A_{mn}^{(K)}\left[1 + \frac{17\pi^2 h^2}{84(1-v)}\left(\frac{m^2}{a^2} + \frac{n^2}{b^2} \right) \right] \tag{10}$$

where the solution for $A_{mn}^{(K)}$ has been given as [18]

$$A_{mn}^{(K)} = \frac{q_0}{\pi^4 D}\left(\frac{m^2}{a^2} + \frac{n^2}{b^2} \right)^{-2}. \tag{11}$$

The ratio of deflection by TSDT to that by CPT is independent from the coordinate position (x, y) of the plate

$$\frac{w^{(R)}}{w^{(K)}} = \frac{A_{mn}^{(R)}}{A_{mn}^{(K)}} = \frac{420a^2b^2(1-\nu)+85\pi^2h^2(a^2n^2+b^2m^2)}{420a^2b^2(1-\nu)+\pi^2h^2(a^2n^2+b^2m^2)}, \tag{12}$$

indicating that an amount of $84\pi^2h^2(a^2n^2+b^2m^2)$ is attributed to the transverse shear deformation according to TSDT. As is well known, the transverse shear deformation reduces as the plate becomes thinner. The second terms on at the numerator and denominator on the RHS of Eq.(12) also suggests that transverse shear deformation reduces as the load is less wavy (i.e. smaller values of m and n). Alternatively, the transverse shear deformation can be reduced by using auxetic materials, as implied in Eq.(12).

3. Results and Discussion

Figure 1(a) shows the plots of Reddy-to-Kirchhoff deflection ratio for square plates $(a=b=l)$ under half sine wave load $(m=n=1)$ for various dimensionless plate thickness h/l. Although it is known that thicker plates encounter greater transverse shear deformation, it is of interest to note that for a thick auxetic plate with $h/l=0.18$ and $\nu\le-0.65$, the transverse shear deformation is lower than that of an incompressible plate ($\nu=0.5$) of lower dimensionless thickness of $h/l=0.1$. In addition, a thick plate of $h/l=0.18$ that is auxetic at $\nu=-0.95$ is mechanically equivalent to a thin plate of $h/l=0.1$ that is conventional with $\nu=0.4$, as the Reddy-to-Kirchhoff deflection ratio for both cases approximate $w^{(R)}/w^{(K)}\approx1.0656$. Similarly, a highly auxetic plate ($\nu=-0.95$) of dimensionless thickness $h/l=0.1$ is mechanically equivalent to a very thin conventional plate $(\nu=0.3, h/l=0.06)$ as both plates exhibit $w^{(R)}/w^{(K)}=1.02$, i.e. a transverse shear deformation that exceeds the bending deformation by 2%.

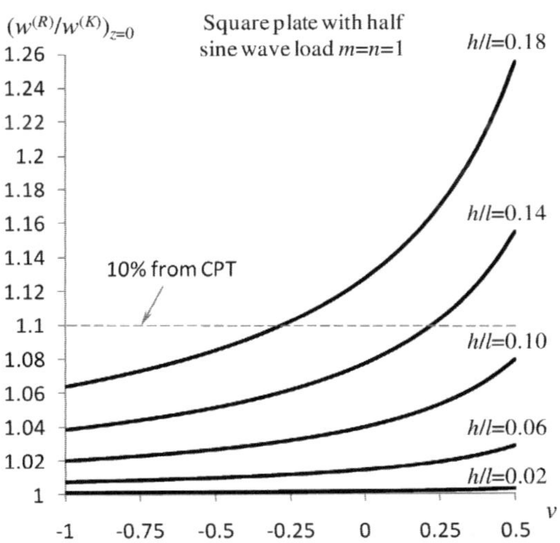

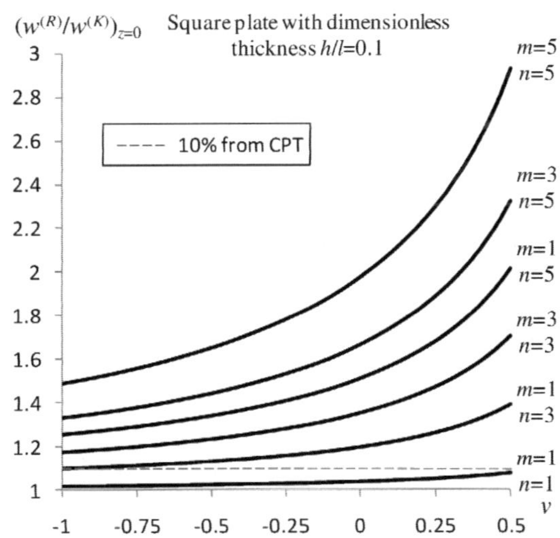

Figure 1. Plots of TSDT-to-CPT deflection ratio for square plates under half sine wave load for various dimensionless thicknesses (left), and at dimensionless thickness of 0.1 for various load waviness (right).

A family of Reddy-to-Kirchhoff deflection ratio for square plates $(a=b=l)$ and dimensionless thickness of $h/l=0.1$ is plotted in figure 1(b) for different patterns of load distribution, in which the extent of load waviness is denoted by the number of m and n, i.e. m and n refer to the waviness along the x and y axes of the plate, respectively. It is readily seen from Eq.(15), and verified from figure 1(b), that the waviness controls the extent of transverse shear deformation via (m^2+n^2) for a square plate or $(b^2m^2 + a^2n^2)$ for a rectangular plate. It can be observed that an auxetic plate under a load of higher waviness deflects as if for a conventional plate under a load of lower waviness. For example, a highly auxetic plate of $\nu = -0.7$ under a very high waviness load pattern of $m = 3, n = 5$ (i.e. $m^2 + n^2 = 34$), a moderately auxetic plate of $\nu = -0.3$ under a moderately high waviness load pattern of $m = 1, n = 5$ (i.e. $m^2 + n^2 = 26$), a conventional plate of $\nu = 0.1$ under a moderately low waviness load pattern of $m = 3, n = 3$ (i.e. $m^2 + n^2 = 18$), and an incompressible plate (i.e. $\nu = 0.5$) under a very low waviness load pattern of $m = 1, n = 3$ (i.e. $m^2 + n^2 = 10$) give equal Reddy-to-Kirchhoff deflection ratio of $w^{(R)} / w^{(K)} = 1.392937$.

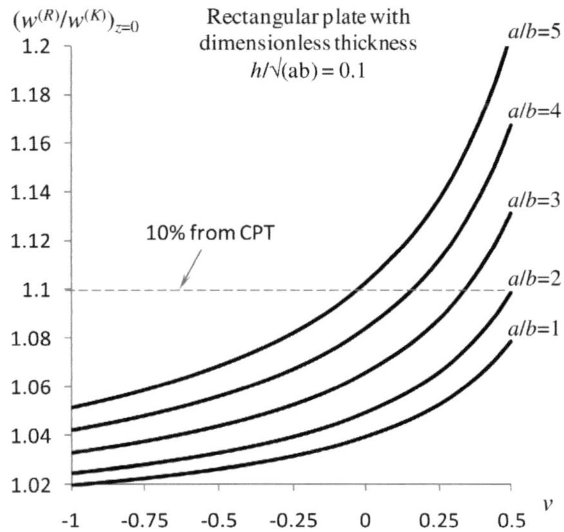

Figure 2. A family of TSDT-to-CPT deflection ratio for rectangular plates with dimensionless plate thickness of 0.1 under various half sine load pattern.

Curves of Reddy-to-Kirchhoff deflection ratio for rectangular plates are displayed in figure 2 for dimensionless plate thickness of $h / \sqrt{ab} = 0.1$ under half sine load pattern (i.e. $m = n = 1$). As before, the Reddy-to-Kirchhoff deflection ratio decreases as the Poisson's ratio of the plate becomes more negative. It is interesting to note that at the Reddy-to-Kirchhoff deflection ratio of $w^{(R)} / w^{(K)} = 1.049319$ is valid for a negative Poisson's ratio ($\nu = -0.7$) plate with aspect ratio of $a / b = 4$, a zero Poisson's ratio plate with aspect ratio $a / b = 2$, and a positive Poisson's ratio plate ($\nu = 0.2$) with aspect ratio $a / b = 1$. It is also noticeable that an equal Reddy-to-Kirchhoff deflection ratio can be obtained at $\nu = \pm 1/4$, i.e. $w^{(R)} / w^{(K)} = 1.052605$ when $a / b = 1$ at $\nu = 1/4$ and when $a / b = 3$ at $\nu = -1/4$; for completeness' sake, the same Reddy-to-Kirchhoff deflection ratio can be obtained when $a / b = 5$ at $\nu = -0.95$.

As a design guideline, a 10% deviation of the more accurate TSDT deflection from the less accurate CPT is shown in figures 1 and 2, such that the CPT remains valid under the dashed lines even for thick and high aspect ratio plates under loads of high waviness pattern.

4. Conclusions

The deflection characteristic of an auxetic plate, vis-à-vis a conventional plate, has been herein investigated using a third-order shear deformation theory (TSDT), and investigated for a simply supported rectangular plates for the entire range of Poisson's ratio of isotropic solids, with variations to the plate thickness, plate aspect ratio and load pattern waviness. Results from the present study imply that, should there be a requirement to use thick plates and/or long narrow plates with very wavy load pattern such that significant the transverse shear deformation takes places, then the choice of auxetic materials will not only reduces the extent of transverse shear deformation with respect to bending deformation, but also justifies the use of classical plate theory (CPT). Hence a geometrically thick plate becomes a mechanically thin plate should the Poisson's ratio of the plate material be sufficiently negative to effectively diminish the extent of shear deformation vis-à-vis bending deformation. Auxetic plates are therefore useful for state of the art pressure sensor plates due to the enhanced in-plane deformation at the surface arising from decreased shearing-to-bending deformation ratio. It is herein suggested that experimental work be performed in future to validate Eq.(12).

References

[1] Kirchhoff G 1850 Über das gleichgewicht und die bewegung einer elastischen scheibe *Journal für die Reine und Angewandte Mathematik* **40** pp 51-88
[2] Love A E H 1888 The small free vibrations and deformation of a thin elastic shell *Philosophical Transactions of the Royal Society of London A* **179** pp 491-546
[3] Reissner E 1945 The effect of transverse shear deformation on the bending of elastic plates *Journal of Applied Mechanics* **12** pp 68-77
[4] Mindlin R D 1951 Influence of rotatory inertia and shear on flexural motions of isotropic, elastic plates *Journal of Applied Mechanics* **18** pp 31-38
[5] Reddy J N and Wang C M 1998 Deflection relationships between classical and third-order plate theories *Acta Mechanica* **130** pp 199-208
[6] Gao X L, Huang J X and Reddy J N 2013 A non-classical third-order shear deformation plate model based on a modified couple stress theory *Acta Mechanica* **224** pp 2699-2718
[7] Gao X L and Zhang G Y 2015 A microstructure- and surface energy-dependent third-order shear deformation beam model *Zeitschrift für Angewandte Mathematik und Physik* **66** pp 1871-1894
[8] Lim T C 2013 Circular auxetic plates *Journal of Mechanics* **29** pp 121-133
[9] Lim T C 2013 Optimal Poisson's ratios for laterally loaded rectangular plates *Journal of Materials: Design and Applications* **227** pp 111-123
[10] Lim T C 2016 Bending stresses in triangular auxetic plates *Journal of Engineering Materials and Technology* **138** article 014501
[11] Lim T C 2016 Simply-supported elliptical auxetic plates *Journal of Mechanics* **32** pp 413-419
[12] Lim T C 2013 Shear deformation in thick auxetic plates *Smart Materials and Structures* **22** article 084001
[13] Lim T C 2014 Shear deformation in rectangular auxetic plates *Journal of Engineering Materials and Technology* **136** article 031007
[14] Lim T C 2014 Elastic stability of thick auxetic plates *Smart Materials and Structures* **24** article 045004
[15] Lim T C 2014 Vibration of thick auxetic plates *Mechanics Research Communications* **61** pp 60-66
[16] Lim T C 2015 *Auxetic Materials and Structures* (Singapore: Springer)
[17] Wang C M, Reddy J N and Lee K H 2000 *Shear Deformable Beams and Plates* (New York: Elsevier)
[18] Timoshenko S P and Woinowsky-Krieger S 1970 *Theory of Plates and Shells* (Singapore: McGraw-Hill)

IOP Conf. Series: Materials Science and Engineering **157** (2016) 012012 doi:10.1088/1757-899X/157/1/012012

Influence of Clearance and Punch Velocity on the Quality of Pure Thin Copper Sheets Blanked Parts

Didin Zakariya Lubis[1] and Muslim Mahardika[1]

[1] Department of Mechanical and Industrial Engineering, Faculty of Engineering, Gadjah Mada University, Jl. Grafika No.02,Yogyakarta, 55281, Indonesia.

Abstract. Research on the influence of clearance and punch velocity to determine the quality of the punched edge were conducted. This study uses pure copper sheet material with the clearance variation of 2.5, 5, 7.5 and 10%. Punch velocity is based on the ability of about Micro Punch CNC machine which is 100 and 2600 mm/min. At highest speed with a clearance of 2.5%, sheared zone is of about 395 μm or 79% of the material thickness. It can be concluded that the punch velocity gives positive influence on the sheared zone in copper. Basically the ideal outcome of the sheared edge of punching result is having rollover and small burr and contain at least 75% of the shear zone. This can be achieved with a clearance of 2.5%.

1. Introduction

The development of science and technology encourages the creation of a new product with good quality. Miniaturization and weight reduction of various industrial products continue to be developed, machines with high accuracy and good quality machining results are needed today [1]. Micromachining is the most efficient technology to produce components with very small size. This technology can accommodate the demand for the products of micro-sized and micro components in various industrial sectors including electronics, medical, optics, biotechnology, and automotive [2].

Shearing process had been observed by some researchers for a variety of material or workpiece and tool conditions to improve the quality of shearing product, especially the quality of material surface. Quality of the punched edge during punching process is evaluated based on part-edge characteristics and dimensional accuracy. Basically, there are a lot of factors that affect the product quality of part-edge characteristic, according to [3-5] and the major factor which affects the product quality of the punched edge are clearance between punch and die and punch velocity.

Based on several experimental studies conducted by FEM [6, 7] clearance has the most significant influence in affecting the accuracy of the punching process. In the simulation also found that the rollover and shear edge increases and fracture edge decreases with decrease in the value of clearance [8]. Mentioned on their research [9-11] when the clearance value is decrease, smooth sheared and punch force will increase. When the clearance value is increasing, rollover depth will also increase, in other hand smooth sheared value will decrease.

In addition, increasing the pace of work in production process is not only closely related to the increase in profitability of the components, improve the working speed can reduce production costs and energy consumption [12]. High-speed blanking will improve the product quality of the punched edge, burr and rollover are in low altitude while the shear zone increases [13]. Thus, further emphasis of this research in addition to optimization of clearance also represents the potential punch velocity as process parameters to influence product quality of tested material feasibility.

2. Experimental conditions

2.1. Micro Punch CNC Machine

One of the products developed is Micro Punch CNC Machine. Micro Punch CNC Machine is a combination of press dies machine with CNC (Computer Numerically Controlled) driving system as shown in figures 1 (a) and (b). Press dies can produce a part with uniform quality with minimum machining time. In addition, the blanking process on press dies system has several advantages, including it can produce a uniform punched edge quality and size products with lower cost in the manufacturing of mass products. The working principle of press dies is using penetration between punch and die.

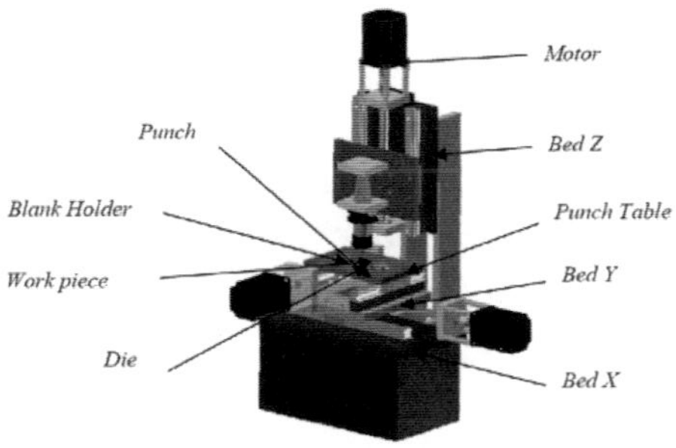

Figure 1. (a) Mechanical Design Micro Punch CNC Machine, (b) Micro Punch CNC Machine

2.2. *Initial tool position and condition*

The alignment of punch and die reach 10 μm. The accuracy alignment measurement uses "Dino-Lite Edge AM 4515 T5" camera, featured with an LED light source that is located on the side of the punch serves as illumination during alignment process. The alignment between the punch and die is very important before doing the experiment, aimed to minimize the tool wear and to obtain a same cutting edge on the circumference of a circle.

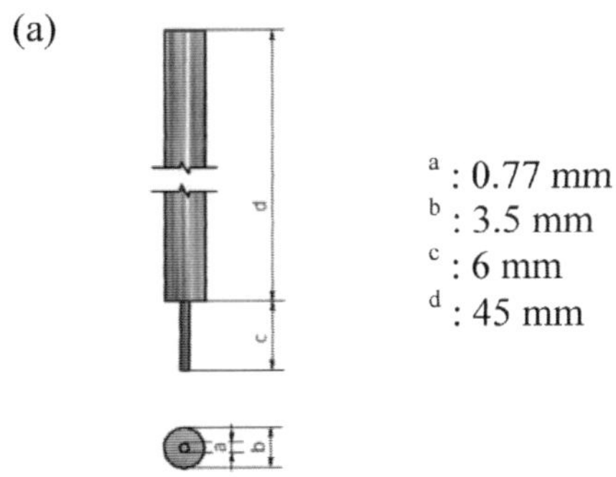

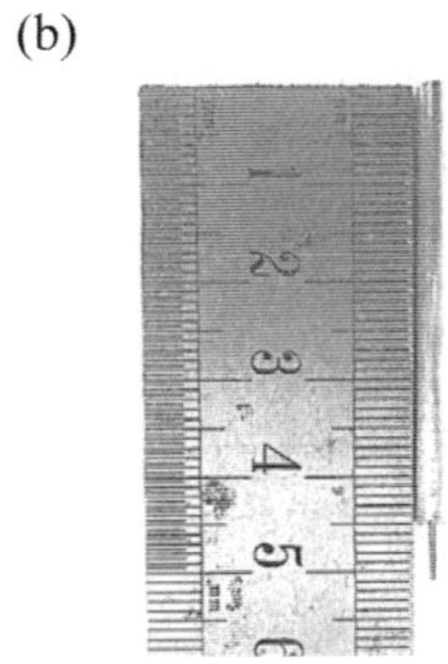

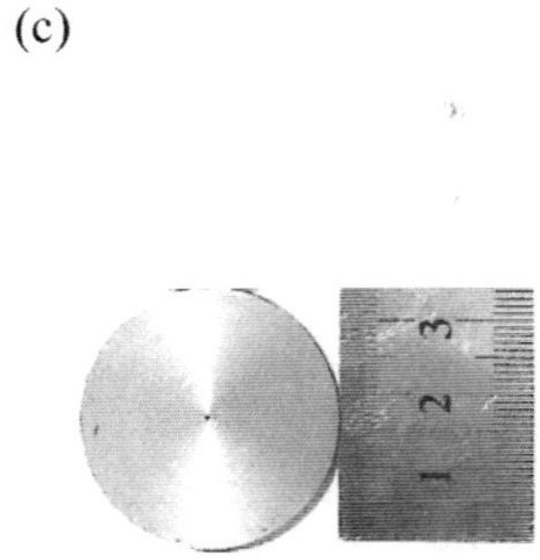

Figure 2. (a) Punch sample, for clearance 2.5%, (b) Punch tool, (c) die

Punch tool using High Speed Steel (HSS) materials type Nachi Standard with a Rockwell C hardness of 64 HRC while the die using Mild Steel (see figure 2). Mechanical properties of experimental setup is shown in Table 1.

Table 1. Mechanical properties of experimental setup

Properties	Material testing	Punch tool (High Speed Steel)	Die (Mild Steel)
Young's Modulus (Mpa)	118 x 10^3	207 x 10^3	220 x 10^3
Poisson's Ratio	0.34	0.27	0.28
Density (kg/m3)	8900	8138	7861
Yield Strength (Mpa)	33.3	3250	207
Tensile Strength (Mpa)	210	172	345

2.3. *Clearance and punch velocity setup*

The experiment was conducted at room temperature. Punch velocity (v) of Micro Punch CNC Machine was 100 (low speed) and 2600 mm/min (high rate). Clearances used for this study were 2.5, 5, 7.5 and 10%. Figure 3 describes clearance punch and die used in the study.

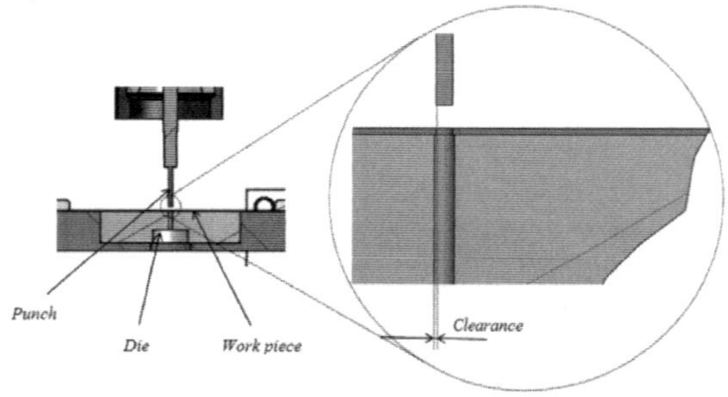

Figure 3. Punch and die Clearance

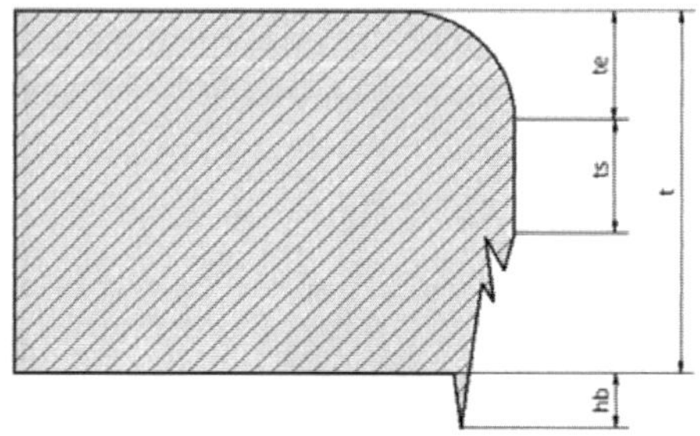

Figure 4. Form errors on sheared workpieces, t is the sheet thickness; te rollover; ts shear zone and hb is burr height

2.4. *Material testing*

This study uses pure copper sheet (99.6%) workpiece with the thickness of 500 μm. On each velocity variation, punching is performed four times to obtain the most valid data, and follow with clearance variation. The equation to measure the total force required to pierce the workpiece with circular section is as follows.

$$P = \pi Dst \tag{1}$$

Known that shear strength (*st*) of the material is 0.7 times the tensile strength (TS) of 210 MPa. The obtained *st* is 147 MPa. It can be obtained that the force (*P*) needed to punching the copper with t: 500 μm is 182.3 N. Based on the calculation of the force required for punching process as compare to optimal punching machines, it can be concluded that the maximum force of 1662.8 N may allow punching process on the copper sheet with the required maximum force of 182.3 N.

The calculation of punch size variations based on the following formula.

$$c = \frac{d_d - d_p}{2t} \cdot 100\% \tag{2}$$

where D_p is the external diameter of the punch, D_d is the internal diameter of the die, and t is the thickness of the workpiece. The punch diameter variations are shown in table 2 below.

Table 2. Variations in diameter punch against clearance

d*ie* Diameter (μm)	*Clearance*	Various punch Diameter (μm) t = 500 μm
790	2.5 %	765
	5%	740
	7.5%	715
	10%	690

2.5. *Measurement method*

The amount of rollover, shear zone and burr height from section of the slug were obtained using an optical microscope camera (Type Olympus C-35AD-4). To observe the validity of rollover, shear zones and burr height, each slug has been measured at 4 different locations around the circumference (see figure 4).

3. Result and discussion

Figure 5 shows the shape of the part-edge characteristic as punching process results. It can be seen that the high rate punching and the smallest clearance shape looked smooth and even glossy. Differ from the shape of the copper with v: 100 mm/min for clearance of 10% figure 5 (d), the shape looks dull and finely fibrous with no reflection of light and with many fine scratch-marks parallel to the punching direction.

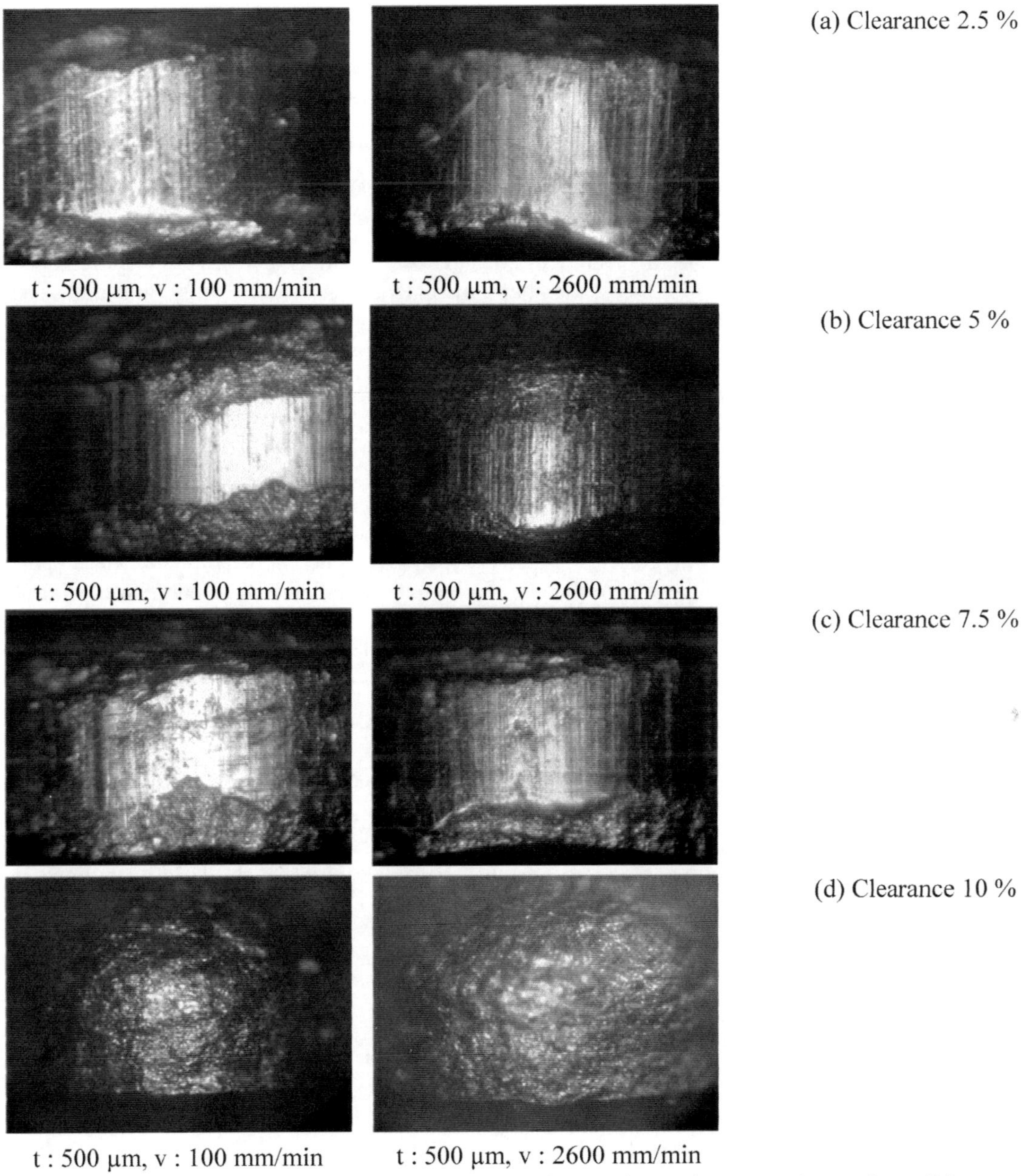

Figure 5. Surface appearance of punched edge surface under various experimental conditions, t : sheet thickness, v: punch velocity

Figure 6 shows the measurement results of a rollover, shear zone and burr height in respect to variations of punch and die clearance for the low speed. Up to 10% clearance, the percentage rollover and burr height of the slugs is constantly increasing. At a clearance of 10% the rollover is about 4 times greater than 2.5% clearance with v: 100 mm/min, while on high rate at a clearance of 10% rollover increase is about 6 times higher (see figure 7).

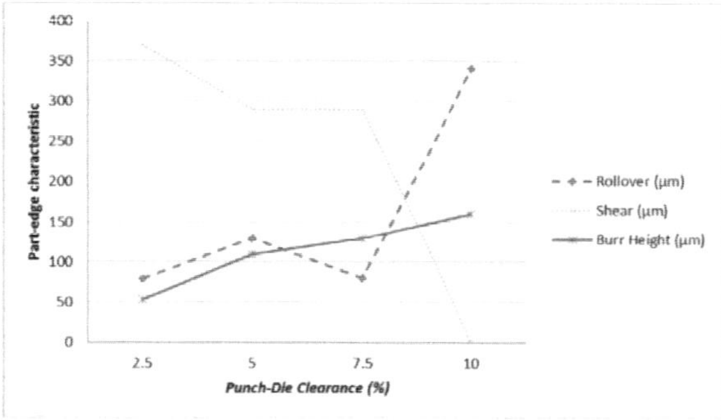

Figure 6. Various punch-die clearance with v : 100 mm/min

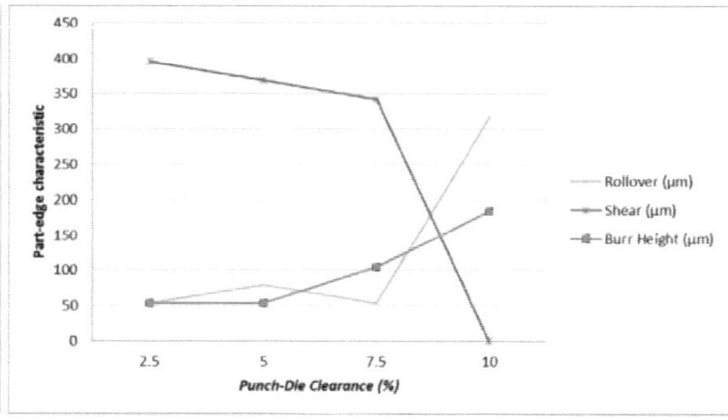

Figure 7. Various punch-die clearance with v : 2600 mm/min

However, with the increasing value of clearance, shear zone is getting lower. At 10% clearance shear zone was not developed see figure 5 (d), the shape of punched edge looks rough and dull. The sheet tends to be pulled into the clearance region, and the perimeter or edges of the shear zone become rougher. Unless such edges are acceptable as produced, secondary operations may be required to make them smoother.

At high speed of v: 2600 mm/min, the shear zone was increase (see figure 7). When punching with small clearance (2.5%) at the high rate a shear zone of up to 395 µm or about 79% of the material thickness could be reached (Table 3). This is because when the high rate punching, the kinetic energy will be greater than the energy required for punching process, so that the heat generated will disappear quickly. Although there is an increase in the percentage of shear when using high rate, it is not enough to compensate the decrease of percentage shear which is caused by an increasing clearance. This means that in order to obtain the shape of the part-edge characteristics of 75% shear, the clearance should be 2.5 to 5%.

Table 3. Percentage of Shape Characteristics

Clearance	Rollover		Shear		Burr Height	
(%)	100 mm/min (%)	2600 mm/min (%)	100 mm/min (%)	2600 mm/min (%)	100 mm/min (%)	2600 mm/min (%)
2.5	15.8	10.6	73.6	79	10.6	10.6
5	26.4	15.8	52.6	73.6	21	10.6
7.5	15.8	10.6	57.8	68.4	26.4	21
10	68.4	63.2	0	0	31.6	36.8

Figure 8 shows the average test results for low speed and high rate. At high rate, burr heights is found to be constant at 2.5 and 5% clearance. When the clearance increase, the height of the burr increases linearly.

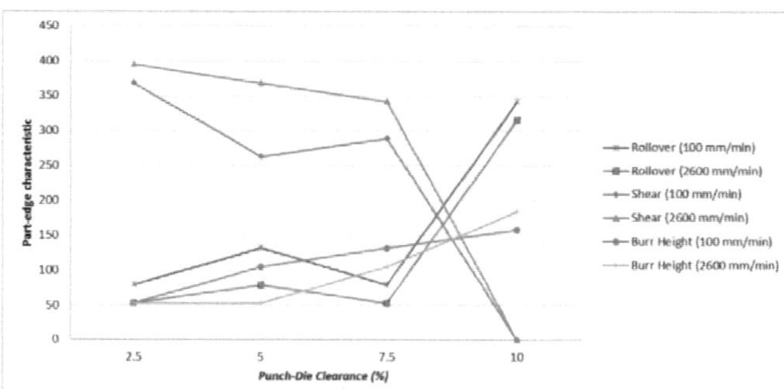

Figure 8. Comparison result between v: 100 mm/min and v: 2600 mm/min

4. Conclusion

In this paper, it can be noted that a positive influence of high rate punching on the quality of the part-edge is more obvious, especially regarding shear zone. At low speed or high rate, it produces similar part-edge characteristics. If the clearance value increases, the shape of the shear zone decreases with the increase in rollover and burr height and vice versa. When clearance value decreases, the shear zone will increase, and it is followed by the increasing height of the rollover and burr. When the clearance value increase, the deformation area becomes larger and shearing zone becomes rough. Rough shearing zone can be accepted as a product, but further machining may be needed to refine the punched edge, and it may increase the production cost.

Basically, the ideal outcome of part-edge characteristics is having small rollover and burr and having at least 75% of shear zone. This can be achieved with clearance value of 2.5%.

Acknowledgment

This study was carried out with the financial support of DIKTI. Work facilities was supported by Gadjah Mada University. Gratefully acknowledgment.

References

[1] Jeswiet J, Geiger M, Engel U et al. 2008 *Metal forming progress since 2000* (CIRP Journal Manufacturing Science Technology 1: Elsevier) pp 2–17

[2] Masuzawa T, 2000 *State of the Art of Micromachining*, CIRP Annals - Manufacturing Technology, Volume 49, Issue 2, 2000, pp 473–488

[3] Lange K 1985 *Handbook of Metal Forming* (Michigan: Mc Graw-Hill, Inc.) chapter 24.19

[4] Suchy I 1998 *Hand Book of Die Design* (New York: Mc Graw-Hill, Inc.)

[5] Kalpakjian S and Schmid S R 2009 *Manufacturing Engineering and Technology Sixth Edition*,. (Singapore: Prentice Hall) p 381

[6] Subramonian S, Altana T, Campbell C 2013 *Determination of forces in high speed blanking using FEM and experiments* (Journal of Materials Processing Technology 213: Elsevier) pp 2184– 2190

[7] Grunbaum M, Breitling J, Altan T 1996 *Influence of high cutting speeds on the quality of blanked parts*. In: Report No. ERC/NSM – S-96-19. Center for Precision Forming (USA: Columbus, Ohio

[8] Husson C, Correia J, Daridon L, et al. 2008 *Finite elements simulations of thin copper sheets blanking: Study of blanking parameters on sheared edge quality* (journal of materials processing technology 1 9 9: Elsevier) pp 74–83

[9] Kibe Y, Okada Y, and Mitsui K 2007 Machining accuracy for shearing process of thin-sheet metals — Development of initial tool position adjustment system (International Journal of Machine Tools & Manufacture 47: Elsevier) pp 1728–1737.

[10] Tekiner Z, Nalbant M, Gurun H 2006 *An experimental study for the effect of different clearances on burr, smooth-sheared and blanking force on aluminium sheet metal* (Materials and Design 27: Elsevier) pp 1134–1138

[11] Groover, P Mikell 2010 *Fundamentals of Modern Manufacturing Materials, Processes and System 4th Edition*. (USA: Jhon Wiley and Son) p 443

[12] Neugebauer R, Bouzakis D K, Denkena B et al., 2011 *a Velocity effects in metal forming and machining processes*, (CIRP Annals - Manufacturing Technology 60: Elsevier) pp 627–650.

[13] Gotoh M, Yamashita M 2000 *A study of high-rate shearing of commercially pure aluminum sheet* (Materials Processing Technology 110: Elsevier) pp 253-264

IOP Conf. Series: Materials Science and Engineering **157** (2016) 012013 doi:10.1088/1757-899X/157/1/012013

Introduction of Enhanced Compressive Residual Stress Profiles in Aerospace Components Using Combined Mechanical Surface Treatments

Abhay Gopinath[1,2], Andre Lim[1,2], Balasubramanian Nagarajan[3], Chow Cher Wong[1], Rajarshi Maiti[1,2], Sylvie Castagne[2,*]

[1] Rolls-Royce Singapore, Advanced Technology Centre, 6 Seletar Aerospace Crescent, Singapore 797575
[2] School of Mechanical and Aerospace Engineering, Nanyang Technological University, 50 Nanyang Avenue, Singapore 639798
[3] Rolls-Royce@NTU Corporate Laboratory, c/o School of Mechanical & Aerospace Engineering, 50 Nanyang Avenue, Nanyang Technological University, Singapore 639798

[*]scastagne@ntu.edu.sg

Abstract. Mechanical surface treatments such as Shot Peening (SP) and Deep Cold Rolling (DCR) are being used to introduce Compressive Residual Stress (CRS) at the surface and subsurface layers of aerospace components, respectively. This paper investigates the feasibility of a combined introduction of both the surface and sub-surface compressive residual stress on Ti6Al4V material through a successive application of the two aforementioned processes, one after the other. CRS profiles between individual processes were compared to that of combination of processes to validate the feasibility. It was found out that shot peening introduces surface compressive residual stress into the already deep cold rolled sample, resulting in both surface and sub-surface compressive residual stresses in the material. However the drawback of such a combination would be the increased surface roughness after shot peening a deep cold rolled sample which can be critical especially in compressor components. Hence, a new technology, Vibro-Peening (VP) may be used as an alternative to SP to introduce surface stress at reduced roughness.

1. Introduction

Crack initiation and propagation is one of the key challenges faced in aerospace industry [1]. High levels of vibrational stresses, frettage fatigue and Foreign Object Damage (FOD) are the main sources of crack initiation in an aerospace component. Introduction of Compressive Residual Stresses (CRS) into the material delays the mechanisms leading to crack initiation and crack propagation thereby improving the fatigue life of the component [2].

Shot Peening (SP) is one of the widely used cold working techniques to induce residual stresses on the component surface [3]. This process uses spherical balls called as shot or projectile to impact the part surface at high velocity, creating plastic deformation through indents on the surface. A larger region of compressive residual stress in the surface zone is introduced as the thin plastic deformation near the surface is resisted (compressed) by the bulk elastic material [4].The key variables in shot peening are the density, size and shape, hardness of the shot, air pressure used for peening, nozzle type, shot velocity, angle of impingement and hardness, geometry and preload of the component.

Deep Cold Rolling (DCR) is a mechanical surface treatment technique that uses a hydrostatically controlled ball or roller tool, to impart a contact stress larger than the yield stress of the component, leading to plastic deformation and a three-dimensional Hertzian stress pattern around the point of contact [5]. Although DCR has its roots in roller burnishing and shares common benefits, the main aim of DCR is to induce plastic deformation which results in compressive residual stress in the surface and sub-surface region of the component, while roller burnishing is mainly used to attain a high quality mirror-like finish on the surface of the component [6].

Figure 1 shows the typical stress distribution through the depth of a component for a surface and a subsurface process. DCR generally introduces sub surface stress of significant compressive level up to a depth of 500 μm to 1 mm, while SP introduces compressive surface stress below 200 μm. This paper investigates the feasibility of combining surface and sub-surface stresses.

One of the characteristics of DCR is that, this process induces directional stresses unlike SP which introduces isotropic stresses. In DCR, stresses that are perpendicular to the rolling direction (crosswise) are higher than the stresses along the direction of rolling (lengthwise). Figure 2 illustrates the crosswise (CW) and lengthwise (LW) direction with respect to the rolling direction for DCR. By combining DCR with SP, we intend to investigate the possibility of decreasing the gap between lengthwise and crosswise stresses as well.

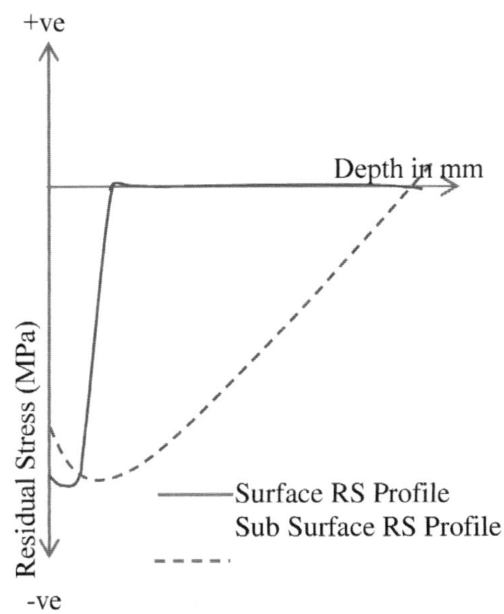

Figure 1. Approximation of Residual Stress profiles for Surface and subsurface treatment.

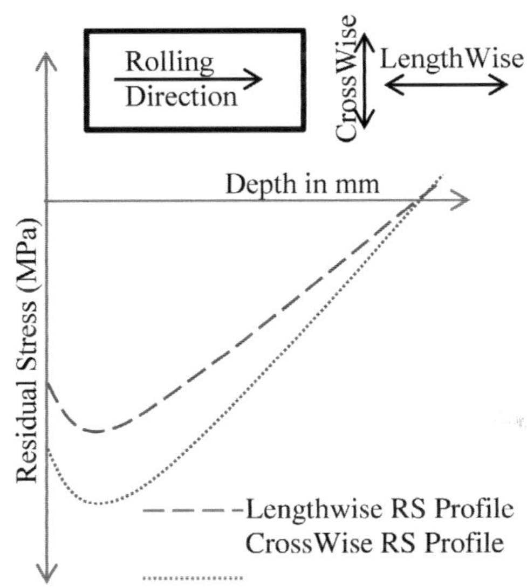

Figure 2. Depiction of lengthwise and crosswise stress profiles for DCR.

2. Measurements and methodology

Three flat coupons (30x30x20 mm^3) of Ti6Al4V (ASTM Grade 5) material were used in this study. The first coupon was deep cold rolled using a 12.6 mm ball, the second was shot peened using 200% coverage using 8A intensity and the third one had DCR followed by SP. The tests were repeated for two sets of coupon for repeatability check and the average of the two sets was taken. The residual stress measurements were conducted by Centre Hole Drilling (CHD) with Vishay Precision Group type EA-06-062RE-120 gauge using the orbital hole drilling procedure. Stresses obtained had a length wise and crosswise component since DCR is a directional technique.

3. Results and discussion

The compressive residual stress profiles are compared in Figure 3 between SP, DCR and DCR + SP based on the following parameters: stresses at the surface, peak residual stress and the depth at which peak stress occurs and the transition depth from compressive to tensile residual stresses. The residual stress profiles have been normalized against the maximum compressive residual stress value measured across all five residual stress depth profiles.

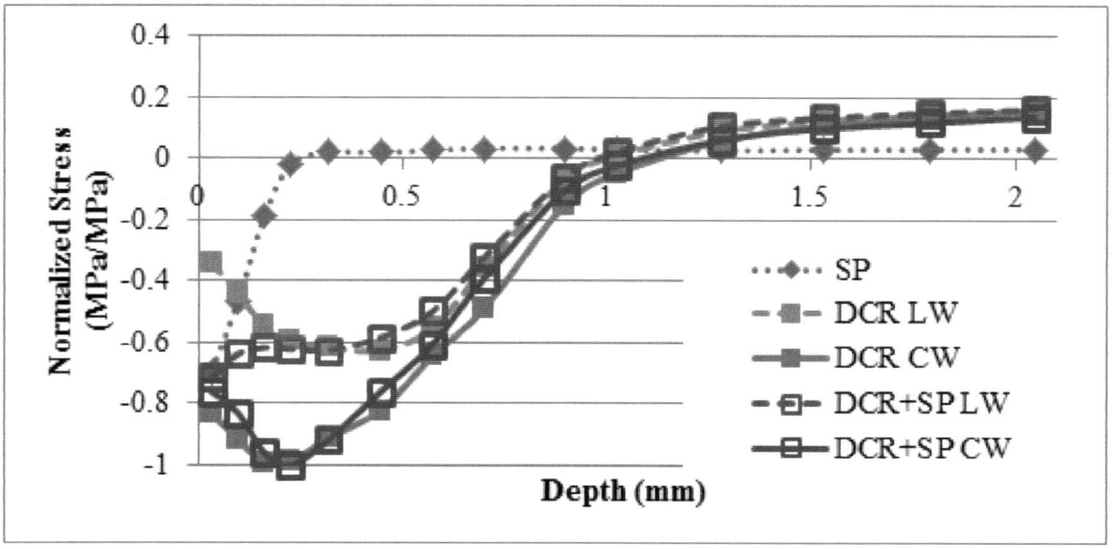

Figure 3: Normalized residual stress profiles for both individual processes and the combination process.

It can be observed that the residual stress depth profile from SP is identical in both directions. However, for the near surface region, up to a depth of about 500 μm, DCR has different residual stress profiles in the lengthwise and crosswise directions as shown in Figure 3. This is due to the mechanism of the DCR process in which the burnishing ball element generates a bow wave in advance of the rolling direction. As a result of this bow wave, material is pushed perpendicular to the rolling direction. More material is pushed perpendicularly to the rolling direction relative to that parallel to the rolling direction. This was reported by Balland et al. [7] through the use of finite element simulation modelling of a single path of the burnishing process. In addition, Lim et al. [8] demonstrated that the net movement of material in the surface region led directly to larger plastic strains and hence more compressive residual stress, through the finite element process modelling of a full DCR treatment.

It can be seen from Figure 3 that the effect of the combined process led to a significant improvement of the compressive surface RS in the lengthwise direction after SP. However, the effect on the crosswise stresses was insignificant as no improvement was observed after SP. This behaviour could be attributed to the fact that the high levels of near-surface stress after DCR had already led to saturation of induced stresses in the crosswise direction, and the plastic deformation in the component was unable to gain the benefits from SP. Overall, this also means that SP post DCR reduces the deviation between crosswise and lengthwise stresses at the surface thereby reducing the directionality of DCR.

4. Discussions and recommendations

Based on the experimental investigations it can be concluded that combination of SP with DCR could introduce the surface stress into Ti6Al4V, thus producing both surface and sub-surface RS into the specimen. However, the introduction of SP after DCR increases the roughness of the surface to about 1.5 μm, negating the surface smoothness produced by DCR (Ra ~ 0.3 μm). This might create minor aerodynamic issues especially if used in compressor components. In order to address this limitation, a new method called Vibro-Peening (VP) is proposed as an alternative to SP [9].

Vibro-Peening (VP) is a process that involves both impact and sliding of media on to a component. With the impact of media on the component being comparable to that of shot peening, the near-surface stresses from VP can be compared to that of SP. The additional sliding of the media on to the surface helps in reducing the surface roughness compared to SP. Hence instead of SP, VP can be combined with DCR to produce both surface and sub-surface CRS into the material and also results in smoother surface than SP. Feldmann et al. [9] compared the roughness between SP and VP in 6 different specimens. It has been proved that the SP surface is rougher than the pre-treated surface while the VP surface has a better finish compared to the pre-treated surface.

In addition to the surface roughness improvement, there are more advantages of using VP as compared to SP, one being the fact that the tool path of SP process consumes a long processing time in order to complete the peening treatment, especially for complex geometry, whereas VP is expected to be faster for larger components. VP can hence be developed for its potential in achieving necessary compressive residual stresses and lower surface roughness as a single step process. Its success would result in considerable reduction of both cost and process time.

5. Conclusion

A feasibility study was performed for combining sub-surface (DCR) and surface (SP) manufacturing techniques to get both sub-surface and surface stresses on a Ti6Al4V flat coupon. The results showed that the lengthwise surface stress induced by DCR was improved by combining it with SP thereby producing surface and sub-surface RS and also made DCR technique more directional. In addition, VP is suggested as an alternative to SP to improve the surface finish.

Acknowledgements

This work was partially conducted within the Rolls-Royce@NTU Corporate Lab with support from the National Research Foundation (NRF) Singapore under the Corp Lab@University Scheme. Andre Lim wishes to acknowledge the support of the Singapore Economic Development Board (EDB) Industrial Post Graduate Programme (IPP).

References

[1] Jata K V, Roy A and Parthasarathy T A 2009 *Encyclopedia of Structural Health Monitoring,* (Hoboken: John Wiley & Sons, Ltd)

[2] Altenberger I 2006 *Shot Peening,* ed L Wagner (Weinheim: Wiley-VCH Verlag GmbH & Co. KGaA) pp 419-34

[3] Schulze V 2006 *Modern mechanical surface treatment: states, stability, effects* (Weinheim: John Wiley & Sons)

[4] Verpoort C and Gerdes C 1989 *Metal Behaviour & Surface Engineering,* , ed S Curioni, *et al.* (Lincoln: IITT International) pp 97-155

[5] Altenberger I 2005 Deep rolling—the past, the present and the future. In: *Proc. 9th Int. Conf. Shot Peening,* ed V Schulze and A Niku-Lari (Paris: IITT-International) pp 144-55

[6] Majzoobi G, Azadikhah K and Nemati J 2009 The effects of deep rolling and shot peening on fretting fatigue resistance of Aluminum-7075-T6 *Mater. Sci. Eng. A-Struct. Mater. Prop. Microstruct. Process.* **516** 235-47

[7] Balland P, Tabourot L, Degre F and Moreau V 2013 Mechanics of the burnishing process *Precis. Eng.-J. Int. Soc. Precis. Eng. Nanotechnol.* **37** 129-34

[8] Lim A, Castagne S and Wong C C 2016 Effect of Deep Cold Rolling on Residual Stress Distributions between the treated and untreated regions on Ti-6Al-4V alloy *J. Manuf. Sci. Eng.-Trans. ASME*

[9] Feldmann G, Wong C C, Wei W and Haubold T 2014 Application of Vibropeening on Aero–Engine Component *Procedia CIRP* **13** 423-8

Features of Creation and Operation of Electric and Hybrid Vehicles in Countries with Difficult Climatic Conditions, for Example, in the Russian Federation

K Karpukhin, A Terenchenko

NAMI Russian State Research Center, 2 Automotornaya st, Moscow, Russia, 125438

Abstract. The trend of increasing fleet of electric or hybrid vehicles and determines the extension of the geographical areas of operation, including the Northern areas with cold winter weather. Practically in all territory of Russia the average winter temperature is negative. With the winter temperatures can be below in Moscow -30°C, in Krasnoyarsk -50°C. Battery system can operate in a wide temperature range, but there are extremes that should be remembered all the time, especially in cold climates like Russia. In the operating instructions of the electric car Tesla Model S indicate that to save the battery don't use at temperatures below -15°C. The paper presents the dependence of the cooling time and heating of the battery cell at different ambient temperatures and provides guidance on allowable cooling time while using and not thermally insulated thermally containers Suggests using the temperature control on the basis of thermoelectric converters Peltier connection from the onboard electrical network of the electric vehicle.

1. Introduction

Personal mobility is a basic need of humans. In 1998, we had a total about 700 million vehicle, but in 2016, we have over 1 billion vehicle in the world. Currently most of the world's automotive countries link the solution of a complex ecological safety and problems of energy saving by regulating CO_2 emissions, as well as increasing adoption in the production of electric vehicles, which can significantly reduce the pollution on the planet. According to forecasts of the European Commission for transport by 2020 in Europe 7 % of all new passenger and light commercial vehicles will be electric, increasing to a massive 31% by 2030. This measure is a forced connection with the sharp deterioration of the situation on the planet associated with air pollution and global warming. The solution of these problems is the creation of electric vehicles, production of which is established in the leading countries of the world.

2. Main part

Modern electric vehicle inferior to the vehicle with the internal combustion engine in the range, the absence of a widely developed infrastructure of charging stations and the initial cost. These disadvantages are compensated, by the advantages of this type of transport. The electric vehicle doesn't emit exhaust gases, noise, doesn't require constant and costly maintenance, charging from a household outlet, can make the path of 150 km or more without recharging and doesn't require refueling with petrol. The production of electric vehicles becomes more widespread and almost all automobile corporations have an electric vehicle in their lineup, as a concept or production model.

Today, is very actively developed, green technology. The use of the vehicle environmental product - electricity, instead of non-renewable natural resource of petroleum do electric vehicles really the future of transportation. Thus, it is likely that the electric vehicle will win in the next decade, the auto market and will take on their leading position. Today, when you create energy-efficient vehicles focusing on reducing energy consumption to overcome the resistance movement, the main of which is the force of rolling resistance, the inertial force of the vehicle when the acceleration and the force of aerodynamic resistance. Given the operating conditions of the electric vehicle in the modern metropolis with high density of traffic with frequent cycles of "acceleration- braking", the availability of high capacity battery is inefficient from the point of view of these losses. Regenerative braking, which is used in modern vehicles, the electric traction, when driving in the conditions described does not provide a balance between energy spent on the acceleration and recharged during braking. Of particular importance in the design of electric vehicle, has a choice of type of storage of electrical energy. Modern batteries have several times the best specific characteristics compared to lead-acid batteries (table 1.), however, electric vehicles are inferior to traditional vehicles with internal combustion engine by the ratio of the total mass and transported payloads (1, 2, 3).

Table 1. Performance characteristics of various battery chemistries.

	Theoretical			Practical		
Battery type	V_{max}	Charge Capacity Ah/g	Energy Density Wh/kg	V_{nom}	Energy Density Wh/kg	Energy Density Wh/L
Lead-acid	2,1	0,12	252	2,0	35	100
NiCd	1,35	0,181	244	1,2	40	90
Ni-MH	1,35	0,178	240	1,2	80	220
Li-NiCo	4,1	0,109	448	3,8	200	420
IMR	3,5	0,122	426	3,0	150	400
LiFePo4	3,65	0,111	405	3,4	115	255
LiTiO	2,8	0,09	252	2,5	75	150
Li-S	2,5	0,341	950	2,15	400	365

As already mentioned, one of the key drawbacks that prevent the widespread use of electric vehicles is the limited mileage (4). The existing problem is difficult to solve the increase in battery capacity, as it leads to increase the cost of the vehicle and the weight gain that, in turn, increases the consumption of energy to overcome forces of resistance. In addition, the bigger battery has a long charge time (up to 7-8 hours). This circumstance is compounded, if the electric car is operated in difficult climatic conditions in countries where winter air temperature is negative from 0 °C to -30 °C.

The climatic conditions of the Russian Federation are among the harshest in the world. For example, in Moscow, winter temperatures can be -30°C, and in Irkutsk -50°C. The Russian Federation has a vast territory and long distances between cities. Lithium-ion battery systems can work in a fairly wide range of temperatures, but there are extremes that should be remembered all the time, especially in countries with challenging climates like Russia. The inherent disadvantages of electric vehicles, superimposed on the climatic features of the countries with a difficult climate, and exacerbate the problems hampering the popularization of electric vehicles.

Negative consumer characteristic of all lithium batteries is that they are very poorly adapted to sub-zero temperatures. As a rule, they allow for discharge at temperatures below -10...-20°C (with reduction of capacity and a given current), and the charge is only at a positive temperature.

Batteries suffer from the cold due to the fact that the electrolyte, which is a conducting substance, the electric battery and provides power transmission between positive and negative pole, in the cold weather thickens, and electrochemical processes in the battery slower. Poor conductivity of the electrolyte increases the internal resistance of the battery system, which produces much less energy (5).

Li-ion batteries are quite well charged with a small positive temperature, and also can be put on a fast charge at a range of temperatures from +5 to +45°C. Process charge and discharge is well to proceed at the highest allowable temperatures (up to +45°C), but this reduces the lifetime of the battery (6, 7). At temperatures below +5°C, the charge current must be lowered. The charge is not valid at temperatures below 0°C, in this case the external changes are observed, but the chemical processes necessary for the correct operation of the battery will be compromised, which could lead to permanent damage to the battery. During charging, at this temperature, the particles of lithium metal can be transferred to the anode, and this coating is not lost during the cycles of charge/discharge (8, 9). The batteries with such a coating become less resilient and can fail under vibrations.

In the NAMI Russian State Research Center for many years works on creation of electric vehicles, adapted to the difficult climate, and explores its nearest competitor. On the roads of avtopoligon NAMI has conducted tests electric vehicle Nissan Leaf (Japan) (10, 11). The aim of experimental researches was to determine the influence of climatic and road conditions on the power reserve of the electric vehicle Nissan Leaf (according to the manufacturer the length of mileage on a single charge of the electric vehicle Nissan Leaf is 160 km). A significant factor influencing the reduction in mileage is higher energy consumption during the winter period of operation, due to the necessity of heating the passenger space and increase the duration of lighting due to short daylight hours. Studies have shown that decreasing ambient temperature from +25 °C to -7°C causes a decrease in the power reserve at 9% and 44% respectively when turned off and on the consumers of energy. With further decrease of the ambient temperature and when the consumers power reserve of electric vehicles was reduced to 50 km. When driving on loose snow with a power reserve of 14% less than when driving on dry asphalt.

Currently, in the NAMI Russian State Research Center created the experimental model of a driverless electric ecovehicle with system energy storage (figure 1 and figure 2).

Figure 1. Driverless electric ecovehicle.

Figure 2. 3D model of system energy storage.

On this vehicle was implemented the experimental system of electric energy storage, intended for validation and investigation of engineering solutions in the field of ensuring of electric energy of the traction module, implementation of algorithms for monitoring and active balancing of single batteries battery adapted to climatic conditions of Russia. The unit for monitoring and balancing is built on a modular principle. Each module allows you to balance up to 12 cells and consists of 2 circuit boards, designed to work with battery from 100 to 400V. BCB LT, which serves all elements relating to the low-voltage parts, and PCB transformers, which connects to the entire capacity of the battery. The system of accumulation of electricity consists of 96 battery cells. Nominal voltage — 317V.

When designing the system of energy storage based on lithium cells was taken into account the negative impact of raising or lowering the temperature of batteries in operation with intensive current impact on the operating period of the battery. In the summer, when the weather is hot, even a simple trip through the city leads to the fact that lithium batteries are heated, and their temperature may exceed the critical value lying in the range +60 ÷ +70°C. In winter the situation is similar, because the temperature below - 5°C affect the mileage of the electric vehicle and the durability of its battery.

Experimental system of accumulation of energy was developed taking into account the use in difficult climatic conditions of the Russian Federation and was equipped with a system termostatics. Calculations were made with the battery in stationary mode with constant current.

Thermal power generated in the battery is calculated by the formula 1:

$$P = I^2 r_{ac} = 12.5^2 \text{ x } 0.0576 = 9 \text{ W} \quad (1)$$

where: I – charging current; r_{ac} – accumulator cells total impedance.

Thus the heat dissipation of the battery during the charge mode will be 9 watts. When charging from an external charging device allowed charging current of 63 A. In this case, the power dissipation of the battery, will be 228.6 W.

Also was calculated the value of the temperature of the battery at the end of the cycle the full charge current of 63 A, on the Duration of this cycle is 1143 seconds. Specific heat capacity of lithium – ion batteries is evaluated with a = 1.1 kJ/kg*^{0}C. The temperature increment for our case is calculated using the following relationship:

$$\Delta T = \frac{P*t}{c*m} \tag{2}$$

where: P - the heating power, is equal to 228.6 W; t – the duration of the charge, t = 1143 sec.;
c - specific heat capacity of lithium – ion batteries; m - the mass of the batteries.

In the result, we find that Δ T = 5^{0}C.Thus, during fast charge temperature battery storage system electricity will increase by 5^{0}C. In both operating modes, i.e. in modes of vehicle movement, the calculation of energy dissipation it is advisable to conduct involving mathematical modeling. For conducting computational research of mathematical simulation of the vehicle model VAZ - 1117, taken as the basis for design of the battery, used the software package AVL CRUISE. The required information can be obtained at a design research integrated current values or the square of the current over the cycle. Figure 3 shows the curve of the integral value of the square of the current of the battery, built on a full cycle, according to the UNECE Regulation No. 83.

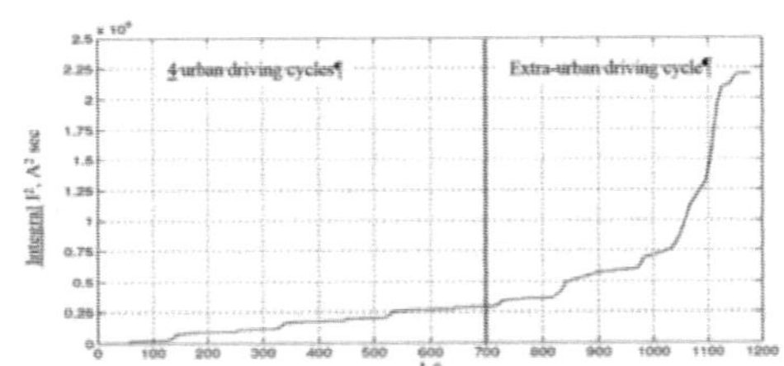

Figure 3. The curve of the squared accumulator battery current integral value on the full cycle, according to the UNECE Regulation # 83.

According to the results of calculations and research, it was decided on the placement of the experimental storage system in insulated container with temperature control system. Temperature control system based on the use of Peltier elements. To ensure compensation of heat transfer was selected for the Peltier module, type TB-127-2.0-1.15 ensuring required temperature in storage tanks at a temperature of exposed surface at +30°C, the main characteristics Peltier module presents on a figure 4.

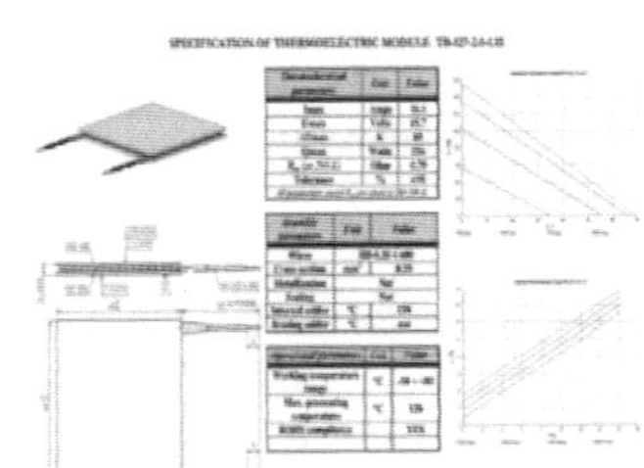

Figure 4. Specification of thermoelectric module TB-127-2.0-1.15.

Using this module allows you to provide temperature control of the battery in this mode of operation level +20 - +25°C. Temperature control system, is powered by the onboard battery voltage of 12 V. In case of temperature increase in the battery above the desired, i.e. above +20°C, the block enable control signal U on. The signal U± is absent. Turn on the fan and the Peltier element in the cooling mode. The inclusion continues until the temperature falling below +20°C. If in conditions of low external temperatures the temperature of the battery decreases below +10°C, the control unit battery are formed from U on; U± and the Peltier element is switched on in heating mode. The fan also turns on. Next, the process continues as before, heating the battery module to achieve the desired temperature. The advantage of Peltier element is small, the absence of any moving parts, gases and liquids. When the direction of the current, possibly as cooling and heating, enabling temperature control at ambient temperatures both above and below the temperature of the temperature control. Were also conducted thermal and gas-dynamic calculations of the application. It is possible to find the optimum layout of the battery module. The problem of gas-dynamic calculation was resolved with the use of a complex of programs allowing to perform solid modeling, and the integration of the air flow by the finite element method. Figure 5 shows how the currents of air near the fans, are distributed without significant turbulence, which suggests that the chosen arrangement minimized the effect of the surge.

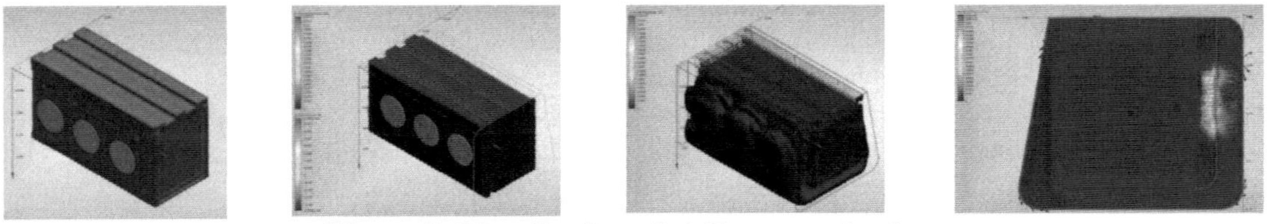

Figure 5. Visualization of airflow near the fans.

As a result, of carried out experimental research into the behavior of two battery modules. In this case, the module number one had an inner insulating coating and an integrated device for temperature control is built on a ventilated thermoelectric Peltier converters, and the module number two hadn't insulation coating. Tests of the two battery modules is presented in figure 6.

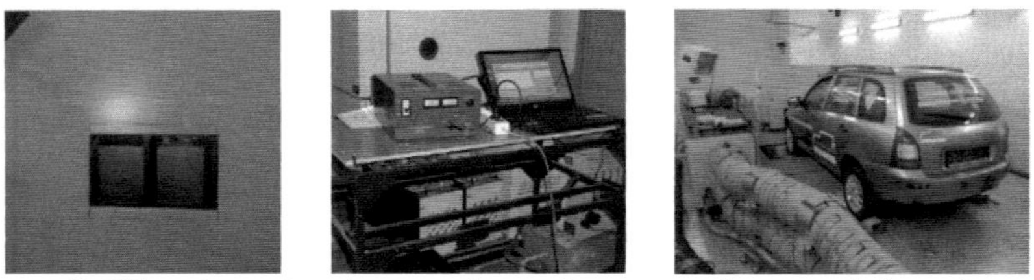

Figure 6. Climatic testing the experimental system of accumulation of energy.

An experimental study showed that the accumulator unit, which had an inner insulating coating and built-in temperature control device was subjected to cooling in a heat chamber at a temperature of -30°C. During the cooling process, temperature control system is included for heating the internal volume of the container. After the temperature is established to a predetermined level, and the cooling process inside the battery module from ambient temperature to a predetermined temperature lasted about seven hours, the power supply system of temperature control was 69 watt, including the energy consumed by the fan. This coincides with the calculated value for the required heating and shows that in the heating mode thermoelectric Peltier work together effectively. The temperature inside an insulated module maintained at +17°C - +20°C, all battery cells are charged evenly at a temperature in the heat chamber – 30°C and applying a charging current of 120 A. In the module that had no insulation coating of any abnormal phenomena in the process of heating or cooling the battery cells not shown, however, when the temperature of the battery module below – 10 °C the charge of the cells was absent. Reviewed the processes of charge at 0°C and +5°C. After a long exposure, about 7 hours, at these temperatures revealed that at 0°C the charge current 5A cells was perceived in different ways, and some of the cells, about 8% of the total, the charge was not accepted. At temperatures above+5°C, all cells charged evenly.

3. Results and Disccusion

The results of the study allow to formulate the main aspects of ensuring the effective functioning of the battery systems of electric vehicles and vehicles with a hybrid powertrain in difficult climatic conditions. Operation not protected against ambient temperatures below +5°C accumulator systems are not desirable, as leading to accelerated degradation of the batteries. Vehicles containing an insulated battery system, allowed to leave at negative external temperatures for a limited time, decreasing with decreasing temperature. Prolonged Parking is permitted in the conditions of heated rooms or when using independent heating. When implementing independent heating battery systems, for example, by using thermoelectric converters Peltier, daily power consumption is at -30°C of not less than 1.6 kWh. Therefore, if the long Parking time is not known, and electric cars with a hybrid powertrain must be connected to a stationary power grid, or in vehicles with a hybrid powertrain to provide periodic startup to warm the engine and recharge the battery system.

4. Conclusion

(1) Claimed by the manufacturers of electric vehicle operational performance, in particular range and required charging time of the batteries, often aren't focused on the country's complex climate, and to the middle European climate zone, excellent road and charging infrastructure.
(2) Operation not protected against ambient temperatures below +5°C and above +50°C battery systems is not desirable, as leading to accelerated degradation of the batteries.
(3) Vehicles containing an insulated battery system, allowed to leave at negative external temperatures for a limited time, decreasing with decreasing temperature. Prolonged Parking is permitted in the conditions of heated rooms or when using independent heating.
(4) When implementing independent heating battery systems, for example, by using thermoelectric converters Peltier, daily power consumption is at -30°C of not less than 1.6 kWh. Therefore, if the long Parking time is not known, hybrid and electric vehicles must be connected to a stationary power grid or hybrid vehicles to provide periodic startup to warm the engine and recharge the battery system.
(5) Developed temperature control system of energy storage devices designed for hybrid and electric vehicles operating in difficult climatic conditions. Practical application of the developed principle of temperature control enables operation of the rechargeable battery in the most favorable temperature range 20÷30°C.

Acknowledgments

The paper was prepared under the agreement # 14.625.21.0006 with the Ministry of Education and Science of the Russian Federation (unique project identifier RFMEFI62514X0006).

References

[1] John M. Miller, 2010. Propulsion systems for hybrid vehicles 2nd Edition *M. UK: The Institution of Engineering and Technology.*
[2] John Wiley & Sons 2014 Encyclopedia of Automotive Engineering *M.* ISBN: 978-0-470-97402-5 Online.
[3] International Energy Agency. Hybrid and Electric Vehicles *M. The Electric Drive Delivers*, 2015.
[4] Selifonov V, Baulina E, Karpukhin K, Electric vehicle very small class with a hybrid power installation 2004 *J. Automotive electrical equipment*, **9** pp.20-22.
[5] Gaines L, Cuenca R, Life-Cycle Costs of Lithium-Ion Vehicle Batteries 2000-01-1483 *J. SAE paper*, p.12.
[6] Karpukhin K, Terenchenko A, Shorin A, Bakhmutov S, Kurmaev R 2015 Temperature control of the battery for hybrid or electric vehicle. *J. Biosciences biotechnology research Asia*, **12 (2)** pp. 1297-1301.
[7] Abdullah AL-Refai, Osamah Rawashdeh, Rami Abousleiman. 2016. An Experimental Survey of Li-Ion Battery Charging Methods. *J. SAE International Journal of Alternative Powertrains*, **5 (1)**.
[8] Slabosritsky R, Khazhmuradov M, Lukyanova V, Analysis and calculation of battery cooling system 2011 *J. Radioelectronics and informatics*, **3** pp.3-8.
[9] Slabosritsky R, Khazhmuradov M, Lukyanova V, Studying of battery cooling system 2012 *J. Radioelectronics and informatics*, **2** pp.5-8.

[10] Terenchenko A, Karpukhin K, Kurmaev R 2015 Features of operation of electromobile transport in the conditions of Russia. *J. Paper of EVS 28 International Electric Vehicle Symposium and Exibition*, KINTEX, Korea.

[11] Kurmaev R, Terenchenko A, Karpukhin K, Struchkov V, Zinov'ev E. 2015 Maintaining the required temperature of high-voltage batteries in electric cars and hybrid vehicles. *J. Russian engineering research*, **35 (9)** pp. 666-669.

IOP Conf. Series: Materials Science and Engineering **157** (2016) 012015 doi:10.1088/1757-899X/157/1/012015

Generic Automated Multi-function Finger Design

M Honarpardaz [1,2], M Tarkian [2], D Sirkett [1], J Ölvander [2], X Feng [1,2], J Elf [2] and R Sjögren [2]

[1] ABB, Corporate Research, Västerås, Sweden.

[2] Machin Design Division, Linköping University, Linköping, Sweden.

Abstract. Multi-function fingers that are able to handle multiple workpieces are crucial in improvement of a robot workcell. Design automation of multi-function fingers is highly demanded by robot industries to overcome the current iterative, time consuming and complex manual design process. However, the existing approaches for the multi-function finger design automation are unable to entirely meet the robot industries' need. This paper proposes a generic approach for design automation of multi-function fingers. The proposed approach completely automates the design process and requires no expert skill. In addition, this approach executes the design process much faster than the current manual process. To validate the approach, multi-function fingers are successfully designed for two case studies. Further, the results are discussed and benchmarked with existing approaches.

1. Introduction

This paper aims to provide a generic and reliable method for design automation of multi-function fingers for industrial robot grippers. Multi-function fingers allow the gripper to handle multiple workpieces with the same fingers.

In most industrial assembly applications, multiple workpieces are required to be handled in the assembly process. Three solutions for multiple workpiece handling are common in the robot industry.

1. To change the gripper (tool) for each assembly task. This incurs a significant cycle time penalty and consequently reduces the throughput of the workcell.

2. To use multi-gripper (or interchangeable) end-effectors [1]. These end-effectors usually have 2-3 grippers that are attached to a rotatory wrist mechanism which activates the relevant gripper for the specific task. This solution is expensive and reduces the graspable payload of the robot considerably due to the high weight of the additional grippers and interchange mechanism [1], [2].

3. To use multi-function fingers.

Designing multi-function fingers enables the robot to handle more than one workpiece without any tool-change operation. As a result, multi-function fingers play a vital role in reducing assembly cycle time and increasing workcell throughput [3]. On the other hand, the manual design of practical multi-function fingers requires a high level of skill and experience. Besides, the current manual design process consists of several time consuming iterations which may take weeks to accomplish. Thus, automating the design procedure of multi-function fingers is highly valuable for robot industries [4].

Existing approaches for multi-function finger design automation are limited to workpieces having specific geometrical properties. Therefore, in this paper, Generic Automated Finger Design 2.0 (GAFD 2.0) is proposed to automate the multi-function finger design independent of the geometrical properties of the workpieces. GAFD 2.0 is a further development of the previous version of the GAFD.

Section 2 states the relevant work in this research area and the overview of the GAFD method is presented in section 3. Section 4 describes the methodology that is utilized in this study and Results of the implementation of the proposed approach are presented in section 5. Section 6 compares the results to existing methods and section 7 concludes the contributions.

2. Relevant Work

A review of the literature reveals that few studies [1]–[3], [5]–[11] consider the importance of multi-functional fingers in robot applications and only a couple of works [7], [11] propose approaches for the design automation of multi-function fingers [4].

In [2], [3], [9] multi-function fingers are consider to be one of the significant methods for increasing throughput of the robot workcell. Causey and Quinn [3] assert that multi-function fingers are cheaper, more effective and reliable than tool changing systems and rotary wrist mechanisms.

Pham et al. [10] present a trial and error based approach to achieve multi-function fingers. Their method begins by categorizing similar workpieces into various groups. Each group contains workpieces with similar geometrical properties (e.g. shape, mass, etc.). Then, a set of fingers are designed for a group of workpieces (e.g. group A). The designed fingers are used to grasp another group of workpieces (e.g. group B). If the fingers can grasp both groups of workpieces, they are considered as the final finger design. Otherwise the process is repeated yet this time, the initial fingers are designed for the group B workpieces and then checked against group A. In the case of failure in both iterations, the algorithm designs a set of fingers for each group.

Velasco and Newman [7] propose a CAD-based approach to design multi-function fingers for two workpieces. In the proposed method, the geometric superposition of the workpieces is determined and assumed to be the only workpiece to be handled. In [7] finger design approach, fingers are assumed to be solid blocks from which the geometry of the workpiece is subtracted. The remaining geometry forms the designed fingers. Velasco and Newman further propose an additional hypothesis for geometrically-similar workpieces.

In this hypothesis, two sections from the grasping sides (left and right) of each workpiece are extracted and a hypothetical workpiece is created by superposing the sections extracted from all workpieces. Then the hypothetical workpiece is subtracted from the finger blocks. The result is considered as the multi-function fingers design.

3. GAFD Method Overview

The current manual procedure for designing fingers requires several exhaustive and time-consuming trial and error iterations. Earlier work [12] proposed Generic Automated Finger Design (GAFD) as a successful method to automate the finger design procedure using the customized design approach. As shown in Fig. 1, the presented design procedure consists of three stages of data arrangement, grasp and finger design.

In the data arrangement stage, two parallel processes prepare the required data for executing the key process in the grasp and finger design stages:

- The first parallel process imports the workpiece CAD model and the gripper properties (i.e. gripper model, maximum finger length, maximum workpiece weight, maximum jaws stroke). Then the surfaces of the workpiece are reproduced and discretized (meshed). Based on the meshed surfaces and the element nodes, a point cloud model of the workpiece is generated. In the final step, surface-normals at each point are extracted.
- In the second parallel process, material properties of the fingers and workpiece are imported to the system. Following this, the friction coefficient (μ) of the contacts are extracted and the contact model is determined based on the value of μ. Then the minimum number of contact points are obtained in regard to the contact model. The grasp stage begins by randomly selecting n points and generating all the possible grasp sets based on the minimum number of contact points. Then the contact wrenches of all the grasp sets are determined and the convex hull of the contact wrenches for each grasp set is calculated. The stability of the grasp sets is

verified by checking the location of the origin of the wrench space [13], [14]. The grasp sets whose origins lie inside the convex hull are considered stable and listed as feasible grasp sets.

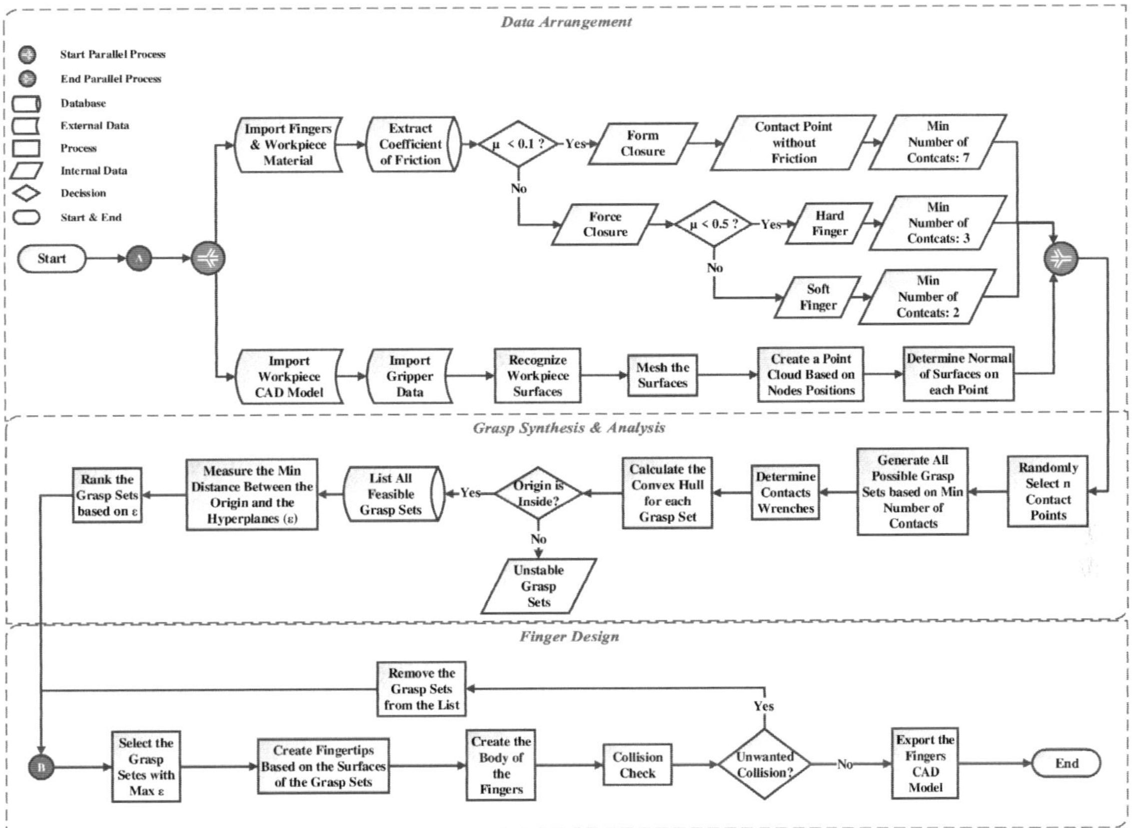

Figure 1. Finger Design Automation Process Flowchart of GAFD.

In the next step, the minimum distance between the origin of the grasp sets and the hyperplanes of the convex hull (ε) are measured. The list of the feasible grasp sets are ranked based on the value of the ε. The larger the value of the ε, the more stable the grasp set [15], [16].

In the final stage of GAFD, finger design, the grasp set with the largest ε is selected and the fingertips are designed. The designed fingertips mimic the contour of the surface of the workpiece at the contact point to increase the contact area and the reliability of the fingers.

Then the bodies of the fingers are designed based on the gripper properties and the grasp types (i.e. internal and external grasp). The feasibility of the designed fingers is verified by detecting whether the grasp approach is collision-free. If the approach is free of unwanted collisions, the designed fingers are considered as feasible and CAD models are exported for the production and experimental verification [12].

4. Method

This section describes the utilized methodology in this study to automate the multi-function finger design for industrial grippers. With reference to Fig. 1, point A represents the beginning of the data arrangement and point B denotes the end of the grasp stage. The data arrangement and grasp stages used in this work are identical to those presented in the earlier version of the GAFD, as described in detail in [12].

Fig. 2 demonstrates the procedure that is employed to design multi-function fingers. This procedure begins by selecting the grasp set with the highest quality and creating fingertips based on the surfaces of the workpiece at the contact points. This feature enables the fingertips to imitate the surface contour of the workpiece. Then, the algorithm counts the number of workpieces that are required to be handled. If the number of workpieces is more than one, the area of the projection of the fingertips' surface (A_p) is measured and the value of the measured area is listed in a database. Then the steps in the data arrangement and grasp stages (point A to point B) are repeated for the next workpiece. The best grasp set for the second workpiece is selected from the ranked list of the feasible grasp sets.

Next, the fingertips are created based on the selected grasp sets and then the A_p of the fingertips are measured and stored in the database. This process is repeated until fingertips are created for all workpieces and the A_p of the fingertips are measured and listed in the database. In the next step, the list of A_p is sorted so the fingertip with the largest A_p stands at the top of the list and the fingertip with the smallest A_p at the bottom. For cases where the areas, A_p of two fingertips are identical, the fingertip with the smaller actual contact area (not projected) (As) stands above the other fingertip in the list.

By sorting all the fingertips, the one with the largest A_p is selected as the main fingertip. The fingertip with the second largest A_p is selected and its contact surface is located on the contact point of the main fingertip. Then the contact surface of the second fingertip is subtracted from the main fingertip. The remaining geometry of the first fingertip is considered as the main fingertip and the second fingertip is removed from the list of the fingertips area. This process is repeated until the list of the fingertips area is empty. The main fingertips are used in the next step of the process to generate the finger bodies.

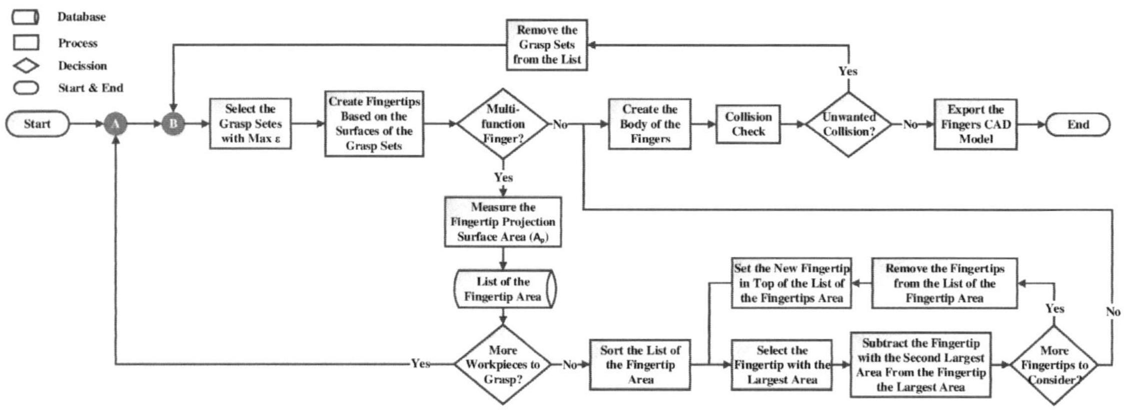

Figure 2. Multi-function Finger Design Automation Process Flowchart of GAFD 2.0.

To check the feasibility of the multi-function fingers produced by the method, the collision detection process verifies whether the grasp approach of the fingers is free from unwanted collisions for every workpiece. If the designed multi-function fingers are found to be collision-free, CAD models of the fingers are exported to be manufactured and tested in practice. Similar to the earlier version of the GAFD, this version (GAFD 2.0) is also implemented in CATIA V5 [17] and automated by implementing the described logic with Visual Basic script [12].

5. Results

In this section, the results of applying the developed version of GAFD (GAFD 2.0) on two workpieces are presented. As shown in Fig. 3, the chosen workpieces are a ball (Fig. 3 (a)) and a rack gear (Fig. 3 (b, c)). To properly demonstrate the capability of the proposed approach, two different case studies are defined and the GAFD 2.0 design process (see Fig. 2) is implemented on each case study. The case study 1 consists of the ball and the gear in its first stable pose (Fig. 3 (a, b)) and in the case study 2, the ball and the gear in its second stable pose are considered (Fig. 3 (a, c)). In each case study, the presented algorithm automatically generates fingers that are able to grasp both workpieces. As stated earlier, the processes carried out in the data arrangement and grasp stages (point A to point B) are not replicated in this work (see [12]). Thus, only the ranked list of the grasp sets is presented here as the result of the grasp stage.

Figure 3. CAD model of the: (a) ball; (b) rack gear in stable pose 1; (c) rack gear in stable pose 2.

A. Fingertip Design

Once the grasp set location is known, the finger design stage commences with the design of the fingertips. In both case studies, the ball is considered as the first workpiece and the gear as the second. However, the order of the workpieces does not affect the result. According to the design process (Fig. 2), the first step is selecting the best grasp set for the ball from the grasp set list (see Fig. 4 (a)). Then the fingertips are created in such a way that the contact surface conforms to the surface contour of the ball (see Fig. 5 (a)). To do so, a preliminarily-designed fingertip is extruded to meet the contact surface of the workpiece. As a result the contact surface of the fingertip mimics the surface contour of the workpiece.

In the next step of the process, the area of the projection of the contact surface (A_p) is measured and stored in a database. Then the second workpiece (i.e. the gear) is imported to CATIA and the ranked list of the grasp sets is obtained from the grasp stage. In a similar manner to the processes applied to the ball, the best grasp set for the gear is selected. Fig. 4 (b) and Fig. 4 (c) represent the best grasp sets for case study 1 and case study 2 respectively. Then the fingertips of the gear are generated and the A_p of the finger tips is measured and stored in the database.

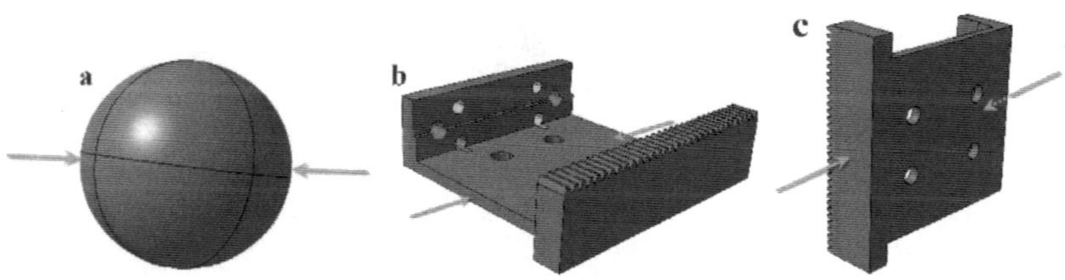

Figure 4. Best grasp set of the: (a) ball in both case studies; (b) gear in case study 1; (c) gear in case study 2.

As shown in Table 1, in the case study 1 the A_p of the ball is larger than of the gear; and in the case study 2, the gear and the ball have the same A_p. Thus, the actual contact surface area (A_s) of the workpieces are measured and the workpiece with the smaller A_s (i.e. the gear) stands at the top of the list.

The first fingertips in the list are selected as the main fingertips which are the ball-gripping fingertips for case study 1 (Fig. 5 (a)) and the gear-gripping fingertips for case study 2 (Fig. 5 (c, e)). Then the contact surface of the second workpiece is coincident with the main fingertips at the contact points. To ensure the contact surfaces of the workpieces do not fully overlap, the contact surface of the second workpiece is slightly translated towards the fingertip base along the grasp axis. Then the contact surface of the second workpiece is subtracted from the main fingertips. Fig. 5 (b) and Fig. 5 (d, f) illustrate the fingertips generated for case study 1 and case study 2, respectively.

Table 1. The area of the projection of the fingertips' surface (A_p) and actual contact area (A_s) for the ball and the gear. Numbers with underlines represent main fingertips.

		Ball		Gear		Selected Main Fingertip
		A_p [mm^2]	A_s [mm^2]	A_p [mm^2]	A_s [mm^2]	
Case Study 1	**Fingertip 1 (right)**	**230**	241	38	38	Ball
	Fingertip 2 (left)	**230**	241	38	38	Ball
Case Study 2	**Fingertip 1 (right)**	169	174	169	**169**	Gear
	Fingertip 2 (left)	160	165	**172**	172	Gear

		Main Fingertip	Final Fingertip Design
Case Study 1	Both Fingers	a 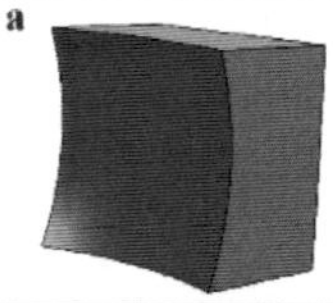	b 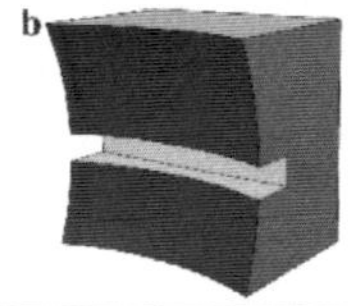
Case Study 2	Finger 1 (right)	c 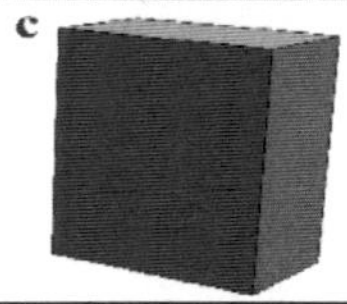	d 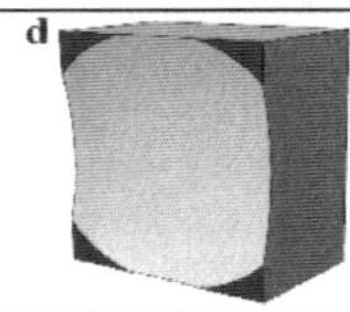
	Finger 2 (left)	e 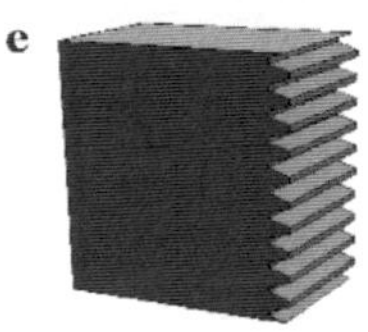	f 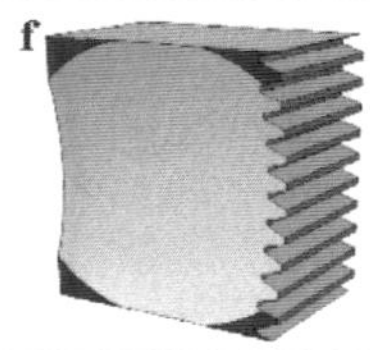

Figure 5. Case study 1: (a) main fingertip for both fingers; (b) final fingertip design for both fingers. Case study 2: (c) main fingertip for finger 1 (right); (d) final fingertip design for finger 1; (e) main fingertip for finger 2; (f) final fingertip's design for finger 2.

B. Finger Body Design

The next step in the finger design process is generating the body of the finger. Considering the fact that the finger body is responsible for properly connecting the fingertip to the gripper jaw, a group of parametric lines are defined as a guideline for designing the finger body. The algorithm automatically adjusts the parameters of the lines to ensure the feasibility of the design while taking the gripper properties (e.g. maximum finger length, maximum jaws stroke, etc.) into the account. Then the Sweep feature (in CATIA) is used to follow the guideline and construct the solid finger body. Fig. 6 (a) and Fig. 6 (b) respectively represent the fingers designed for case study 1 and case study 2.

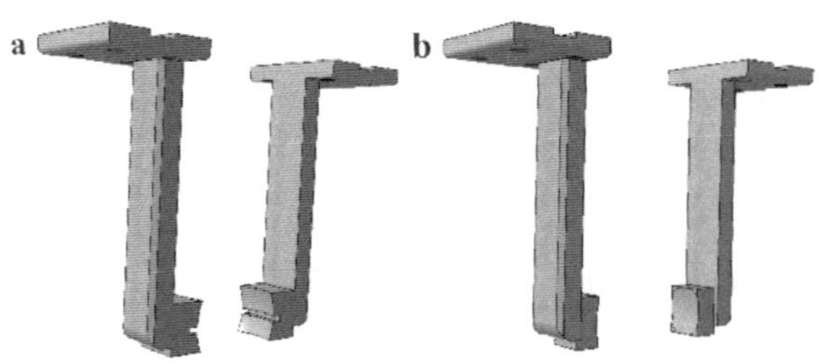

Figure 6. Designed multi-function fingers for the: (a) case study 1; (b) case study 2.

C.Collision Detection

The final step in the proposed design process is verifying the feasibility of the generated fingers. The feasible multi-function fingers should be able to grasp both workpieces without incurring unwanted collisions with the workpieces. Therefore, the inbuilt collision detection module of CATIA (Clash Analysis) is employed to this end. As shown in Fig. 7, the designed multi-function fingers for both case study 1 (Fig. 7 (a, b)) and case study 2 (Fig. 7 (c, d)) are collision free and feasible. As a result, the proposed approach shows good potential for successfully designing multi-function fingers for handling multiple workpieces. The execution time to accomplish the design process for the case study 1 and 2 is respectively 1670 and 1840 seconds on a 2.7 GHz Intel i7 with 16 GB RAM running Windows.

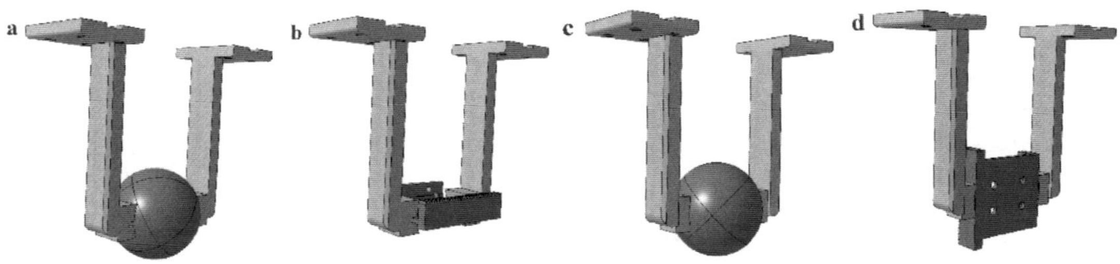

Figure 7. Collision detection of the designed fingers for the: (a, b) case study 1; (c, d) case study 2.

6. Discussion

This section discusses the pros and cons of the few existing approaches for the multi-function finger design automation. Further the results that are obtained in this work are benchmarks with the existing approaches.

Pham et al. [10] propose an iterative approach for producing multi-function fingers. The algorithm uses a trial and error approach to detect whether the designed fingers for a group of workpieces are able to grasp another group. While this approach is simple and easy to implement, it has the following weaknesses and limitations:

- The approach is effective only for similar groups of workpieces.
- The approach is unable to ensure a reliable grasp as the fingertips do not fully match the surface of the workpiece.
- The efficiency of the approach falls considerably with increasing number of workpieces.

In contrast to [10] method, [7] present a CAD-based approach for designing multi-function fingers. Their proposed approach consists of two complementary methods. The first method geometrically superposes the workpieces that are required to be handled and considers the superposition result as the only workpiece to be grasped. Then the superposed workpiece is subtracted from the finger blanks. The outcome of the fingers geometry should be able to grasp all the workpieces. However, this method is inapplicable for similar workpieces with different dimensions (e.g. two spheres with different radii). As a consequence, the second method is utilized to design fingers for workpieces with similar geometries. In the second method, two sections (from the right and left side) of each workpiece are extracted. Then, all the sections of the workpieces are superposed and a hypothetical object is created. The method uses the hypothetical object and subtracts it from the finger blanks. The remaining geometries are considered as the multi-function fingers. Although [7] approach is more generic than the one presented by [10], the approach suffers the following shortcomings:

- The approach encounters difficulties finding a solution for workpieces with complex geometries.
- The presented approach lacks a logical method for superposing the extracted section.
- The approach is unable to guarantee a feasible solution. For instance, a cube and a sphere with equal length and radius, respectively.

In consonance with the results of this study, the developed GAFD process for designing multi-function fingers overcomes the limitations of the existing approaches in the literature. The elaborated design process consists of the succeeding advantages:

- The approach is generic and independent of the geometrical complexity of the workpieces.
- The approach can ensure the stability of the grasp by utilizing grasp synthesis and analysis.
- The algorithm completely automates the design process.
- The approach verifies the designed multi-function fingers using a collision detection process.
- The approach has the potential to be quicker than the current manual design process.

It should be noticed that some of time-consuming processes in the manual design procedure (i.e. fingers manufacture and experimental verification) are not considered in this benchmark, thus comparing designs lead-time cannot be precise.

7. Conclusion

This paper presents the development of Generic Automated Finger Design (GAFD) algorithm for designing multi-function fingers for industrial robot grippers. The proposed approach automates the design process for multi-function fingers to overcome the complexity and lengthy lead time of the current procedure and is able to handle workpieces with complex geometries. In addition, the approach can ensure the stability of the grasp and produces fingers with a smaller foot-print in comparison to other approaches that are therefore more suitable for assembly applications.
The future work suggestions are pointed out as follows:

- Experimental verification of the designed multi-function fingers using GAFD 2.0.
- Benchmarking the performance of the multi-function fingers designed by GAFD 2.0 with other approaches.
- Optimizing the GAFD 2.0 design process to reduce lead time.
- Developing the algorithm to be able to design multi-function fingers for the combination of internal and external grasps.

Acknowledgement

The research leading to these results has received funding from the European Community's Framework Program Horizon 2020 – under grant agreement No 644938 – SARAFun.

References

[1] Hodge Jenkins, "Design of Robotic End Effectors," in *Robotics and Automation Handbook*, 0 vols., CRC Press, 2004.
[2] G. C. Causey, "Elements of agility in manufacturing," 1999.
[3] G. C. Causey and R. D. Quinn, "Gripper design guidelines for modular manufacturing," in *1998 IEEE International Conference on Robotics and Automation, 1998. Proceedings*, 1998, vol. 2, pp. 1453–1458 vol.2.
[4] M. Honarpardaz, M. Tarkian, J. Ölvander, and X. Feng, "Finger Design Automation for Industrial Robot Grippers: A Review," *Submitted*, 2015.
[5] V. B. Velasco, Jr., and W. S. Newman, *An Approach to Automated Gripper Customization Using Rapid Prototyping Technology*. 1996.
[6] V. B. Velasco, W. S. Newman, Y. Zheng, and S. Choi, "Automated Gripper and Fixture Customization via Rapid Prototyping," 1998.
[7] J. Velasco V. B. and W. S. Newman, "Computer-assisted gripper and fixture customization using rapid-prototyping technology," in *1998 IEEE International Conference on Robotics and Automation, 1998. Proceedings*, 1998, vol. 4, pp. 3658–3664 vol.4.
[8] D. T. Pham and S. H. Yeo, "A knowledge-based system for robot gripper selection: criteria for choosing grippers and surfaces for gripping," *Int. J. Mach. Tools Manuf.*, vol. 28, no. 4, pp. 301–313, Jan. 1988.
[9] Greg Causey, "Guidelines for the design of robotic gripping systems," *Assem. Autom.*, vol. 23, no. 1, pp. 18–28, Mar. 2003.

[10] D. T. Pham, N. S. Gourashi, and E. E. Eldukhri, "Intelligent conceptual design of robot grippers for assembly tasks," in *Intelligent Production Machines and Systems: First I*PROMS Virtual Conference, 4-15 July 2005*, Amsterdam, 2005, pp. 321–326.
[11] D. T. Pham, N. S. Gourashi, and E. E. Eldukhri, "Automated configuration of gripper systems for assembly tasks," *Proc. Inst. Mech. Eng. Part B J. Eng. Manuf.*, vol. 221, no. 11, pp. 1643–1649, Nov. 2007.
[12] M. Honarpardaz, M. Tarkian, J. Ölvander, X. Feng, and D. Sirkett, "Generic Automated Finger Design," in *submitted*, 2016.
[13] J. K. Salisbury, "Kinematic and Force Analysis of Articulate Hands," 1985.
[14] B. Mishra, J. T. Schwartz, and M. Sharir, "On the existence and synthesis of multifinger positive grips," *Algorithmica*, vol. 2, no. 1–4, pp. 541–558, Nov. 1987.
[15] C. Ferrari and J. Canny, "Planning optimal grasps," in , *1992 IEEE International Conference on Robotics and Automation, 1992. Proceedings*, 1992, pp. 2290–2295 vol.3.
[16] J. C. Trinkle, "A Quantitative Test For Form Closure Grasps," in , *Proceedings of the 1992 IEEE/RSJ International Conference on Intelligent Robots and Systems, 1992*, 1992, vol. 3, pp. 1670–1677.
[17] CATIA, "CATIA," 2016. [Online]. Available: http://www.3ds.com/. [Accessed: 02-Feb-2016].

Power-based Shift Schedule for Pure Electric Vehicle with a Two-speed Automatic Transmission

Jiaqi Wang[1], Yanfang Liu[1, *], Qiang Liu[1] and Xiangyang Xu[1]

[1] School of Transportation Science and Engneering, Beihang University

* Correspondence: liuyf@buaa.edu.cn

Abstract. This paper introduces a comprehensive shift schedule for a two-speed automatic transmission of pure electric vehicle. Considering about driving ability and efficiency performance of electric vehicles, the power-based shift schedule is proposed with three principles. This comprehensive shift schedule regards the vehicle current speed and motor load power as input parameters to satisfy the vehicle driving power demand with lowest energy consumption. A simulation model has been established to verify the dynamic and economic performance of comprehensive shift schedule. Compared with traditional dynamic and economic shift schedules, simulation results indicate that the power-based shift schedule is superior to traditional shift schedules.

1. Introduction

Electric vehicles (EVs) with a multi-speed automatic transmission have become more attractive in recent years on vehicle marketing [1, 2]. Shift schedule is a significant designing factor for EVs equipped with automatic transmission [3]. It will directly affect the drivability and economy of vehicles [4]. Traditional shift schedule is often classified as dynamic shift schedule and economic shift schedule, both considering about only one factor [5]. In order to achieve the driver's power demand with low energy consumption, a comprehensive shift schedule should be widely used in EVs.

There are some optimal methods to optimize the gear selection of shift schedule to improve the vehicle performance [6, 7]. Hou Rui has come up with a dynamic programming (DP) method to optimize shift schedules by changing different control weight. The economic benefit can be increased 0.94-2.28%. However, the dynamic characteristic may be reduced by 1.55-3.52% [8]. Yu Xiang proposed a DP algorithm to map the optimal gear operational points. Simulation result of calculating the battery state of charge (SOC) indicates that the energy consumption economy of EV could be increased 5.5%. But the dynamic characteristic is not mentioned [9]. Lulu Guo presents that both model predictive control (MPC) strategy and DP algorithms could insure the 2-speed EV dynamic performance. MPC could better improve economy performance [10]. In addition, Lulu Guo considers about the road grade and improves the energy efficiency about 6.57% [11].

Some researches just consider about the economic characteristics, but neglect the dynamic performance of EVs. In this paper, the comprehensive shift schedule simultaneously considers the dynamic and economic characteristics. Principles to design the power-based comprehensive shift schedule are proposed. Simulations to compare about three kinds of shift schedule are shown in this paper.

2. The two-speed automatic transmission

The research objective of this paper is a pure electric vehicle with a two-speed automatic transmission. Table 1 is the shift logic and gear ratios. Table 2 shows the design parameters of the EV.

Table 1. The shift logic and gear ratios of the 2-speed automatic transmission.

Gear	Ratio	Shift Clutch	
		C1	C2
1st gear	9.185	engaged	disengaged
2nd gear	5.609	disengaged	engaged

Table 2. The main design parameters of aim vehicle.

	Parameter	Value	
	Maximum mess m (kg)	2256	
	Wind area A (m2)	2.2	
	Coefficient of air resistance C_D	0.35	
	Coefficient of rolling resistance f	0.015	
	Wheel rolling radius r_w (m)	0.3217	
Dynamic	Maximum speed (km/h)	≥150(instantaneous)	≥130(continuous)
	0-100km/h Accelerate time (s)	≤13	
	Maximum grad ability (%)	≥30%	
Economic	Limited driving distance(km)	NEDC ≥100	60km/h ≥140
	100km energy consumption(kWh)	NEDC ≤18	60km/h ≤15

3. Dynamic model of electric vehicle

3.1. *Drive motor model*

The drive motor output torque T_M is a piecewise function of the accelerator pedal opening α and the motor speed ω_M, separated by base speed ω_{Mb} as:

$$T_M = \begin{cases} \alpha T_{M\max}, & 0 \le \omega_M \le \omega_{Mb} \\ \frac{\alpha P_{M\max}}{\omega_M}, & \omega_M > \omega_{Mb} \end{cases}, \tag{1}$$

where α ranges from 0 to 1, T_{Mmax} is the maximum motor torque, P_{Mmax} is the maximum motor power. Figure 1(a) shows the experimental characteristic curve for the drive motor with different accelerator pedal opening. The following shift schedule in this paper is based on these motor experimental data. Figure 1(b) is the motor efficiency contour and external characteristic curve.

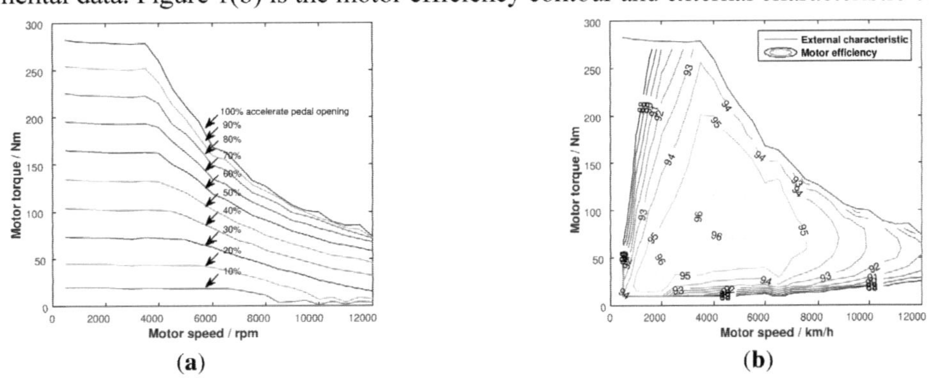

Figure 1. (a) The torque-speed characteristic curve of drive motor; **(b)** The motor efficiency contour and external characteristic curve.

3.2. *Automatic transmission model*

The function of automatic transmission is selecting a proper gear to change the speed ratio of powertrain. The speed relationship between the input and output of transmission is:

$$\omega_{S2} = \frac{\omega_{S1}}{i_k \cdot i_0}, \tag{2}$$

where ω_{S1} and ω_{S2} are the speed of input and output, i_k and i_0 are the ratio of normally engaged gears and the final reduction gears. When k=1, $i_k \cdot i_0$ is the total ratio of the first gear. When k=2, $i_k \cdot i_0$ is the total ratio of the second gear.

3.3. *Vehicle dynamic model*

This model is to calculate the vehicle driving resistance torque. The vehicle speed has a relationship with the transmission output speed ω_{S2} as:

$$v_v = \omega_{S2} \cdot r_W, \tag{3}$$

where v_v is the vehicle speed and r_W is the rolling radius of wheels. The vehicle driving resistance force can be calculated, containing the air F_D, the rolling F_R, and the gradient resistance forces F_G:

$$\begin{aligned} F_D &= \tfrac{1}{2} C_D A \rho (v_v - v_a)^2 \\ F_R &= fmg\cos\beta \\ F_G &= mg\sin\beta \end{aligned} , \tag{4}$$

where C_D, A, f and m are shown in Table 2. ρ is the dentist of air, v_a is the speed of wind, g is the gravity coefficient, β is the gradient. Then the vehicle driving resistance torque is:

$$T_{DR} = (F_D + F_R + F_G) \cdot r_W. \tag{5}$$

4. Shift schedule

4.1. *Dynamic shift schedule*

For a specific vehicle speed v and a specific accelerator pedal opening α, the relationship curve between traction torque and vehicle speed can be obtained with different ratio [12]. The dynamic shift curve can be plotted by connecting intersections for different pedal opening as shown in Figure 2. In order to void cyclic shift, the downshift curve should be delayed 5-10 km/h.

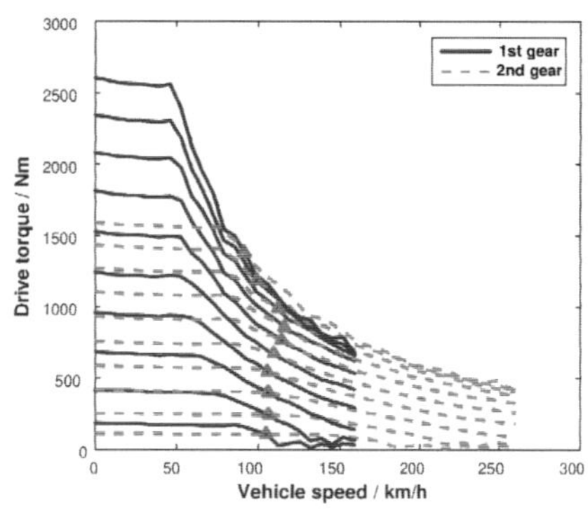

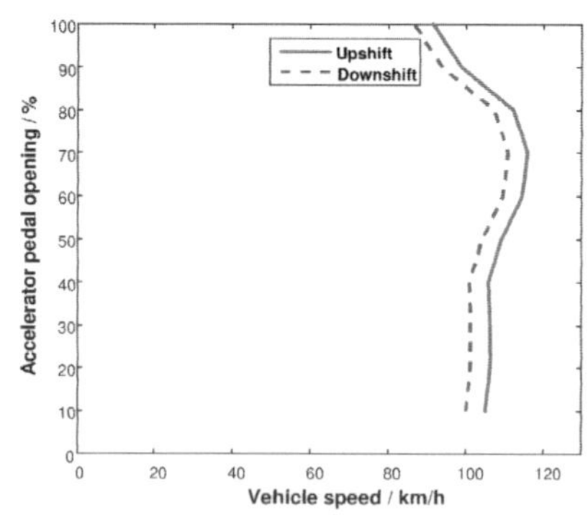

Figure 2. (a) Intersections of drive torque for different pedal opening; **(b)** Dynamic shift curve.

4.2. *Economic shift schedule*

Figure 3(a) shows the relationship curve between motor efficiency and vehicle speed in different pedal opening. The economic shift curve is obtained by connecting efficiency intersections [13]. Figure 3(b) shows the economic shift curve. The downshift curve is also delayed 5-10 km/h.

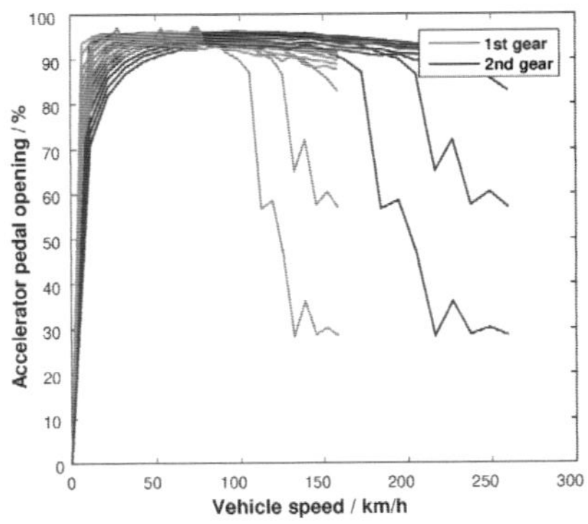

(a)

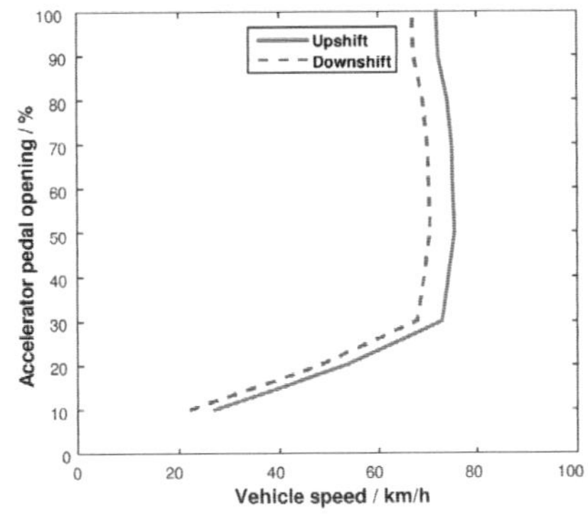

(b)

Figure 3. (a) Intersections of motor efficiency for different pedal opening; **(b)** Economic shift curve.

4.3. Power-based shift schedule

The above shift schedules have their limitations due to the only one-factor consideration about the vehicle dynamic or economic performance. This paper introduces a method to draw a power-based shift map to firstly meet the dynamic demand and secondly ensure the motor working in a higher efficiency area. The power-based shift schedule follows these three principles.

4.3.1. Maximum motor speed principle. In this paper, the motor working speed can achieve no more than 7000rpm. The motor speed can be reduced by a certain vehicle speed v and two gear ratios as:

$$n_k = \frac{v \cdot i_k}{0.377 r_w}, \tag{6}$$

where n_k is the k-gear motor speed. Since $i_1 > i_2$, n_1 is larger than n_2. Thus, the second gear should be the current gear once n_1 exceeds 7000 rpm.

4.3.2. Maximum motor torque principle. For a given vehicle speed v and a given power demand P, the torque request can be counted as:

$$T_{kd} = 9550\frac{P}{n_k}, \tag{7}$$

where T_{kd} is the k-gear motor torque demand. The motor actual torque $T_k = \min\{T_{kd}, T_{Mmax}\}$. Then the power motor provided is known and the gear with larger power should be selected as current gear:

$$P_k = \frac{T_k \cdot n_k}{9550}. \tag{8}$$

4.3.3. The lowest energy consumption principle. In addition to the above two conditions, for a specific vehicle speed and power demand, the motor efficiency is known according to Figure 1(b). Current gear should be the one with higher efficiency to ensure lower energy consumption.

According to these three principles, the power-based shift map can be drawn as shown in Figure 4.

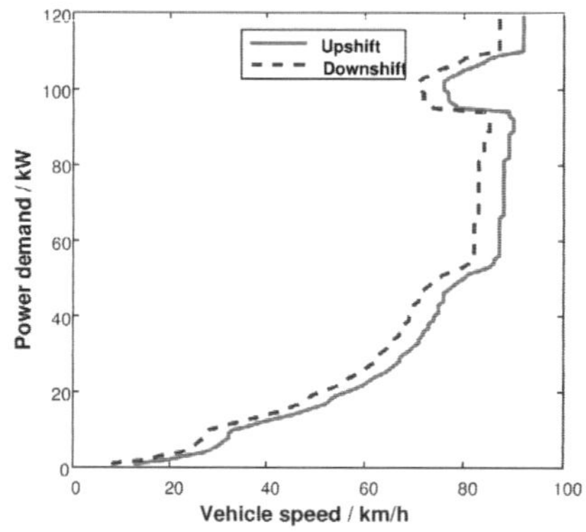

Figure 4. The power-based shift curve.

5. Simulation results and analysis

5.1. Simulation for dynamic performance

Under the 100% accelerator pedal opening, the accelerate time for 0-50 km/h, 50-80 km/h, 0-100 km/h without shift continue time is listed in Table 3. As can be seen, the economic shift schedule accelerates rapidly for 0-50 km/h, while for 50-80 km/h and 0-100 km/h, the drive ability of dynamic schedule and power-based schedule are almost the same, both are superior to economic schedule. The reason of the simulation result is both the dynamic shift and power-based shift schedule firstly consider about the traction performance. While the economic shift schedule ignores the vehicle dynamic ability.

Table 3. Accelerate time for 0-50 km/h, 50-80 km/h and 0-100 km/h.

Shift schedule	0-50 km/h (s)	50-80 km/h (s)	0-100 km/h (s)
Dynamic shift	4.206	3.452	11.088
Economic shift	4.207	3.548	11.214
Power-based shift	4.215	3.450	11.095

5.2. Simulation for economic performance

Comparing to traditional gasoline-powered automotive, the biggest superiority of electric vehicle is the economic performance with low energy consumption. A forward modelling method is adopted in this paper to infer the practical accelerator pedal opening. The energy consumption E_M in accordance with NEDC driving cycles is calculated by integrating the motor output power as:

$$E_{con} = \int \frac{n_M \cdot T_M(n_M, \alpha)}{9550} \cdot \frac{1}{\eta(n_M, \alpha)} dt, \tag{9}$$

where η is the transmission efficiency and other parameters have the same meaning as above.

To satisfy the design demand, 100-km energy consumption is also calculated for uniform velocity 60 km/h. The simulation results are shown in Table 4. Due to the dependence of the accelerator pedal, the power-based shift schedule has a less energy consumption than traditional dynamic and economic shift schedules.

Table 4. 100-km Energy consumption of three kinds of shift schedule.

Shift schedule	NEDC (kWh)	60km/h(kWh)
Dynamic shift	14.46	14.08
Economic shift	14.35	13.59
Power-based shift	14.10	13.59

6. Conclusion

This paper introduces a method to generate the power-based shift curve. It is a comprehensive shift schedule by considering both dynamic and economic characteristics of electric vehicles. Three principles to develop the power-based shift schedule are proposed. As the simulation result illustrates, the comprehensive power-based shift schedule has a fairly driving ability with traditional dynamic shift type. Power-based shift schedule has a strong accelerate ability in 50-80 km/h. Moreover, the energy consumption of comprehensive shift type is the lowest. In NEDC driving cycles, the 100-km energy consumption of power-based shift schedule is less than the traditional economic shift schedule by 2%. The power-based shift schedule regards the vehicle tractive ability and economy performance as the input parameters of shifting schedule. Thus, the motor load power and working efficiency can be improved simultaneously. Therefore, the comprehensive power-based shift schedule should be widely employed in electric vehicles.

References

[1] Yu, Zhe, S. Li, and L. Tong. "Market dynamics and indirect network effects in electric vehicle diffusion." *Transportation Research Part D Transport & Environment* 47(2016):336-356.

[2] D., A., Crolla, and Morris. "Effect of Transmission Design on Electric Vehicle (EV) Performance." *Vehicle Power and Propulsion Conference, 2009. VPPC '09. IEEE* 2009:46-51.

[3] Roberts, Simon. "Multispeed transmission for electric vehicles." *Atz Worldwide* 114.4(2012):8-11.

[4] Torrelli, Dr. Ing. Claudio. "Transmissions for Electric Vehicles." *Atz Worldwide* 115.2(2013):10-13.

[5] Qin, Da Tong, et al. "Parameters design of powertrain system of electric vehicle with two-speed gearbox." *Journal of Chongqing University*34.1(2011):1-6.

[6] Hofman, T., S. Ebbesen, and L. Guzzella. "Topology Optimization for Hybrid Electric Vehicles With Automated Transmissions." *IEEE Transactions on Vehicular Technology* 61.6(2012):2442-2451.

[7] Ngo, V., et al. "Optimal control of the gearshift command for hybrid electric vehicles." *IEEE Transactions on Vehicular Technology*61.61(2012):3531-3543.

[8] Hou, Rui, J. Zhang, and C. Lin. "New Approach of Comprehensive Shift Schedule Based on Electric Drive Automated Mechanical Transmission ☆." *Energy Procedia* 88(2016):945-949.

[9] Xiang, Yu, et al. "A Study on Gear Shifting Schedule for 2-Speed Electric Vehicle Using Dynamic Programming." *Chinese Control and Decision Conference* IEEE, 2013:3805-3809.

[10] Guo, Lulu, et al. "Shift schedule optimization of 2-speed electric vehicle using model predictive control." *Control Conference* IEEE, 2014:156-161.

[11] Guo, Lulu, B. Gao, and H. Chen. "On-line Shift Schedule Optimization of 2-Speed Electric Vehicle Using Moving Horizon Strategy.".

[12] Zhu, Bo, et al. "Gear shift schedule design for multi-speed pure electric vehicles." *Proceedings of the Institution of Mechanical Engineers Part D Journal of Automobile Engineering* 229.1(2014):70-82.

[13] Li, Wei, X. Yin, and X. Wu. "Shift Schedule of Electrical Vehicle based on the Optimization of Motor Efficiency." *Journal of Mechanical Transmission* (2014).

A new approach to tolerance analysis method based onthe screw and the Lie Algebra of Lie Group

X C Zhai[1],Q G Du[1],W X Wang[2], Q Wen[1], B S Liu[1] and Z Q Sun[1]

[1]School of Mechanical& Automotive Engineering, SCUT Wushan RD, Tianhe District, Guangzhou 510641, P.R.China
[2]School of International Studies,SYSU Tangqi RD, Tangjiawan Town, Xiangzhou Dis trict, Zhuhai 519000, P.R.China

zhaixiaochen@126.com

Abstract. Tolerance analysis refers to the process of establishing mapping relations between tolerance features and the target feature along the dimension chain. Traditional models for tolerance analysis are all based on rigid body kinematics, and they adopt the Homogeneous Transformation Matrix to describe feature variation and accumulation. However, those models can hardly reveal the nature of feature variations. This paper proposes a new tolerance analysis method based on the screw and the Lie Algebra of Lie Group, which describes feature variation as the screw motion, and completely maps the twist, an element of the Lie Algebra, to the Lie Group that represents the feature configuration space. Thus, the analysis can be conducted in a more succinct and direct way. In the end, the method is applied in an example and proven to be robust and effective.

1. Introduction

In the product design phase, the geometric variations of a feature relative to its nominal size and state in the tolerance zone are known as errors. The introduction of GD&T specification made it possible to analyze feature variations in the three-dimensional space. Therefore, tolerance analysis should consider transmission and accumulation of errors in three-dimensional space.

Main models of tolerance analysis include the Matrix Method proposed by Desrochers et al.[1]in 1997; the State Transition Method put forward by Mantripragada et al.[2]in 1999; and the Jacobian Torsor Method presented by Desrochers et al.[3]in 2003. The above models all adopt Homogeneous Transform Matrix (HTM) methods or methods derived from HTM[1][2, 3]to express the variation and transmission of features, which means that they all describe coordinate systems in a macroscopic way. However, such models lack microscopic analysis which is necessary for exploring the nature of feature variations. Moreover, the direction cosine matrix of HTM shows deficiency in describing special configuration of an object, for it can hardly express the position and orientation changes in a direct way. Therefore, this paper puts forward a tolerance analysis method based on the screw and the Lie Algebra of Lie Group, which describes Chasles' motion using the matrix exponential of the Lie Algebra, and further describes feature variations in the tolerance zone using Chasles' motion. What's more, instead of establishing sophisticated local coordinate systems, this method can express position and orientation variations of the functional feature simply by setting up the twist axis of each tolerance feature. Thus, the solution process is more succinct and direct.

2. Method based on the Screw and the Lie Algebra of Lie Group

2.1. Introduction to the method

The Lie Group is one of the cores of modern geometry, and the Lie Algebra is an algebra tool used for study of the Lie Group, differentiable manifolds and so on. There is a correspondence of the Lie Group and the Lie Algebra[4]. According to ASME Y14.5 Standards, the tolerance zone exerts constraints on the variations of toleranced features which form the Lie subgroup. Using the geometric algebraic theory, configuration variations of a rigid body in the space can be visually described, on which some scholars have already conducted studies. In 1999, Gou et al.[5]found that a toleranced feature demonstrates a symmetry subgroup under the action of the Lie Group, and the configuration space of a tolerance feature can be considered as the homogeneous space of the Lie Group. With the help of exponential mapping, mathematic models can be obtained for verification of form, profile and orientation tolerances. In 2000, Gou et al.[6]further pointed out that the establishment of datum reference frame and the tolerance verification in the position tolerance is similar to the sequential location problem in multi-featured workpieces, and a datum feature also exhibits a symmetry subgroup under the action of the Lie Group.

The above studies have proven the practicability of the Lie Algebra of Lie Group, but they mainly focused on the tolerance verification, instead of proposing a complete model for tolerance representation and analysis. Based on the outcomes of previous studies, this paper further introduces the screw theory[4, 7]in which the exponential matrix is surjective, and it possesses merits such as clear geometrical concepts, neat expression form and convenient algebraic operation.

2.2. Formula derivation

As is stated by the Chasles' theorem[4], any rigid body movement can be achieved by screw motion which is a comprehensive one including rotating around and translating along an axis. The rigid body movement is equivalent to the screw motion. The infinitely small quantity of screw motion, twist $\hat{\xi}$, is also an element of the Lie Algebra. This concept combines the Lie Algebra of Lie Group, the screw and the screw motion together. As is shown in figure 1., in the screw motion, $\omega \in R^3$ is the unit vector on the direction of the revolution axis, $\theta \in R$ is the rotation angle around the revolution axis, and the $\hat{\xi}$ expression is manifested in figure 1.

$$\hat{\xi} = (v, \hat{\omega}) = \begin{bmatrix} \hat{\omega} & v \\ 0 & 0 \end{bmatrix} \tag{1}$$

Where: $\hat{\omega}$ is the skewsymmetric matrix of the direction vector ω ;

$v = -\omega \times q + h\omega$, $q \in R^3$ is the coordinate of any point on the axis in the global coordinate system; $h = d/\theta$, d stands for the translation amount along the twist axis. The twist represents the pure rotational motion when h=0, the pure translational motion when h=∞ , and a mix of both rotational and translational motion when h is a finite value.

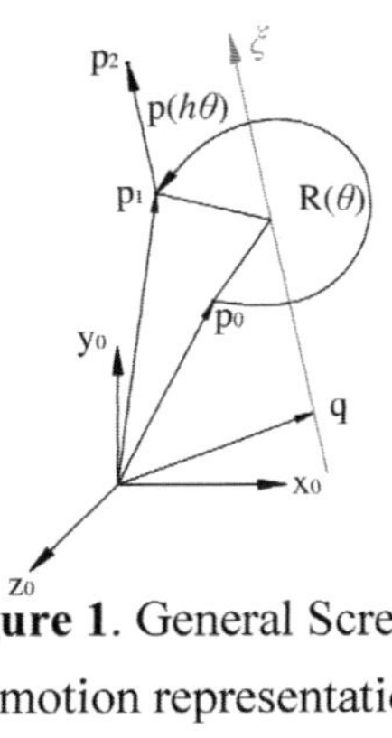

Figure 1. General Screw Motion

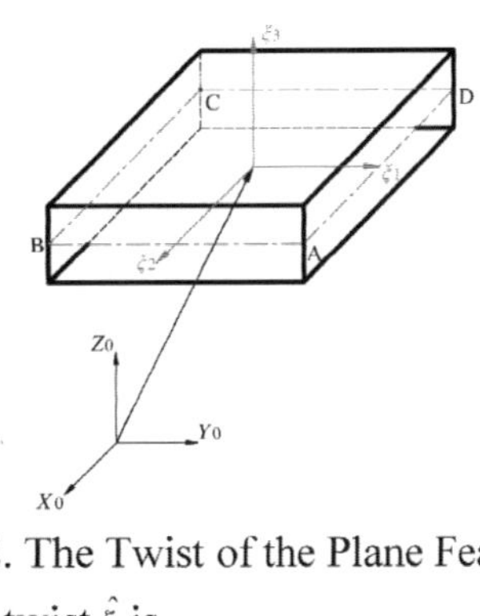

Figure 2. The Twist of the Plane Feature

So the screw motion representation determined by the twist $\hat{\xi}$ is

$$g = e^{\hat{\xi}\theta} = \begin{bmatrix} e^{\hat{\omega}\theta} & \left(I - e^{\hat{\omega}\theta}\right)q + h\theta\omega \\ 0 & 1 \end{bmatrix} \tag{2}$$

Therefore, therigid body movement can be described by mapping the elements in $\hat{\xi} \in se(3)$ to $g \in SE(3)$. Since the exponential mapping is surjective[7], any rigid body movement can be represented as an exponent of a certain twist. When there is more than one rigid body, the rigid body movements of each kinematic pair can be combined, and such combination of movements is expressed by the mapping g_{st}

$$g_{st}(\theta) = e^{\hat{\xi}_1\theta_1} e^{\hat{\xi}_2\theta_2} \cdots e^{\hat{\xi}_n\theta_n} g_{st}(0) \tag{3}$$

$g_{st}(0)$ and $g_{st}(\theta)$ respectively stand for the initial configuration and the instantaneous configuration of coordinate system *T* relative to the datum coordinate system *S*.

Therefore, the screw can express any rigid body movements as Chasles' motion, while the exponential matrix can express them as screw matrix operator. Detailed derivation process is shown in the References[7].

3. Tolerance Representation

Tolerance representation is the first step of tolerancing. This paper takes the plane feature shown in figure 2.as an example to illustrate the tolerance representation of the screw. In figure 2., the red chain lines show the ideal position of the plane feature. In order to describe the screw motion, the reference coordinate system and revolution axis should be determined in the first place. Although the position of the reference coordinate system and the configuration of the revolution axis can be selected arbitrarily, the most convenient way is make the direction of the axis be in line with that of the reference coordinate system according to degree of freedom (DOF) of the features, as the revolution axes ξ_1,ξ_2 and ξ_3 of the plane feature in figure 2. The DOF of a feature is basically determined as (6-*n*) according to its degree of invariance *n*[1]. The number of DOF is influenced by the structure of the tolerance zone as well as feature, mating and datum types[8].

Features with multi degree of freedom can be seen as combinations of revolute pair and prismatic pair. Take the plane feature in figure 2. again as an example. Its three DOF in the tolerance zone include two revolute DOF and one translational DOF. Hence, the three twists, $\hat{\xi}_1$, $\hat{\xi}_2$ and $\hat{\xi}_3$, can be established, and the mapping of the plane feature can be expressed as:

$$g = e^{\hat{\xi}_1\theta_1} e^{\hat{\xi}_2\theta_3} e^{\hat{\xi}_3\theta_3} = e^{\sum_{i=1}^{3}\hat{\xi}_i\theta_i} \tag{4}$$

Where: twists $\hat{\xi}_1$ and $\hat{\xi}_2$ are pure revolute variations whereas $\hat{\xi}_3$ is pure translational variation. θ_1 and θ_2 stand for rotation angles, and θ_3 is the translation amount. Through changing the value of the three parameters, the plane feature can traverse the whole tolerance zone.

What's more, along the DOF, feature variations are also constrained by the boundaries of the tolerance zone. Therefore, feature variations in the tolerance zone can be expressed as products of exponentials (POE) of the twist, through which the complete geometric representation and parameterization of the feature variations can be realized. The tolerances of another six types of features in Reference[1]can be obtained through the similar way as well.

4. Tolerance Analysis

Tolerance analysis is in nature a repositioning of nominal features with errors being considered, and a description of various feature positions and orientations in the tolerance zone. Since the Lie Algebra of Lie Group involves the concept of set of variations, the method based on it is quite suitable for tolerance analysis.

The new method uses exponential mapping to represent the configuration of each feature, and expresses the variation of each feature in the tolerance zone as the exponent of its corresponding twist.

Hence, each feature can be mapped from its nominal position to the position with errors. When the feature j has more than one DOF, the matrix exponential of the screw should be determined based on the DOF of the feature and the screw type.

$$g_j(\theta) = e^{\sum_{i=1}^{6} \hat{\xi}_{ji}\theta_{ji}} \tag{5}$$

Hence, mapping of the target feature configuration along the whole dimension chain can be expressed as

$$g_{st}(\theta) = e^{\sum_{i=1}^{6} \hat{\xi}_{1i}\theta_{1i}} e^{\sum_{i=1}^{6} \hat{\xi}_{2i}\theta_{2i}} \cdots e^{\sum_{i=1}^{6} \hat{\xi}_{ni}\theta_{ni}} \cdot g_{st}(0) = \prod_{j=1}^{n} e^{\sum_{i=1}^{6} \hat{\xi}_{ji}\theta_{ji}} \cdot g_{st}(0) \tag{6}$$

Where: $g_{st}(0)$ is the nominal configuration of the target feature when every feature is in the ideal position; j stands for the sequence number of variable features along the dimension chain from the datum reference system; and i is the number of DOF of the j_{th} feature. Therefore, the entire tolerance analysis based on the screw and Lie Algebra of Lie Group can be realized.

5. Numerical Examples

The centering pin[9]is taken as an example to prove practicability of the new tolerance analysis method. As is shown in figure 3., the assembly is comprised of(a)Base,(b)Pin and(c)Block (see figure 4. for their detailed sizes and tolerance requirements). Based on the tolerance specification in Table 1., the functional requirement (FR) refers to the variable from the Pin point Q to the upper surface of the Base.

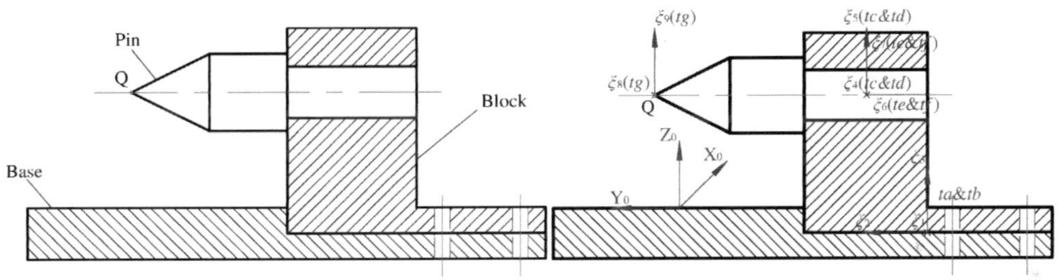

(a) Detail Assembly for Centering Pin Mechanism (b) Twist

Figure 3. Centering Pin Mechanism

Table 1. Tolerance Specifications Proposed

Tolerance specifications	ta	tb	tc	td	te	tf	tg
Tolerance Value	0.2	0.1	0.2	0.1	H11:0.00/0.13	H8:-0.033/0.000	0.1

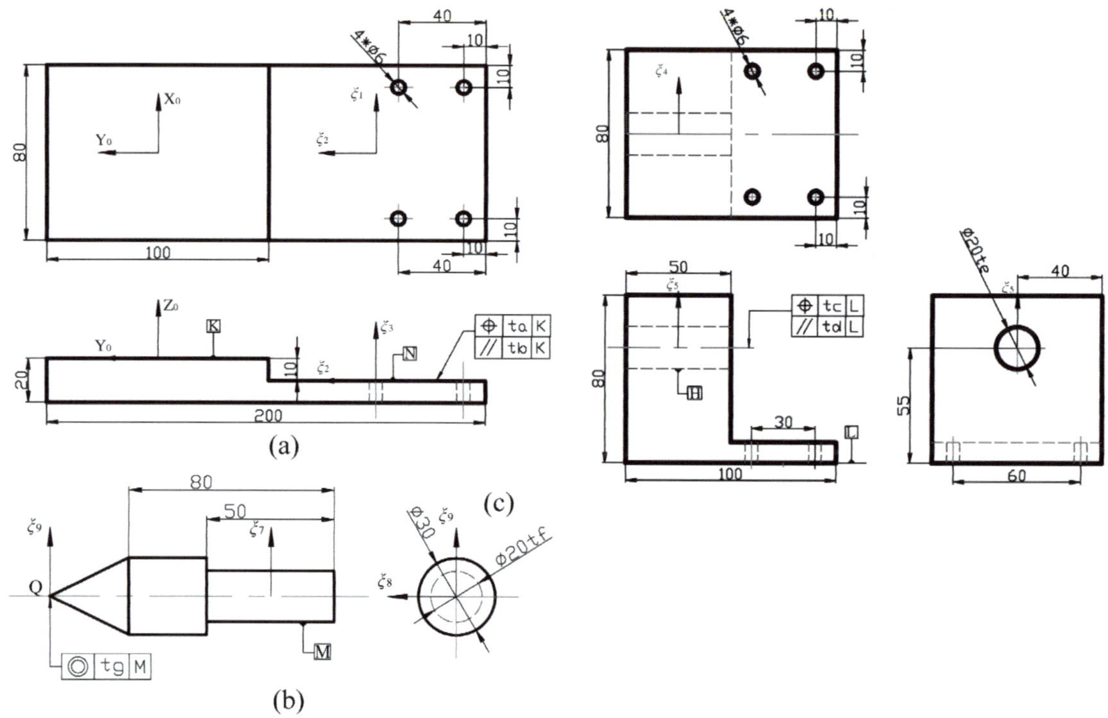

Figure 4. Detailed Sizes and Tolerance Requirements

As is shown in figure 3. (b), the screw based on the reference coordinate system should be first established. Then the screws ξ_1, ξ_2 and ξ_3 of the tolerance feature N, ξ_4 and ξ_5 of H, as well as ξ_6 and ξ_7 of M should be set up, respectively. The screws ξ_8 and ξ_9 of the target feature Q should also established. Where: the two screws, ξ_1 and ξ_2, represent the pure rotational motion, and their corresponding exponential matrix is

$$e^{\hat{\xi}_i\theta_i} = \begin{bmatrix} e^{\hat{\omega}_i\theta_i} & \left(I - e^{\hat{\omega}_i\theta_i}\right)q_i \\ 0 & 1 \end{bmatrix} \tag{7}$$

Where: i=1,2.

The screw ξ_3 represents the pure translational motion, and its corresponding exponential matrix is

$$e^{\hat{\xi}_i\theta_i} = \begin{bmatrix} I & \theta_i v_i \\ 0 & 1 \end{bmatrix} \tag{8}$$

Where: i=3.

The screw ξ_4, ξ_5, ξ_6, ξ_7, ξ_8, and ξ_9 contain both rotational and translational motions, and their corresponding exponential matrix is

$$e^{\hat{\xi}_i\theta_i} = \begin{bmatrix} e^{\hat{\omega}_i\theta_i} & \left(I - e^{\hat{\omega}_i\theta_i}\right) * q_i + h_i\theta_i\omega_i \\ 0 & 1 \end{bmatrix} \tag{9}$$

Where: $h_i = \dfrac{d_i}{\theta_i}$ i=4,5,6,7,8,9.

Then the coordinate of the point q_i on the direction vectors ω_i and revolute axis of each screw can be obtained, and the coordinate of the screw can further be acquired.

At last, through multiplying those exponential matrices, the ultimate mapping matrix g_{st} can be obtained. Since the homogeneous coordinate of the Pin point in the reference coordinate system is known as $p(0)=[0 \quad 40 \quad 45 \quad 1]^T$, through multiplying it by g_{st} using the formula (6), we can acquire the following expression:

$$p(\theta)=g_{st}(\theta)p(0) \tag{10}$$

Hence, the expression containing variation parameters θ_i and d_i can be obtained. Taking into consideration the inequality constraint provided by the tolerance zone, and using the genetic algorithm toolbox of MATLAB software, we can get the variation range of the target feature as (-0.579433, 0.647757)

6. Conclusion

The new tolerance analysis method proposed in this paper demonstrates its merits in the following three aspects: first, this method has clear geometric concepts, defined physical significance, and simple expression form. It is also easy for describing the spatial movement of a rigid body from the perspective of geometry. Revolution axes rather than local coordinate systems can be constructed according to different types of features, with each screw having at most two parameters; second, this method has convenient and efficient algebraic operation. It describes the rigid body movement in global coordinate systems, avoiding the singularity that may be brought about by the local coordinate systems; third, this method has a wide scope of application. It can easily establish relationships with vector, matrix, and influence coefficient methods. To sum up, this new method can be well applied to tolerance analysis.

7. References

[1] Desrochers A andRivière A 1997A matrix approach to the representation of tolerance zones and clearances*International Journal of Advanced Manufacturing Technology***13**630-6
[2] Mantripragada R and Whitney D E1999Modeling and controlling variation propagation in mechanical assemblies using state transition models*Robotics and Automation*, IEEE Transactions on**15**124-40
[3] Laperrière L, Ghie W and Desrochers A2003Application of a unified Jacobian - Torsor model for tolerance analysis*Journal of Computing and Information Science in Engineering***3**2-14
[4] Dai J S 2014*Geometrical foundations and screw algebra for mechanisms and robotics*(Beijing: Higher Education Press)
[5] Gou J B, Chu Y X and Li Z X 1999A geometric theory of form, profile, and orientation tolerances*Precision Engineering***23**79-93
[6] Gou J B, Chu Y X, Xiong Z H, et al2001A geometric method for computation of datum reference frames*IEEE Transactions on Robotics & Automation***16**797-806
[7] Murray R M, Li Z X and Sastry S S1994*A mathematical introduction to robotic manipulation* (Boca Raton:CRC Press)
[8] Luo J M2006Study on the structural synthesis about the partly-DOF parallel robot mechanisms and constraint chains(China:Northestern University)
[9] Ghie W2009Statistical analysis tolerance using jacobian torsor model based on uncertainty propagation method*The International Journal of Multiphysics***3**11-30

IOP Conf. Series: Materials Science and Engineering **157** (2016) 012018 doi:10.1088/1757-899X/157/1/012018

Modeling and Simulation of Metallurgical Process Based on Hybrid Petri Net

Yujuan Ren[1] and Hong Bao[1]

[1]School of Computer & Communication Engineering, University of Science and Technology Beijing, Beijing, China

Email:baohong@ies.ustb.edu.cn; ustb_ryj@163.com

Abstract. In order to achieve the goals of energy saving and emission reduction of iron and steel enterprises, an increasing number of modeling and simulation technologies are used to research and analyse metallurgical production process. In this paper, the basic principle of Hybrid Petri net is used to model and analyse the Metallurgical Process. Firstly, the definition of Hybrid Petri Net System of Metallurgical Process (MPHPNS) and its modeling theory are proposed. Secondly, the model of MPHPNS based on material flow is constructed. The dynamic flow of materials and the real-time change of each technological state in metallurgical process are simulated vividly by using this model. The simulation process can implement interaction between the continuous event dynamic system and the discrete event dynamic system at the same level, and play a positive role in the production decision.

1. Introduction

Recently, the high speed development of the metallurgical industry brought enormous impact on resources, energy and environment [1]. Therefore, development of the new generation of Iron and steel production process has become the primary problem of metallurgical industry. Due to the high complexity, high cost and high energy consumption of metallurgical process, modeling and simulation technology has become an important way to research and analyze the metallurgical process. At the present stage, there are many methods of metallurgical process modeling. The traditional standard material flow diagram [2], which only comprises the input and output flow of iron material, lacks analysis of all the production material and the dynamic evolution process of various materials. For the more, the Petri Net of Iron and Steel Process Based on Technological Indicators (ISPPNSTI) [3] is short of modeling the discrete event dynamic system in the metallurgical process, and it can't show the interaction between the dynamic evolution of material and the change of technological state.

Metallurgical production process is a typical hybrid process which contains both continuous and discrete event systems [4]. In metallurgical production process, the Coking, Sintering and some other processes are continuous production process, which can be seen as continuous event dynamic system; the change of each technological state (running or idle), which is independent of each other, can be regarded as discrete event dynamic system. In order to simulate the metallurgical production process more accurately, an appropriate model of the process is necessary to reflect the discrete parts and continuous parts of metallurgical process, and achieve interaction between them.

Hybrid Petri net [5] is an extension of the traditional Petri net [6,7], which not only inherits the advantages of traditional Petri net in modeling, but also can achieve the interaction between discrete event dynamic system and continuous event dynamic system at the same level. A model of Hybrid Petri net is composed of discrete parts and continuous parts, and the two parts are connected to an Arc between a discrete node and a continuous node. So that, the model can realize mutual influence between discrete portions and continuous portion in dynamic system. Therefore, the characteristics and modeling methods of Hybrid Petri net can be just used to analyze metallurgical process.

In this paper, we propose the definition and modeling theory of MPHPNS, and build the Hybrid Petri net system model of metallurgical process based on material flow by using the modeling algorithms. Through this model, we can clearly show the static structure of metallurgical process. Then we simulate the metallurgical production process by using this model, and analyze the simulation process, which can not only achieve the dynamic simulation of the whole production process, but also directly show the state of each technics.

2. Basic Concepts

2.1. Definition of Hybrid Petri Net System of Metallurgical Process

In MPHPNS, the continuous parts represent the flow of material resources; the discrete parts represent the change of technological state. The continuous places (P^c) represent storage places of the material resources; the continuous transitions (T^c) represent productive technics; Arcs ($F = \{(P^c \times T^c)\}$) represent the flow-direction of material resources. The discrete places (T^d) represent the state of technics (running or idle); the discrete transitions and the Arcs ($F = \{(P^d \times T)\}$) lead to the change of technological state; the initial marking (M_0) represents the initial distribution of resources of places; the capacity function (K) represents the maximum of resources that the place can hold; the Arc labels ($W = \langle Pre, Post\rangle$) indicate the number of input or output resources in the reaction process. Accordingly, this model can be defined as a 6-tuple $Q_{ME} = \langle P, T; F, M_0, K, W\rangle$, where

- $P = \{p_1, p_2, \cdots p_n\} = P^c \cup P^d \ (n \geq 0)$, a finite, not empty, set of places;
- $T = \{t_1, t_2, \cdots t_m\} = T^c \cup T^d \ (m \geq 0)$, a finite ,not empty, set of transitions;
- $F = \{(P \times T) \cup (T \times P)\}$, a set of resources flow, $\forall\ (p_i \in P^c)$, $(F = \{(p_i \times t_j)\})$ need to satisfy $(t_j \in T^c)$;
- $M_0 : P \rightarrow R^+ or\ N$, a set of the initial distribution of resources (initial marking);
- $K : P \rightarrow \{1, 2, \cdots, k\}\ k \in N^+$, capacity function;
- $W = \langle Pre, Post\rangle$, weight function;
- $Pre : P \times T \rightarrow R^+ or\ N$, weight between input places and transitions;
- $Post : P \times T \rightarrow R^+ or\ N$, weight between transitions and output places;

In the definitions of Pre, $Post$ and M_0, N corresponds to the case where $(p_i \in P^d)$,and R^+ corresponds to the case where $(p_i \in P^c)$. Pre and $Post$ function must meet the following criterion: if p_i and t_j are a place and a transition such that $(p_i \in P^d)$ and$(t_j \in T^c)$, then $pre(p_i, t_j) = Post(p_i, t_j)$ must be satisfied [6].

3. Modeling Theory

3.1. Modeling algorithms

The basic operations of Petri net [8] include insertion operation, deletion operation, and composition operation, which are the foundation of the model building. According to the modeling algorithms of Petri net and the definition of MPHPNS, the modeling algorithms of MPHPNS are established, which mainly include the insertion operation, deletion operation, and composition operation.

Note: The definition of inserting operation, deleting operation and composition operations we proposed above can be found in my master's thesis (c .f [9]).

3.2. Transition Trigger Rules

The fires of transition lead to the change of resources distribution of place, which plays a crucial role in analysing the dynamic evolution of the material and grasping the state of the technics in the actual production process.

Theorem 1 The trigger condition of the transition can be described as follow, $\forall t \in T, ?\ M[t>$ on the condition of:

- $$\begin{cases} M(p) \geq Pre(p,t) & \forall p \in {}^{\bullet}t - t^{\bullet} \\ M(p) + Post(p,t) - Pre(p,t) \leq K(p) & \forall p \in {}^{\bullet}t \cap t^{\bullet} \\ M(p) + Post(p,t) \leq K(p)? & \forall p \in t^{\bullet} - {}^{\bullet}t \end{cases} \quad (1)$$

Theorem 2 The trigger result of the transition can be described as follow, if $M[t > M'$, $\forall p \in P$ obtains:

- $$M'(p) = \begin{cases} M(p) - Pre(p,t) & \forall p \in {}^{\bullet}t - t^{\bullet} \\ M(p) + Post(p,t) & p \in t^{\bullet} - {}^{\bullet}t \\ M(p) + Post(p,t) - Pre(p,t) & p \in {}^{\bullet}t \cap t^{\bullet} \\ M(p) & others \end{cases} \quad (2)$$

Note: ${}^{\bullet}t$ represents the input position of (t) when transition are triggering; $t^{\bullet}$ represents the output position of (t) when transition triggered; $M[t>$ indicates that the transition (t) can be fired in marking (M); $M[t > M'$ describes the result of a fired transition (t) in a new distribution of resources (M').

4. System Modeling

The traditional metallurgical process [10] is generally composed of Coking, Sintering, Blast furnace, Desulphurization pretreatment, Dephosphorization converter, Decarburization converter, Continuous casting, Hot-rolling and Storing. The idea of building the model of MPHPNS can be described as follows: firstly, each stage of productive technics is abstracted into a 6-tuple; secondly, build every stage model by using insertion and deletion operations; finally, integrate every stage model into a traditional metallurgical process by using composition operation.

According to the modeling process above, we can obtain the Hybrid Petri net system model of metallurgical process (as shown in Figure 1).

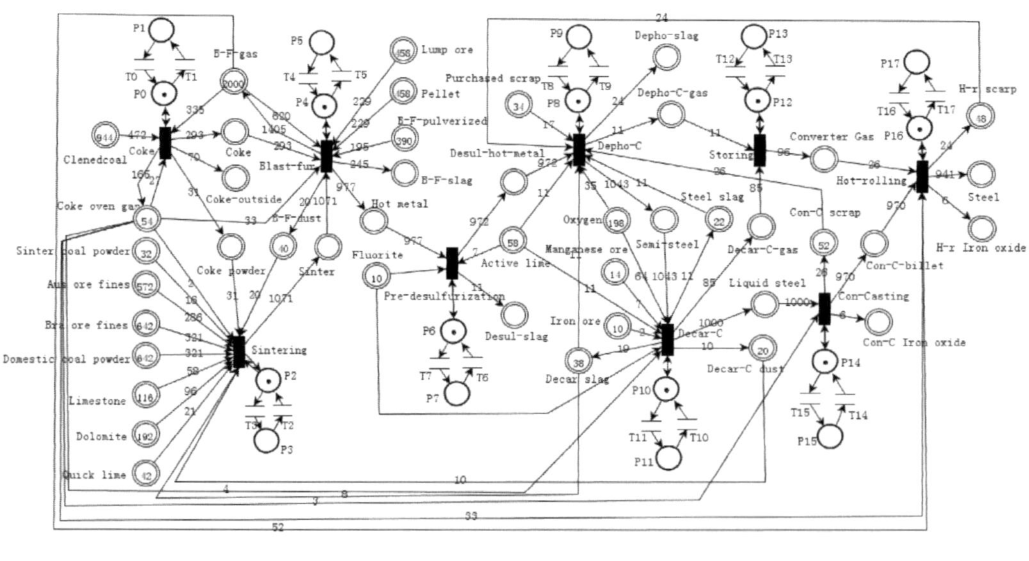

Figure 1. The Hybrid Petri net model of metallurgical process

Note: In the model above, the solid rectangle represents the continuous transitions; the double line represents the discrete transitions; the discrete places are indicated by circle; the continuous places are expressed by double circle. In addition, the sub-technics model can be seen in my master's thesis (c. f [9]).

5. Process Simulation and Analysis

The model of MPHPNS we built above consists of two parts. One part is the continuous Petri net which is used to simulate the dynamic flow of materials resources, and the other part is the discrete Petri net which is used to simulate the change of the technological state in simulation process.

5.1. Process Simulation

Using this model, the simulation process is as follow: Firstly, all of the production technics are running. We can make sure that only the process of Coking can be fired (as shown in Figure 2) by analyzing the trigger condition of each transition in initial marking. Then, the process of Coking fires and leads to the change of resources distribution, which makes the only process of Sintering fire. It follows the process of Blast furnace, Desulphurization pretreatment, Dephosphorization converter, Decarburization converter, Continuous casting, Hot-rolling and Storing fires sequentially. Finally, the output is the corresponding steel, and then the process of Coking switches from running state to idle state because of the number of reactants are not abundant.

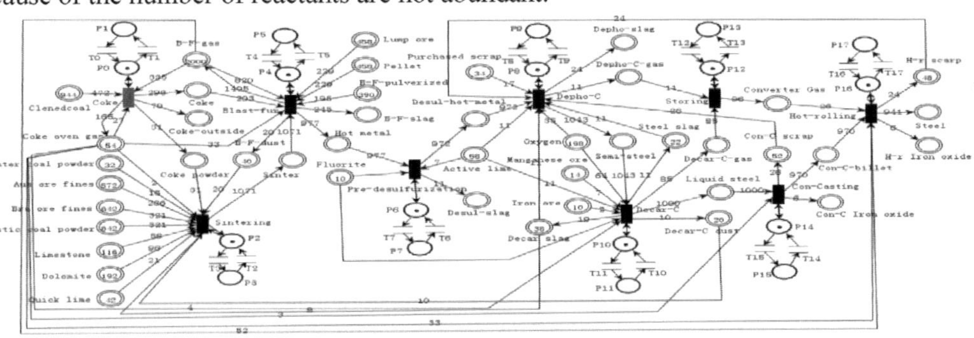

Figure 2. The firing state of Coking

The description of simulation process above can fully reflect the actual production of metallurgical process. It can clearly show the dynamic evolution process of material in each technics process, and describe the technological state at a certain point. At the same time, it can realize the cycle simulation of the whole production process under the condition of abundant material.

Note: In the simulated process, black represents that transitions can't be triggered; red represents that transitions can be fired.

5.2. Process Analysis

The simulation process of the MPHPNS can complete the continuous flow of material and the change of technological state interaction at the same level. This interaction is mainly manifested in two parts. One side is the influence of the continuous part on the discrete part, which mainly performance in: when the initial material of a technics is not sufficient or a material space is insufficient, the technics will switch from running state to the idle state. The other side is the effects of the discrete part on the continuous part, which mainly performance in: when the technics is in idle state, the technics can't be triggered.

Take Coking as an example. At first, the process of Coking fires, and changes the distribution of resources (as shown in Figure.3 (a)). Continue to perform process simulation, we can see that the Coking process is switched to the idle state (as shown in Figure.3 (b)) because of the lack of the number of reactant. At the same time, the process of Coking can't be triggered.

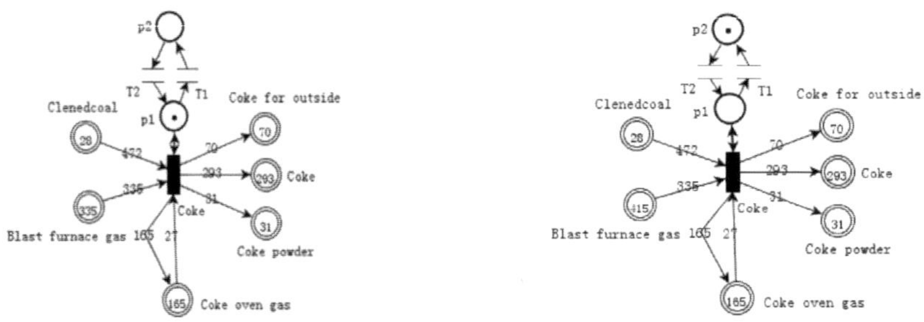

(a) (b)

Figure 3. (a) The firing of Coking (b)Coking equipment switches to idle state

When the material space is not enough to hold more generated materials, technics can also be switched to idle state. In the cycle simulation of production process, it can avoid to compete for the space of material places by the interaction between continuous and discrete parts. We can grasp the state of each technics in the first time through observing the discrete parts of model. So that we can more quickly and accurately find problems in the production process, which plays a positive role to the production decision.

6. Conclusions

The model of Hybrid Petri Net System of Metallurgical Process can exhibit the static structure of metallurgical process. The simulation process fully reflects the actual production of metallurgical process, and implements interaction between the dynamic continuous evolution of the material and the change of discrete state of technics at the same level. The modeling theory of Hybrid Petri Net System of Metallurgical Process provides a feasible and controllable method for modeling. However, the analysis of the model's properties still needs further research and discussion.

7. References

[1] Song Shuqin 2007 Vigorously develop the circular economy to speed up the fundamental change of iron and steel power. *Metallurgy management*. **6** pp 49-51.

[2] Du T, Cai J, Lu Z and et al 2002. Influence of material flow in steel production on atmosphere environmental load. *Iron and Steel(China)* **37(6)** pp 59-63.

[3] Wang Y. and Bao H. 2012. Petri Net System of Iron and Steel Process Based on Technological Indicators. *In 2012 International Conference on Computer Science and Service System* **33** pp 98-102

[4] Ye-qing Z 2014. Research on energy flow network model in iron and steel enterprises based on hybrid Petri net. *Metallurgical Industry Automation*, **5** p 007.

[5] David R and Alla H 2001. On hybrid Petri nets. *Discrete Event Dynamic Systems*, **11(1-2)** pp 9-40.

[6] Wu Zhehui 1989. *Petri net theory and system simulation*. (Xuzhou: China University of Mining and Technology Press) pp 69-72.

[7] Yuan Chongyi 2005. *Principles and application of Petri net* (Beijing: Publishing House Of Electronics Industy) pp 8-11.

[8] Wu Zhehui 2006. *Introduction of Petri net* (Beijing: China Machine Press) pp 101-119.

[9] Ren Yujuan. Study on modeling of metallurgical process based on Hybrid Petri net. unpublished.

[10] Yin Ruiyu 2004. *Metallurgical process engineering*. (Beijing: Metallurgical industry press) pp109-112.

IOP Conf. Series: Materials Science and Engineering **157** (2016) 012019 doi:10.1088/1757-899X/157/1/012019

Study of Photovoltaic Cells Engineering Mathematical Model

Jun Zhou[1,a], Zhengping Yu[1,b], Zhengyi Lu[1], Chenhui Li[2] and Ruilan Zhang[3]

[1]Electrical Engineering College of Northeast Dianli University, Jilin, Jilin, China, 132012;
[2]State Grid Xinyuan Baishan Power Plant, Jilin, Jilin, China, 132013;
[3]Beihua University, Jilin, Jilin, 132013;

[a]jlzhoujun@126.com [b]yzpwxy2010@163.com

Abstract: The characteristic curve of photovoltaic cells is the theoretical basis of PV Power, which simplifies the existing mathematical model, eventually, obtains a mathematical model used in engineering. The characteristic curve of photovoltaic cells contains both exponential and logarithmic calculation. The exponential and logarithmic spread out through Taylor series, which includes only four arithmetic and use single chip microcontroller as the control center. The result shows that: the use of single chip microcontroller for calculating exponential and logarithmic functions, simplifies mathematical model of PV curve, also can meet the specific conditions' requirement for engineering applications.

1. Introduction

Photovoltaic cell output characteristic curve is the foundation of PV theoretical research [1-3]. Accurate PV curve mathematical model plays a crucial role for optimization of the entire PV system. Therefore, establishing an accurate and simple mathematical model of PV characteristic curve is the basic research of photovoltaic power generation. As the PV characteristic curve is known, this study simplifies the mathematical model, and studies how to build mathematical models in single chip microcontroller on the basis of the known photovoltaic cell mathematical model. In the design of photovoltaic cell simulator, we need to consider how to make the simulator output stably of the solar cell output characteristic curve [4]. Through calculation error of simplified mathematical model and true mathematical model, optimization without sacrificing accuracy, minimizes the amount of calculation and make sure the system to calculate the required value quickly and accurately. Finally, achieves analog and output known PV curve finally.

2.Taylor series expansion of curve model

2.1.Output characteristic curve of the photovoltaic cell

IOP Conf. Series: Materials Science and Engineering **157** (2016) 012019 doi:10.1088/1757-899X/157/1/012019

PV simulator is based on the photovoltaic cell mathematical and physical model [5]. The photovoltaic cells characteristic curve:

$$a = 0.0012 \times I_{sc} \tag{1}$$

$$b = 0.0005 \times U_{oc} \tag{2}$$

$$C_1 = \left(1 - \frac{I_m}{I_{sc}}\right) e^{\frac{-U_m}{C_2 \times U_{oc}}} \tag{3}$$

$$C_2 = \left(\frac{U_m}{U_{oc}} - 1\right)\left[\ln\left(1 - \frac{I_m}{I_{sc}}\right)\right] \tag{4}$$

$$DI = a \times \frac{S}{S_{ref}} \times Dt + \left(\frac{S}{S_{ref}} - 1\right) \times I_{sc} \tag{5}$$

$$DV = -b \times Dt - R_s \times DI \tag{6}$$

$$Dt = T - T_{ref} \tag{7}$$

$$I = I_{sc}\left[1 - C_1\left(e^{\frac{U - DV}{C_2 \times U_{oc}}} - 1\right)\right] + DI \tag{8}$$

Obviously, mathematical model contains exponential and the logarithm.In the formula, U_{OC} , I_{SC} , U_M , I_M and R_S are the output parameters of photovoltaic cells, S and T are environmental parameters, $S_{ref} = 1000\ W/m^2$, $T_{ref} = 25\ °C$, U as independent variables, I as the dependent variable.

Equation 1 to Equation 4 only contain the output parameters of the photovoltaic cell, which only need to be calculated when the output characteristics of the photovoltaic cell parameter changes. Equation 5 to Equation 7, which include output parameters of photovoltaic cells, but also the environmental parameters, so they must be recalculated when arbitrary parameters from two parameters change. Formula 8 is the main function. When any parameter changes, it should be recalculated. Argument U varies from 0V to 60V, stepping by 0.1V. As shown in figure 1, it is recalculated function logic diagram.

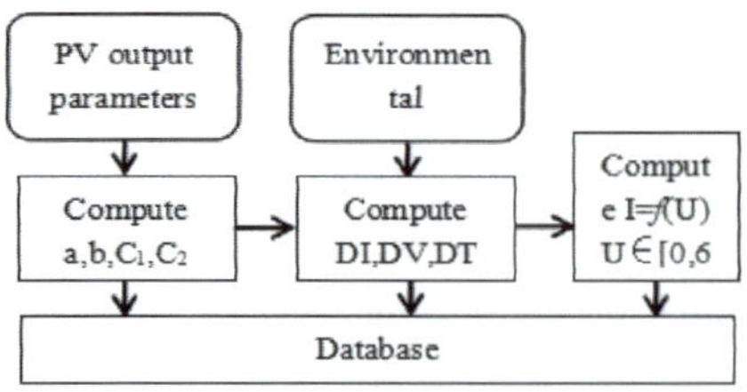

Figure 1.Function recomputation logic.

2.2.Taylor series expansion to exponential and logarithmic functions

Exponential function with a Taylor expansion is:

$$e^x = \sum_{n=0}^{\infty} \frac{x^n}{n!}, \forall x \quad (9)$$

According to Taylor formula expandable depicts the block diagram in figure 2.

Logarithmic function with a Taylor expansion is:

$$\ln(x+1) = \sum_{n=1}^{\infty} \frac{(-1)^{(n+1)}}{n} x^n, \forall x > -1 \quad (10)$$

According to Taylor formula expandable block diagram depicted in figure 3.

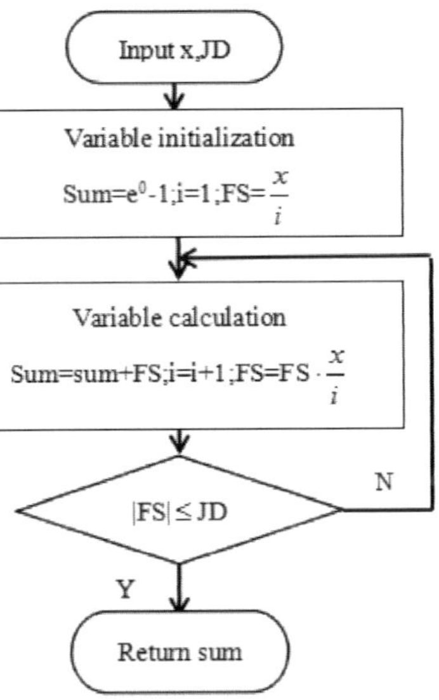

Figure 2. Exponential function calculate process.

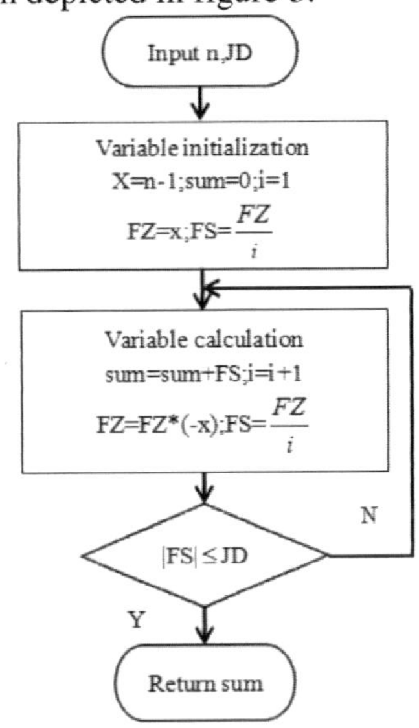

Figure 3. Logarithmic function calculation process.

Among them, by controlling the variable comparison JD to control the accuracy, when JD is smaller, the precision gets higher, corresponding calculation speed will slow down. In figure 2, FS is the term $\frac{x^n}{n!}$ in formula 9, i is the variable n in formula 9. In figure 3, FS was the term $\frac{(-1)^{(n+1)}}{n} x^n$ in formula 10, I was the variables n in formula 10.

3.Optimizated PV curve Experiment

3.1.MK60 introduction

Freescale's chip production MK60DN512, is a 32-bit single chip microcontroller, build-in hardware multiplier, clocked at up to 200MHz. It can fully meet the characteristic curve of single chip microcontroller performance requirements, additionally can meet the Taylor's expansion formula of programming. The microcontroller has 256KB of RAM, can fully store 48KB characteristic curve table.

3.2.The use of single chip microcontroller calculating exponential and logarithmic functions

To approximate the ideal computing speed, prevent the influence of the other block for speed, design a number of programs which are designed for testing single chip microcontroller exponential and logarithmic calculation, program flow shown in figure 4.

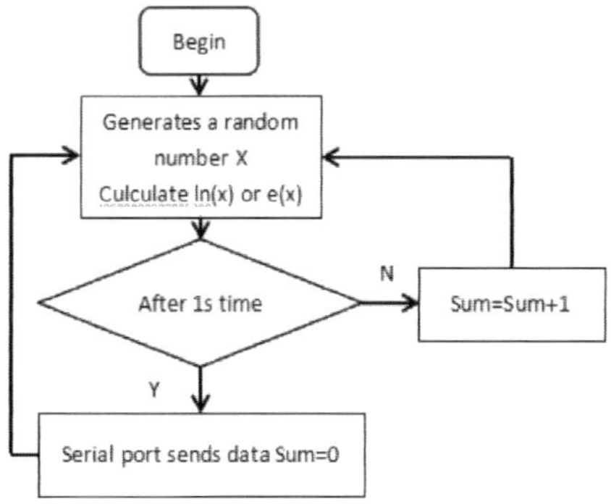

Figure 4.Program flow testing exponential and logarithmic.

The program circulatingly generates a random number x, and calculates the exponential or logarithmic fpr x. During this process, we will continue to count the number of calculations per second, and the calculated number of times through the serial port return to the PC.

After calculating, the obtained results are shown in figure 5, 6. From the below results figures we can see: when JD is smaller, the number of exponential calculations maintain during 800 to 810; the number of logarithmic calculations maintain at 10600.

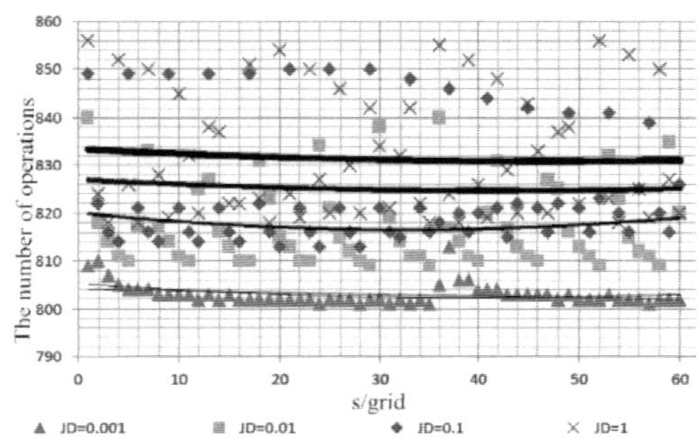

Figure 5.Exponential calculations result.

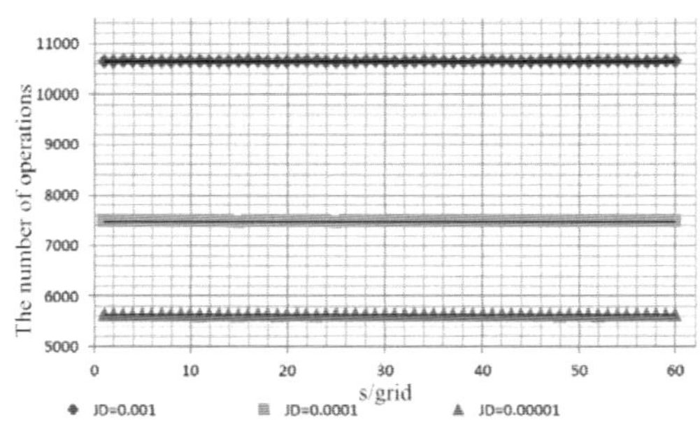

Figure 6.Logarithmic calculations result.

3.3.Using single chip microcontroller drawing photovoltaic cell output characteristic curve

To achieve analoging output characteristics of the photovoltaic cell, first we need to calculate PV output characteristic curve. Since the study of exponential and logarithmic calculations is calculated by Taylor series expansion, existing deviation, as shown in figure 7: actual output characteristic curve of the photovoltaic cell and approximate output characteristic curve of the photovoltaic cell.

System simulated photovoltaic curve shown in figure 8, the solid line is the mathematical model's calculated curve. Various discrete points are system simulator operating point under different load conditions, by comparing the solid lines and the discrete points, you can clearly see the relevant parameters.

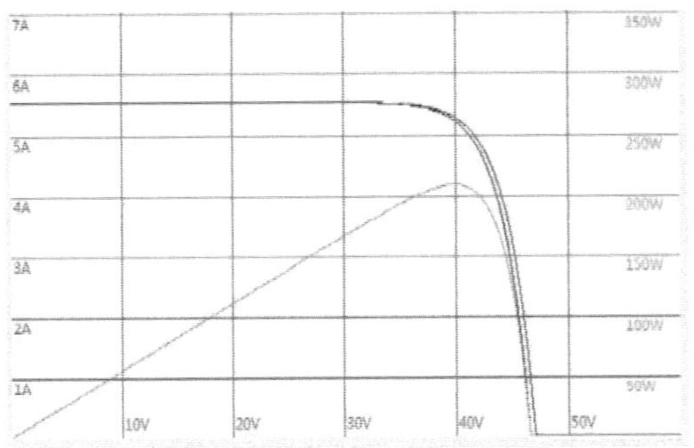

Figure 7.Output characteristic curve.

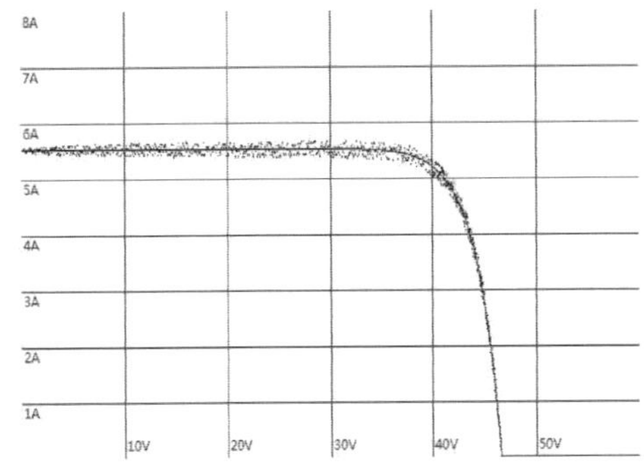

Figure 8.System output characteristics

4. Verify the accuracy of the PV characteristic curve

In order to verify the photovoltaic cell output characteristic curve drew by single chip microcontroller can meet the engineering requirements, now analysis the error rate between the working points of real photovoltaic cell output characteristic curve and the working points of the approximate output characteristic curve. The results are shown in Table 1.

Table 1. List of error rate.

Voltage value	Actual current value	Approximate current value	Error rate
37.00	5.48	5.50	0.36%
37.50	5.47	5.49	0.37%
38.00	5.44	5.46	0.37%
38.50	5.42	5.45	0.55%
39.00	5.40	5.43	0.56%
39.50	5.38	5.42	0.74%
40.00	5.35	5.40	0.93%
40.50	5.31	5.34	0.56%
41.00	5.30	5.32	0.38%
41.50	5.12	5.21	1.16%
42.00	5.00	5.17	3.40%
42.50	4.81	5.00	3.95%
43.00	4.65	4.80	3.23%
43.50	4.42	4.50	1.81%
44.00	4.00	4.22	5.50%
44.50	3.80	4.00	5.26%
45.00	3.62	3.78	4.42%
45.50	2.10	2.22	5.71%

We can see from the analysis results of error rate: working point error rate remains 0.36% to 5.71%, meet the engineering requirements: under 10%, it shows that the use of single chip microcontroller drawing photovoltaic cell output characteristic curve is feasible.

5. Conclusions

The study is based on the mathematical model of general PV array simulator, expand by exponential and logarithmic and simplifies calculation of the mathematical model. It, how to establish mathematical models in microcontroller, can accurately simulate the output characteristic of photovoltaic cells. And analysis the calculation error between calculation mathematical model and the real mathematical model in detail, enabling the system to accurately calculate the required value. Finally, settles the use of single chip microcontroller to calculate the exponential and logarithmic calculations, and achieves the appropriate indicators, PV output characteristics curves meet the project criteria.

6. Acknowledgments

This work was supported by The Jilin Province Science and Technology Development Funds Projects(20130102027JC).

7. References

[1] Azzopardi B, Mutale J, Kirschen D 2008 Cost boundaries for future PV solar cell modules sustainable energy technologies, *Proceedings of the IEEE Internation Conference on Sustainable Energy Technologies*, 589-594

[2] Yazdani A, Dash P 2009 A control methodology and characterization of dynamics for a photovoltaic system interfaced with a distribution network, *IEEE Transactions on Power Delivery,* 1538-1551

[3] Rauschenbach H 1980 Solar cell array design handbook, *Litton Educational Publishing Inc*

[4] Zhou Jun, Li Shuguang, Zhang Yemao, Zheng Hongxia and Li Dianwen 2015 Reserch of the solar cell simulator based on hybrid control algorithm, *Acta Energiae Solaris Sinica*

[5] Wang Fusheng, Wen Jiayan and Zhang Xing 2012 An investigation on digital photovoltaic array simulator, *Acta Energiae Solaris Sinica*, 494-499

IOP Conf. Series: Materials Science and Engineering **157** (2016) 012020 doi:10.1088/1757-899X/157/1/012020

New ENT Laser Micromanipulator Design

LiNing[1], Zhang Ruilan[1], Liu Jian[2],Liu Yingjiu[1]

[1]Basic medical of beihua university, Jilin, 132013; 2. Jilin People's Congress, Jilin,132000

Zhang Ruilan, corresponding author,Email: bhdxzrl@163.com

Abstract: It describes the shortcomings of traditional ENT laser micromanipulator, in order to improve the quality and reliability of output spot, using a reflective structure on the laser micromanipulator. With He-Ne laser as the indication beam, CO_2 laser as the working beam. We have optimized from coincidence, size of the spots and working distance of this device. So that the laser system has been greatly improved in terms of a predetermined distance within the spots coincidence, beam uniformity, stability and reliability, to meet the requirements of long-term care.

1. Introduction

Development of laser medical equipment earlier in seventies in China wasn't forming a market and get little application. As the laser medical equipment, especially manufacturing technology of He-Ne laser acupuncture treatment instrument and CO_2 laser beauty treatment machine was relative simple and inexpensive in the earlier market. Low requirements on medical laser performance, at the time development of laser technology was more suitable for this energy power and performance requirements. Laser medical equipment has becoming increased in demand for health care reforming, laser technology progression, market demand expandition. Such as laser instantaneous burst, short duration, small spot, nondestroy hair follicles, nondamage normal skin, no scars. People understand its advantages at the same time, medical safety is also of great concern.

For improving the quality and reliability of laser micromanipulator continuously, we should design a new kind of medical laser system meeting medical clinic needs. This system is particularly suitable for the throat, ears and other parts of minimally invasive surgery. We got an ideal multi-purpose medical laser system after clinical results.

2. Design ideas

Most biological tissues contain water, the interaction of laser with biological tissue is depending on the interaction of the laser with water to a large extent. Effective intensity of laser on tissue is directly related to the size of energy density, output power density, wavelength, tissue absorption coefficient and other characteristics of laser, also these characeristics decide depth of penetration. The wavelength of CO_2 laser is 10.6μm, so skin has a high absorption on it, and its simple structure, lower price, good performance, stability, easy maintenance, these decide its broadly applications in medical field. He-Ne laser has much advantage, such as good direction (divergence angle less than 10^{-3} rad), good monochrome line width less than 20 Hz, simple structure, long life, small size, light weight, low cost, and easy to use. Besides the output power and wavelength can be controlled very stable. It is widely used for precision measuring, testing, collimation, orientation and other aspects [1]. Based on these characteristics, this design uses CO_2 as the working laser beam, He-Ne laser as the indicator [2].

3 Design and operating results

3.1 Design

3.1.1 Technical requirements

Working distance: 200mm~400mm;
Relationship of working distance and spot size is:
200mm~400mm correspondence 0.16mm~4.6mm;
Working laser and indication laser complete coincidence.

3.1.2 Key considerations in design. At present, the successful development of micromanipulator laser optical system is refractive. Because light beam is different from indication beam at wavelength. Lead to two wavelength of laser light produce great chromatic aberration when crossing the optical system. To correct chromatic aberration make configuration of system becoming very complicated. But it does not make sure the light beam and indication beam complete coincidence, so it affects treatment accuracy.

3.1.3 Selective configuration of reflective system. The reflective optical system has much characteristic such as no colour, light weight, compact structure etc. After reflective surface coating, high reflectivity, from far-infrared to X-rays over a wide spectral range. Reflective system will affect energy efficiency by obscuration of the restrictions, and then affect the therapeutic effect. So this design should consider to reduce obscuration ratio.

Total reflective optical system is more common, such as Newtonian type, Cassegrain type, Gregorian style, Schwarzschild system. Newtonian system is the easiest type of reflective system, only a concave mirror, others belonging to the two mirrors system. The differences between them are bending directions of the reflective surfaces, and secondary mirror is relative to the position of primary mirror's focus. This article hopes to use spherical design and get minimize obscuration. The following discussion is correction primary aberration for spherical system:

Aperture provided on the main mirror, take the normalization condition: $h_1 = 1, u_1 = 0, f' = 1, u_3 = 1$,

the height of principal ray incident on the main mirror h_{1z}=0, incident angle u_z=-1, Lage invariant J=1. Refer to two variables:

$$\alpha = \frac{h_2}{h_1} 、 \beta = \frac{l_2'}{l_2} = \frac{u_2}{u_3} \tag{1}$$

α is aperture ratio of two mirrors, β is magnification of secondary mirror. According to Gaussian optics and geometric relationship between every parameters, we get relations between the parameter and α, β. Index of refraction $n_1 = 1$、$n_2 = -1$.The incident angle of first paraxial ray on the secondary mirror $u_2 = \beta$， incident height $h_2 = \alpha$, the incident height of principal ray on the secondary mirror:

$$h_{2z} = \frac{\alpha - 1}{\beta} \tag{2}$$

radius of curvature of two reflection mirror respectively:

$$r_1 = \frac{2}{\beta} \text{、} r_2 = \frac{2\alpha}{1+\beta} \tag{3}$$

their interval:

$$d = \frac{1-\alpha}{\beta} \tag{4}$$

Use PW method to get the primary aberration coefficients:

$$\begin{cases} \alpha = 4.236 \\ \beta = 1.618 \end{cases} \tag{5}$$

$$\begin{cases} \alpha = -0.236 \\ \beta = 0.618 \end{cases} \tag{6}$$

By (5) and (6) we obtained

$$\bar{s}_1 = \bar{s}_2 = \bar{s}_3 = 0 \tag{7}$$

When

$$\begin{cases} \alpha = 3.56, -0.56 \\ \beta = 0.39, -0.64 \end{cases} \tag{8}$$

$$\bar{s}_3 = \bar{s}_4 = \bar{s}_5 = 0 \tag{9}$$

In this situation, when correcting two aberration , the third one is just zero.

When -1<α<0, β<0, we get a virtual image, the system is divergent, not suitable for laser micromanipulator optical system. In addition, when α=1, the interval of two mirrors is 0, So it should be removed from these solutions. It can clearly be seen the remaining solutions consist Schwarzschild system. Thus can be derived, Cassegrain and Gregorian system can adjust one kind of spherical aberration, the Schwarzschild system can correct up to three aberrations. When α>1, 0<β<1, the primary mirror and the secondary mirror is required openings in Schwarzschild system. Effect of stray light and obscuration are more serious, so we adopt the structure of α>1,β>1 in design.

It is known the incident light needs pass through secondary mirror to reach the primary mirror. So the secondary mirror requires opening. After reflected from the primary mirror toward the secondary mirror, only a part on the secondary mirror, and the rest through the hole on the secondary mirror does not participate in the imaging[3]. This is the first obscuration. Additionally, the secondary mirror reflected light, there may be blocking by the primary mirror to form a second obscuration. Therefore, in order to improve energy efficiency, should be reduced caliber of primary mirror. It is discovery the obscuration is least over optimization when the stop on the main mirror.

3.1.4 200mm～400mm working distance. Working distance of this design need to achieve 200 ~ 400mm adjustable. Regardless of internal or external focus system will bring difficulty on aberration correction in design. The multiple design features in ZEMAX namely Multi- Configuration is a very practical function. It is based on a variety of patterns or structures used for debugging and analysis. For multi-structure analysis, its application involves many aspects, such as the zoom lens, magnification lens, eliminate temperature designing and the scanning lens. It allows the position of optical assembly to make a change with focus or magnification as function. It is also very easy to achieve, therefore, this design consider using multi-functional structure to realize working distance adjustable.

3.2 Operating results

The initial structure of the objective lens shown in Table 1. For debugging, aberration correction, analysis and evaluation the structural parameters of the initial objective with optical design software ZEMAX [4].

Table 1. Initial objective structural parameters

Surf : Type		Comment	Radius		Thickness		Glass		Semi -Diameter
OBJ	Standard		Infinity		Infinity				0.0000
1	Standard		Infinity		25.0000				1.5000
STO	Standard		25.8927		-24.0000	V	MIRROR		1.5000
3	Standard		50.0000		78.0860	V	MIRROR		4.2739
IMA	Standard		Infinity		-				8.32E-005

Setting a single system is very similar to the multiple structure. Multiple Structure Editor is used to indicate those design parameters between different configuration parameters [5]. For this design, we will need three configurations: The first configuration is used to define the working distance 200mm. The second and the third, respectively, will be used to define the working distance of 300mm and 400mm. Table 2 is the multi-system configuration parameters. Using above parameters we got the structure of objective lens. Figure 1 is the structure when working distance is 200 ~ 400mm. It meets the design requirements.

Table 2. multiple structure parameters

Active : 1/3		Config 1*		Config 2		Config 3	
1: THIC	2	-9.174100	V	-6.872908	V	-5.774110	V
2: THIC	3	209.174124	V	306.872911	V	405.774096	V

IOP Conf. Series: Materials Science and Engineering **157** (2016) 012020 doi:10.1088/1757-899X/157/1/012020

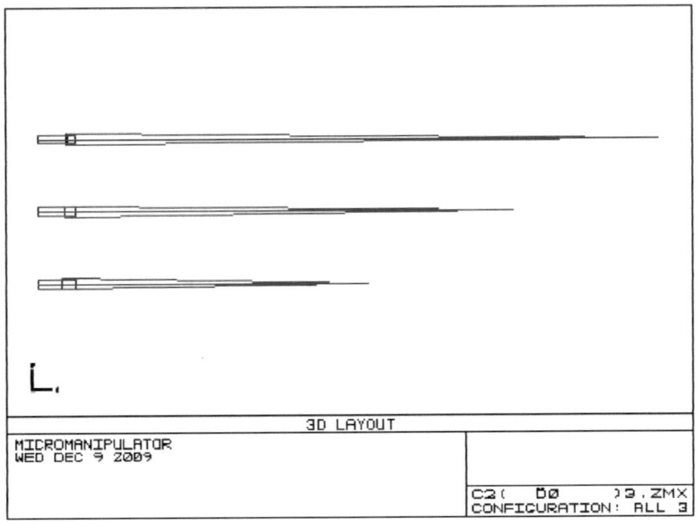

Figure 1. structure of objective lens

After debugging, we got an optical system the working distance can be adjusted. And the spot size is smaller than the required value in their working distance. As shown in Table 3 is spot diagram of 200mm～400mm work distance.

Table 3. Spot diagram of 200mm～400mm work distance

work distance（mm）	200	250	300	350	400
required value（mm）	<2.8	<3.2	<3.7	<4.1	<4.6
geometric radius（μm）	0.595	9.495	17.166	24.131	30.659

4. Conclusion

In this paper, we use the characteristics of Schwarzschild reflective system, combined with the characteristics of the laser micromanipulator. Designed a new ENT laser micromanipulator which direction beam and working beam are complete coincidence and has small spot size. Meet the design requirements, to improve the accuracy of treatment. It is foreseeable that laser microscopy techniques will become increasingly widespread in the medical area. Laser micromanipulator quality and reliability will gradually increase.

5. References

[1] Cai S M, Yang J F,CHENG Y X et[a].Optical design of two wavelength laser therapy system[J]. *Laser and Infrared* ,2003,4(2):104-105.(in chinese)

[2] LI Y X. The progress and development trend of laser medical device[J]. *Medical Device Information* ,2004 19(11):0001-04(in chinese).

[3] LI Y H, LI ZH J, ZHU CH H, et[a]. The Characteristic research for small-size double wave length YAG medical laser[J]. *Optronics Laser*, 1996,7(5):317-319(in chinese).

[4] YUAN X C.Optical Design[M]. Beijing Science Press,1983(in chinese).

[5] David L. Shealy, Shao-Hua Chao. Geometric optics-based design of laser beam shapers[J]. Opt. Eng. 2003,42(11):3223-3124.

IOP Conf. Series: Materials Science and Engineering **157** (2016) 012021 doi:10.1088/1757-899X/157/1/012021

Effect of TiN powder mixed in Electrical Discharge Machining

A Muttamara[1] and J Mesee[1]

[1]Faculty of Engineering, Thammasat University, Pathum thani, Thailand.

Abstract. Many trials were studied about powder mixed in Electrical Discharge Machining (EDM). The experiments were carried for improving surface characteristics and related to the surface modification. The experiment was carried out using a copper tool electrode and EDMed in titanium nitride (TiN) powder mixed in dielectric fluid. In this research, to obtain the even modified layer, the effects of EDMed conditions were investigated. The EDMed surfaces were observed by SEM. Under the suitable discharge conditions in TiN powder mixed kerosene, the stable thick TiN layer adhered on the workpiece surface. The microcrack length per unit area treated in TiN mixed kerosene was greater than that treated in normal kerosene. Titanium carbon nitride (TiCN) was found on the modified layer by XRD analysis. The effect of the diffusion of carbon during cooling on the characteristics of the recast layer was discussed.

1. Introduction

Tool steel Type JIS SKD11 is widely used in applications requiring hardness and good elevated temperature strength. Because of its properties, the SKD11 is good to fabricate mold and die using electrical discharge machining (EDM). EDM is based on the erosion of materials by electrical sparking. Solid, liquid or gas particles can be removed [1,2]. The EDM process generates a recast layer cover the surface after the process. The recast layer is called a white layer due to its colour after etching. This comprises a spattered surface layer and a layer that is removed during the spark process. The next layer is a new layer that is formed by melting but it is not removed from the material and is referred to as a recast layer. This also results in metal is deposited on the workpiece surface as a coating layer. The study of the surface modification using EDM began about a decade ago [3,4]. Furutani et al. [5] used titanium powder in kerosene dielectric and obtained titanium carbide layer on a steel with a copper rotating disk electrode and they kept the powder concentration high in the machining area on carbon steel.This occurs because titanium and carbon from the degradation of the dielectric (a hydrocarbon composition) combine during the spark process with the material surface while it is cooling [2,6]. Yan et al. [7] obtained TiN on the titanium's workpiece using EDM in urea solution. EDMed surfaces have defects such as microcracks and pores that form because of the temperature gradient that is created near the surface, which induces a substantial modification of the thermally affected layer during the machining process. This work aimed to study surface modification and the completeness of the recast layer onto a surface by electrical discharged machining.

2. Experimental procedure

SKD11 (in JIS specification) was used as the workpiece material. Table 1 shows the chemical composition of SKD11. Carbon 1.5 % is important to affect to heat treatment. Figure 1 shows a schematic diagram of experimental apparatus. Experiments are carried out using an electrical discharge machine. Machining conditions were set as shown in Table 2. The tool electrode is a copper cylinder of 3 mm in diameter. As machining fluids, kerosene and titanium nitride powder mixed fluid were used at a concentration rate of 15 g/l.The particle size of TiN powder is not over 2 μm in diameter. In EDM phenomena, polarity of electrode plays important role for machining, the positive (+) electrode polarity is chosen for the machining [4,8]. On the contrary, the negative polarity (-) often uses for the modified technology. Figure 2 shows generation mechanism of modified layer. Carbon from the powder and/or resolved carbon from the kerosene combine with the material of the electrode during the discharge phenomenon, and so a new compound material is formed on the EDMed surface. In our setup, the polarity of the electrode is set to negative in order to make wear of electrode.

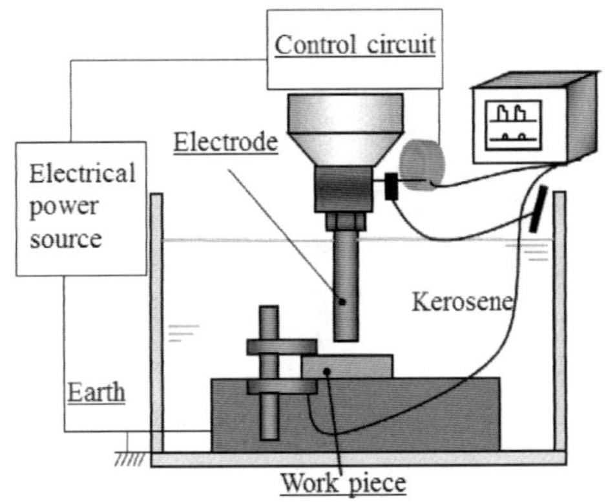

Figure 1. Schematic diagram of experimental apparatus.

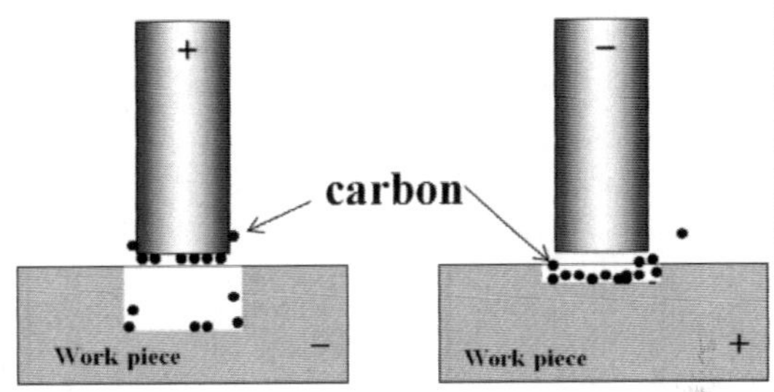

Figure 2. Generation mechanism of modified layer.

Table 1. Chemical Compositions of SKD11 (wt.%)

	Fe	C	Si	Mn	Mo	Cr	Ni	V	Co
AISI 316L	bal.	1.5	0.3	0.3	1.0	12	0.3	0.8	2.1

Table 2. EDMed Conditions

	Levels
Current, I (A)	6 and 21
Discharge duration, t_e (μs)	50 and 500
Duty factor (%)	50

3. Results and Discussions

Figure 3 shows a SEM micrograph of EDMed surface using discharge current 6 and 21 A and on time 50 and 500 μs. All figures are compared with EDM in normal kerosene and TiN powder mixed kerosene. Crack and pores are found on the EDMed surface. When EDM in TiN mixed kerosene, crack can be reduced with covering of the TiN powder. The thickness was measured by observing the cross section. As shown in the figure, the thickness increases with on time. The effect of TiN powder reduces layer thickness. The white parts seen on the micrographs are the coated layers, Besides, the layer contains large cracks and pores. On the other hand, in high carbon powder concentration machining, the undulation is very small and the thickness is approximately uniform. Also, the layer does not contain cracks and pores. This is supposed to be the effect of the more appropriately dispersed electrical discharges. Figure 4 shows the relationships between the recast layer and various EDMed conditions. Higher discharge current (21A) makes thicker recast layer compared with lower current (6 A). This is due to the large amount of heat generated by the discharges, which causes melting of the material, followed by rapid cooling. When TiN powder mixed in kerosene, the thickness of the recast layer decreases because discharge gap is increased with floated powder in the gap related to reducing of discharge energy. Figure 5 shows relationship between surface roughness of EDM and on time in kerosene and TiN mixed kerosene for current 6, 21 A. The surface roughness increases with the increase of discharge current and on time. The increase of surface roughness is due to the increase of spark energy, which causes the surface pits resulting from higher removal of material.

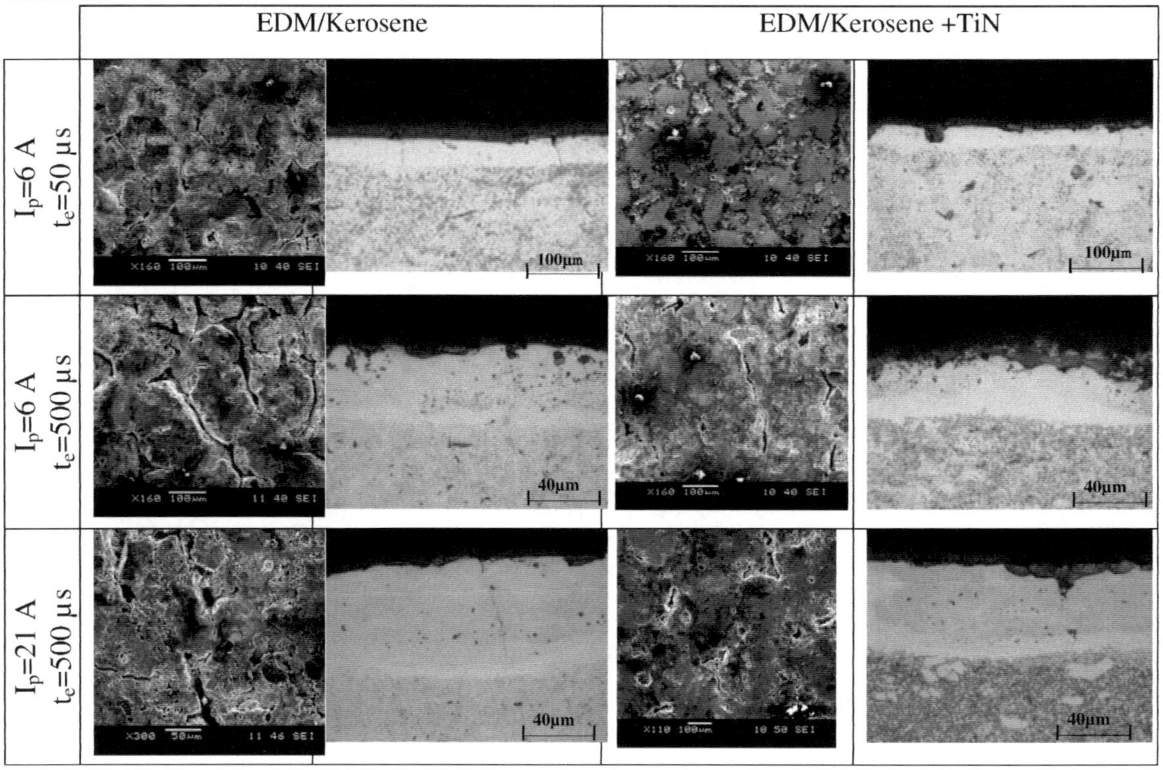

Figure 3. SEM of EDMed surface and cross section images of the modified thickness layer for various conditions.

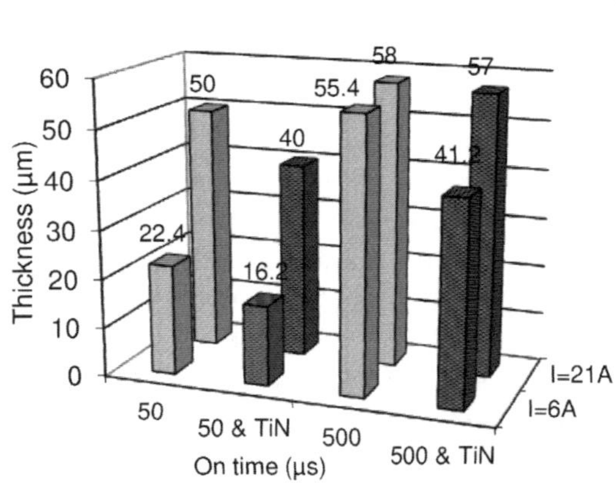

Figure 4. Thickness of the recast layer compared with on time using kerosene and TiN mixed kerosene for current 6, 21 A.

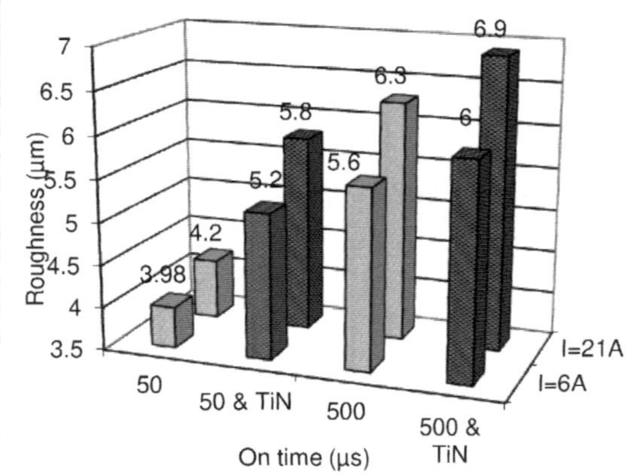

Figure 5. Surface roughness of EDM compared using kerosene and TiN mixed kerosene for current 6, 21 A.

IOP Conf. Series: Materials Science and Engineering **157** (2016) 012021 doi:10.1088/1757-899X/157/1/012021

In order to observe TiN on the recast layer, EDS analysis is discussed. Figure 6 shows the elemental composition of EDS. As can be seen from the figure: Ti, C and Fe elements images of machined surface analysed by EDS. It is also confirmed from the component images that components Ti and C exist in this layer, while component N is rarely detected on the layer. Afterwards, the composition of the layer was analysed using XRD (X-Ray Diffraction) as shown in figure 7. According to this figure, besides the components of the base material such as Fe, the component TiCN also exists on the EDMed surface. Carbon from kerosene reacts to the TiN become to TiCN on the layer. Fe_7C_3 also can be found on the layer. TiN cannot be found due to there is much carbon in the working medium fluid.

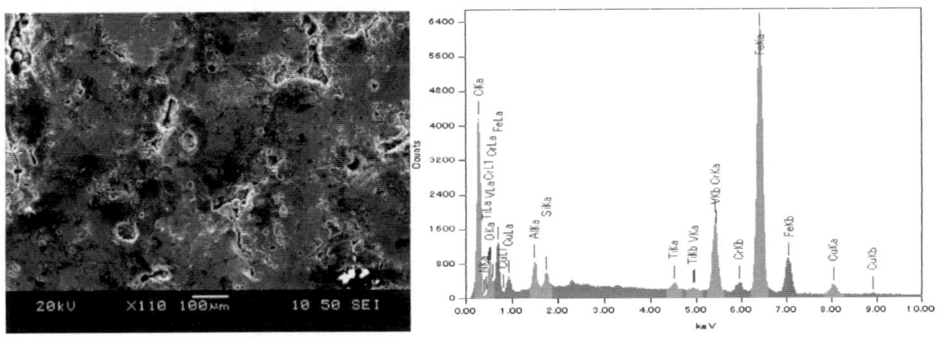

Figure 6. Elemental composition of EDS.

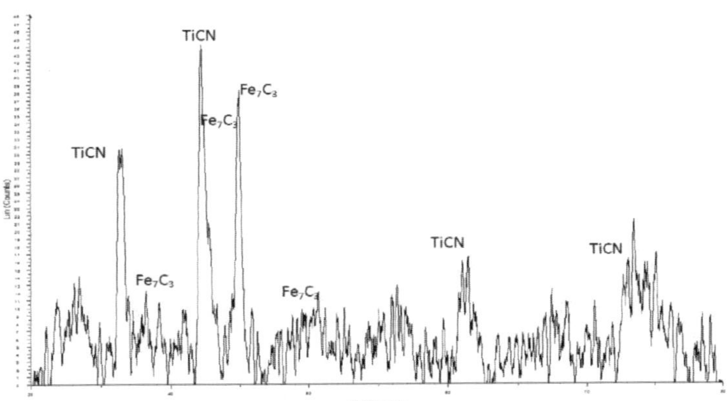

Figure 7. XRD analysis result of EDM with TiN powder electrode.

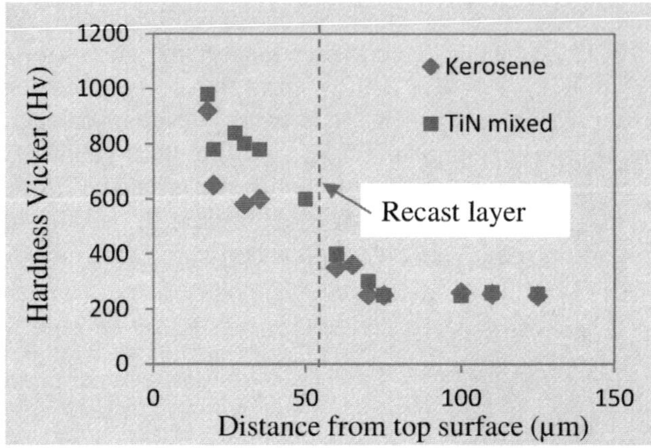

Figure 8. Vickers hardness values of the machined surfaces in kerosene and TiN mixed kerosene (Ip=21A, te=500 μs)

Figure 8 shows the distribution of Vickers hardness on the cross sections of surfaces machined by EDM in kerosene and TiN mixed kerosene. Hardness of layer was examined using load at 10gf. The experiment was conducted on the recast layer, HAZ and matrix material. Both EDM and EDM with TiN powder are produced a high hardened layer, From the figure, it can be seen that, the hardness of the recast layer is much higher than that of base material (SKD61), which indicates that this method hardness of TiCN and Fe_7C_3.

4. Conclusions

In this study, a surface modification technique using EDM with TiN mixed fluid is proposed. Results were summarized as follows:
1. The resolidified layer containing TiCN can be generated on EDMed surface.
2. The EDMed surface with TiN mixed kerosene is rougher than that in EDM with kerosene type fluid.
3. The modified layer thickness grew up to 57 μm. There were few micro cracks and void detected on this thick layer.
4. The hardness of modified surface with TiN mixed kerosene came to 980 HV which is harder than that EDMed with pure kerosene.
5. A hard layer containing TiCN can be formed on the machined surface by EDM.
6. By mixing TiN powder with the dielectric kerosene, smaller hardness value with a little higher surface roughness values can be obtained.

5. Acknowledgments

The authors gratefully acknowledge the financial support provided by Thammasat University Research Fund under TU Innovative Scholar and the National Research University Project of Thailand Office of Higher Education Commission.

6. References

[1] Narumiya H et al 1989 EDM by powder suspended working fluid *Proceeding of ISEM-IX* .pp .5-8
[2] Kruth, J.-P et al 1998 In-process alloying of the white layer of a workpiece machined by Die-sinking EDM *International Journal of Electrical Machining* .3. pp 33-38
[3] Kumar S et al 2009 Surface modification by electrical discharge machining: a review *Journal of Materials Processing Technology* .209(8) .pp 3675–87
[4] Mohri N et al 2003 Layer generation process on work-piece in electrical discharge machining *Annals of the CIRP* .52. 1. pp 161-164

[5] Furutania K et al 2001 Accretion of titanium carbide by electrical discharge machining with powder suspended in working fluid *Journal of the International Societies for Precision Engineering and Nanotechnology* .25. pp 138–144

[6] Tzeng YF and Lee CY 2001 Effect of powder characteristics on electro discharge machining efficiency International *Journal of Advanced Manufacturing Technology* .17. pp 586–592

[7] Yan BH et al 2005 The effect in EDM of a dielectric of a urea solution in water on modifying the surface of titanium *International Journal of Machine Tools and Manufacture* .45 . pp 194–200

[8] Muttamara A and Fukuzawa Y 2012 Influence of carbon in iron on characteristics of surface modification by EDM in liquid nitrogen *American Transactions on Engineering & Applied Sciences* .pp 41-55

Modeling and Timing Simulation of Micro Turbine engine in Starting Process

Libing Huang[1], Feng Gu[2], Ying Zhang[1] and Yuqing He[2]

[1] Information Department, Shenyang Jianzhu University, Shenyang 110016, China
[2] State Key Laboratory of Robotics, Shenyang Institute of Automation, Chinese Academy Sciences, Shenyang 110016, China

huanglibing@sia.cn

Abstract. The stability of start-up process of turbine engine is the foundation of its operation. The process of starting performance optimization is very complex, and the experiment is dangerous, so it is very necessary to make the simulation study on the turbine engine starting process. In this paper, the mathematical model of turbine engine in starting process was established, according to the JetCat-P400 turbine engine starting process principle, using the experimental data of engine ground operation and the actuator voltage data. The simulation results showed the validity of the model.

1. Introduction

Turbine engine is a nonlinear, time-varying and complex pneumatic thermodynamic system, and the longevity of engine is influenced by the performance of the starting process. It is difficult to get the component characteristics of turbine engine at low rotate speed, particularly characteristics of combustion chamber, compressor and turbine. In [1-2], a component-level ground start-up model of turbine engine is established, but it is difficult to obtain the characteristics of the components. In [3-4], a Simple Three Segment Integral Method is put forward to build the model of turbine engine in start-up process. According to the working process of turbine engine ground starting, the starting process includes three phases: engine rotor is driven by starter motor only, engine rotor is accelerated by the both starter motor and turbine together, and the turbine works alone. In [5], Radial Basis Function(RBF) neural network is proposed to build the model of turbine engine in start-up process. However, its physical meaning is not clear. Above all, in the process of starting performance optimization, these three models cannot effectively describe the impact to the turbine engine' performance with the change of actuator state.

Research of this paper was based on the JetCat-P400 turbine engine system. Firstly, the turbine engine starting process sequence was analysed in detail, and then the rotor speed mathematic model of turbine engine starting process was established based on analysing its load condition. Finally, the accuracy of the model was verified by timing simulation of turbine engine in starting process.

2. Ground operation of a turbine engine and analysis of the starting process

The starting process of turbine engine involves five actuators: starting motor, ignition, pump, main fuel solenoid and starting fuel solenoid, which work together. The starting process of engine is carried out by controlling the actuator in accordance with the timing sequence, thus, how to obtain the time sequence of actuators is the foundation of modeling and analysing the start-up process. In this paper, the engine ground operation, aiming at acquiring experimental data, was executed firstly. Then, the actuator state of each stage was obtained through analysing the experimental data. Finally, the time sequence of start-up process was detailed analysed according to the actuator state.

2.1. Turbine engine experiment platform

The ground operation of the turbine engine experiment platform was based on JetCat-P400 turbine engine, shown as Figure 1. The system consists of two parts, one part is the JetCat-P400 turbine engine system, including turbine engine, ground support unit (GCU), electronic control unit (ECU), fuel pump, battery, the other part is a voltage monitoring system [6-7], in which a simple and cheap development board with stm32 is used to monitor the voltage of the four actuators, the starter motor, ignition, the main fuel solenoid and the starting fuel solenoid.

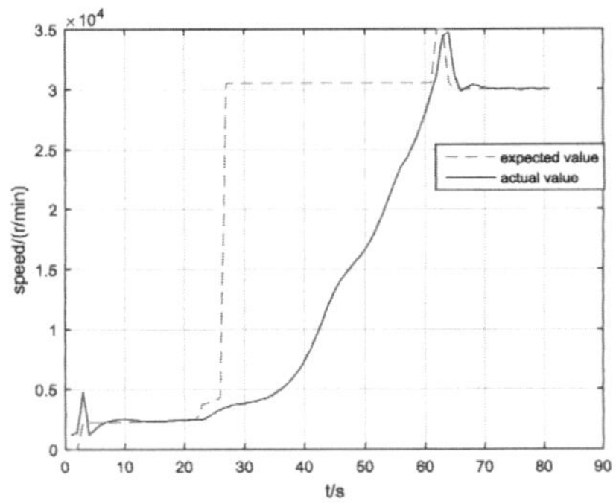

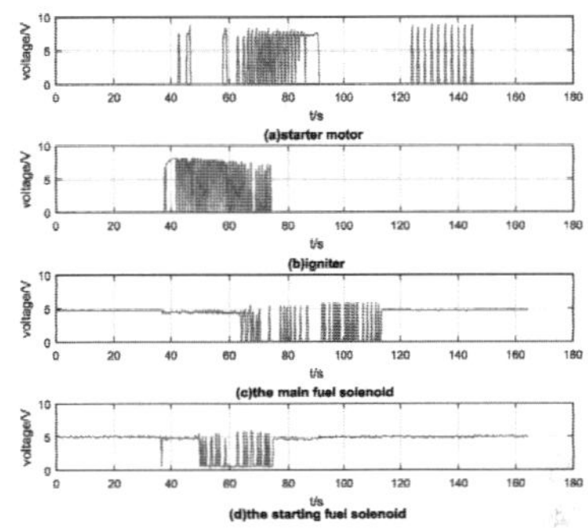

Figure 1. Experiment platform **Figure 2.** Engine speed curve **Figure 3.** Actuator voltage

2.2. Turbine engine ground operation experiment

The aim of the ground operation experiment is to obtain the experimental data of the turbine engine actuator. Data has two sources. One part of data comes from the turbine engine ECU, shown as Figure 2, in which dashed lines are used for the expected value of turbine speed and solid lines are used for its actual measured value. And the other part of data comes from the voltage monitoring system, from which the voltage of the actuator, the starter motor, ignition, the main fuel solenoid and the starting fuel solenoid can be obtained, as shown in Figure 3. In Figure 3 a set of turbine engine successful data is listed and the engine starts from 30 seconds. Figure 3(a) shows the voltage value of starter motor and the value of 0 express the starter motor not works. From the voltage value, the status of the motor can be known. The status of the ignition is the same as the starter motor, however, the main fuel solenoid and the starting fuel solenoid are just the opposite.

3. Modeling of micro turbine engine in starting process

The ground starting process of turbine engine musts rely on the outside motive power, because that the ignition will only cause the engine damage, does not make accelerate, when the engine speed is zero. When going up to a certain rotate speed and temperature, the fuel can be pumped and ignited. The gas created by combustion chamber makes the turbine work, and increases the engine speed. With increasing of the engine speed, the resistance of the turbine goes up. In conclusion, starter motor and gas created by combustion chamber does work to the rotor, thus generates a positive acceleration. On the other side, the resistance of the turbine itself makes the engine speed decrease. So the acceleration of engine rotor can be expressed as:

$$a = a_q + a_T - a_C \tag{1}$$

where the a_q is the acceleration from starter motor, and its value is 0 when motor turn off. The a_T is the acceleration from oil inflame, and its value is 0 when the main fuel solenoid closed. The a_C is the reverse acceleration from the friction or air resistance.

Thus, the engine speed model of turbine engine can be described by the following equation:

$$n = n_0 + \int_0^t a\,dt \tag{2}$$

where the n_0 is initial speed of engine, in the condition of ground experiment, the engine speed is 0.

3.1. Calculation of a_C

When the ground operation experiment being finished, the starter motor, oil pump and oil valve is turned off, the reverse acceleration of engine rotor comes from the frictions and air resistance. So a_C can be identified using the data from engine shut down process. When the engine works with high speed, its mainly resistance comes from the air. Ignoring the driving accessories registers and frictional resistance, then a_C could be considered directly as proportional to the square of the engine speed [8]. It can be described as follows:

$$a_C = C \cdot n^2 \tag{3}$$

where the C is a constant related to motor type.

The experiment data is obtained from the stage of automatic stop in turbine engine ground operation experiment, shown as Figure 4. In automatic stop stage, the starter motor and fuel are not doing any work to engine, and the engine speed reduces slowly. The acceleration of the turbine rotor can be calculated by doing derivation to the engine speed. Finally, the constant can be calculated:

$$C = 2.6921 \times 10^{-5} \tag{4}$$

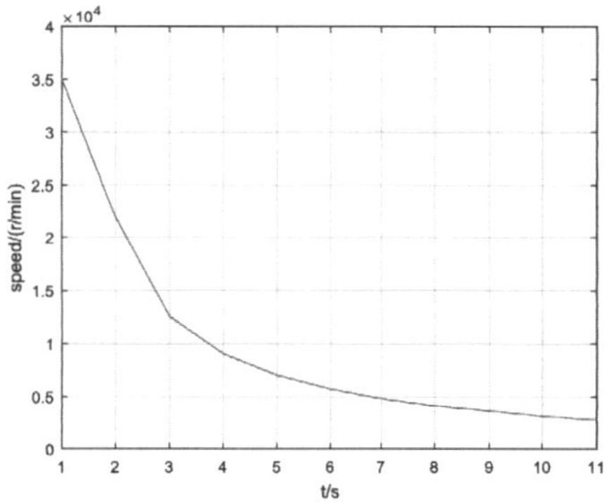

Figure 4. Engine speed curve

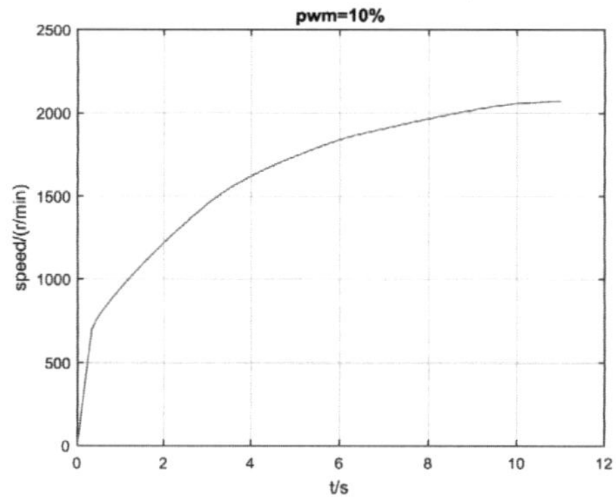

Figure 5. Engine speed curve

3.2. Calculation of a_q

The JetCat-P400 turbine engine uses a direct-current (DC) motor for starter motor. DC motor has a small armature resistance. When DC motor being in an energized state, the motor will generate a large starting current which will bring the motor spins and generates the reverse electromotive force. Then, the current and the motor torque start to decrease, so does the corresponding acceleration. Finally, it will be stable in a certain rotating speed. The common DC motor equation shows as follow [9]:

$$T_m \frac{dw_m(t)}{dt} + w_m(t) = K_m u_a(t) - K_c M_c(t) \tag{5}$$

where the $w_m(t)$ is the motor speed and $w_m(t) = n$. The T_m is the time constant of motor. The K_m is a constant related to the motor type. The $u_a(t)$ is the motor voltage. The $K_c M_c(t)$ is the motor load torque.

Acceleration given by load torque can be calculated as follows:

$$\frac{K_c}{T_m} M_c(t) = C \cdot n^2 \tag{6}$$

The motor voltage is calculated as:

$$u_a(t) = U \cdot pwm\_motor \tag{7}$$

where $U = 9.9V$, which pwm_motor is the motor driver board voltage. The pwm_motor is the duty ratio of motor which varies between 0 and 1.
Combining Eq. (6), (7) with Eq. (8) yields can be obtained:

$$\frac{dn}{dt} = \frac{K_m}{T_m} \cdot U \cdot pwm\_motor - \frac{n}{T_m} - C \cdot n^2 \tag{8}$$

The acceleration given by motor can be expressed as:

$$a_q = \frac{K_m}{T_m} \cdot U \cdot pwm\_motor - \frac{n}{T_m} \tag{9}$$

In order to obtain the parameters of starter motor, the turbine engine ground operation experiment is carried out. In the condition of zero motor speed, the pwm_motor is 10%, when motor speed keeps going up, record the current data, show as Figure 5. Make derivation to motor speed, find $\frac{dn}{dt}$, then the correlation coefficient can be obtained as follows:

$$\frac{K_m}{T_m} \cdot U = 8130 \quad \frac{1}{T_m} = 0.05 \tag{10}$$

3.3. Calculation of a_T

For simplify the problem, a linear relationship between the turbine acceleration and rotate speed is considered as follow [8]:

$$a_T = \frac{C \cdot n_r^2}{(n_r - n_1)} \cdot (n - n_1) \tag{11}$$

where n_1 is the engine speed before the ignition, n_r is the balanced speed of engine between the turbine and the compressor power in contemporary oil supply condition.

After turbine engine being ignited, the stable speed of engine in different fuel injection conditions are different. Based on the analysis on experimental data, the stable speed of engine has a linear relationship with oil pump, and the relationship can be expressed as:

$$n_r = k \cdot U_{pump} \tag{12}$$

where the k is constant. The U_{pump} is voltage of oil pump.

Through the turbine engine ground operation experiment, the stable rpm corresponding to oil pump can be extracted. In the situation of voltage keeping stable, each rpm is processed by average filter. The relationship between the engine speed and oil pump is shown in Figure 6. Using MATLAB Fitting toolbox, we get the parameter as:

$$k = 2.708 \times 10^4 \tag{13}$$

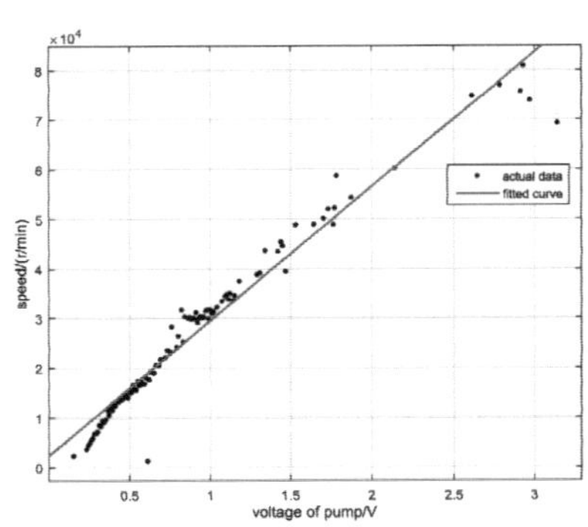

Figure 6. Steady speed relations with oil pump voltage

IOP Conf. Series: Materials Science and Engineering **157** (2016) 012022 doi:10.1088/1757-899X/157/1/012022

4. Timing simulation of micro turbine engine in starting process

The starting process of turbine engine is done strictly according to the timing. From the above analysis we can see, the starting process is not only power on the starter motor and fuel pump, but also includes a few steps in chronological. In different stages, the status of actuator is not the same. Even in the same stage, the starter motor and fuel pump voltage at different time is also different. So the start process of turbine engine should take the time sequence as consideration.

In this paper, the simulation of turbine engine in starting process was achieved based on the software MATLAB/Stateflow [10-11]. Stateflow is good at realizing the logical transfer & event-driven FSM (finite state machine). Such graphical utility can improve the efficiency of turbine engine in starting

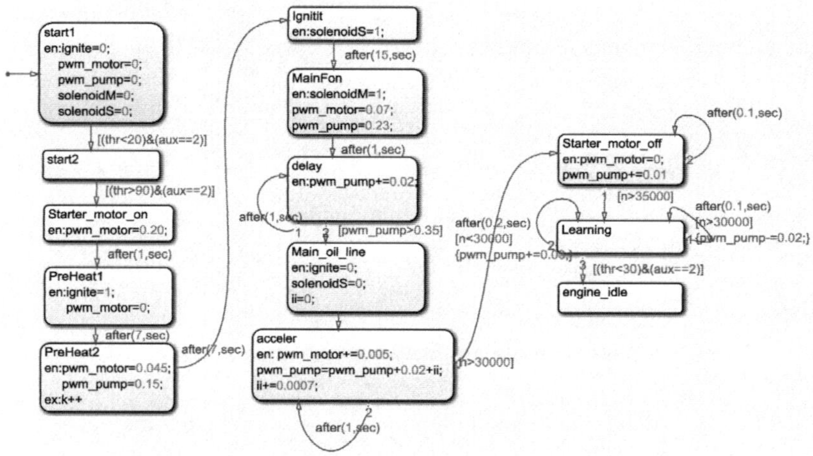

Figure 7. Stateflow chart of turbine engine controller

Figure 7 is the subsystem of the "Turbine engine controller". It as can be seen from the Figure 8, when simulation begins, the system enters into the stage of "start1", and all the actuators will be turned off. Then, waiting for the start-up signal, the throttle stick is moved from minimum to maximum and the auxiliary stick keeps in the start state. After the start signal has been received, the starter motor is activated within one second. With the model of turbine engine in starting process, the engine speed of the moment can be calculated. Then, the starter motor is turned off, the engine steps into the next state after 7 seconds, the engine speed of this moment can also be calculated, and so on, the turbine engine rotor speed can be calculated. The simulation curve is shown in Figure 8.

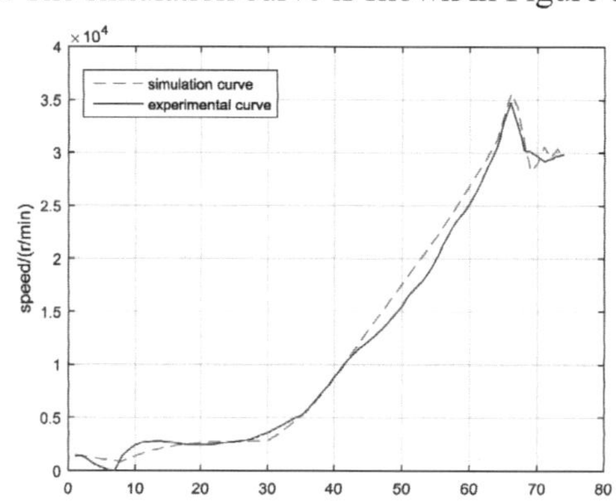

Figure 8. Engine speed curve in starting process simulation

In Figure 8, the solid line represents the turbine engine ground experiment data, while the dotted line shows the predictive value. In this paper, there are two sets of experimental data, the first is used to build the model of turbine engine in starting process, and the second is used to validate the correction of the model. From this figure, the simulation data is well in accordance with the experimental data, and the trend goes to the same direction. Thus, the advantage of modeling according to the timing sequence can be embodied.

5. Conclusion

In this paper, an analytical study about JetCat-P400 turbine engine's starting process sequence was carried out based on the experimental data of engine ground operation. Firstly, the mechanical properties of turbine engine rotor were studied. Then, the start process mathematical model was established. Meanwhile, relevant parameters were identified through analysing the experimental data. Finally, a timing sequence simulation of turbine engine was put forward.

Simulation results showed that the simulation data suited experimental data well. The turbine engine mathematical model and timing sequence simulation method adopted in this paper has be of significance in research guidance of the turbine engine control system.

6. Acknowledgments

The research work is supported by the state key laboratory of robotic, Shenyang Institute of Automation(SIA), Chinese Academy of Sciences(CAS). Which provides the experimental equipment, experimental site and technical support. It will also give much appreciation to my colleagues' contributing to this work. Special thanks to Liying Yang, Bingbing Li and Xiangdong Meng.

7. References

[1] M. Badami, P. Nuccio and A. Signoretto, "Experimental and numerical analysis of a small-scale turbojet engine," Energy Conversion and Management, vol. 76, pp. 225-233, 2013.

[2] R.K. Agrawal, and M. Yunis "A Generalized Mathematical Model to Estimate Gas Turbine Starting Characteristics," Phys.rev.b, 1981, 104:1(1):5668-5687.

[3] Y. Fan, F. Ding, P. Kai, and W. Wangyuan, "Modeling of an Aero-engine in Starting Process Based on the Test Data[J]," *Journal of Air Force Engineering University (Natural Science Edition)*,2013,06:1-4.

[4] K. Xiangxing, W. Xi, Z. Shaoji, Y. Changzhi, and W. Xiang-yu, "Research and experiment on improved sectional combination control scheme for civil turbofan engine's startup process[J]," *Journal of Aerospace Power*,2014,12:2924-2929

[5] J. Tao, L. Yinghong, "A Dynamic Identification Model of Aero engine Starting Process Based on the RBF Network [J]," *Journal of Aerospace Power*,2002,03:381-384.

[6] M. Dub, J. Bajer and M. Stepanek, "Electronic starting control unit for small jet engine," Military Technologies (ICMT), 2015 International Conference on, Brno, 2015, pp. 1-4.

[7] Z. Zheng, L. Fushui, D. Wei, W. Pei, Z. Qinglong and Y. Wang, "Research on fuel injection strategy of start process for the electronic diesel engine," Transportation Electrification Asia-Pacific (ITEC Asia-Pacific), 2014 IEEE Conference and Expo, Beijing, 2014, pp. 1-5.

[8] L. Jianxun, L. Yinghong and Z. Dongfang, "The research progress of numerical simulation of aero engine in starting process [C]," The sixth power conference proceedings of Chinese Society of Aeronautics and Astronautics, 2006:162-166.

[9] H. Shousong. "Automatic Control Theory[M]," Beijing Science and Technology Press.2005.

[10] H. LI, "The Application Study of Experimental Modeling for a Micro Turbine Engine [D]," Nanjing University of Aeronautics and Astronautics, 2010.

[11] W. Jian, Z. Guigang, S. Haixia, Y. Qi, and H. Jiayang, "Simulation and fault diagnosis for aviation engine starting system based on SIMULINK," Chinese Automation Congress (CAC), 2015, Wuhan, 2015, pp. 1834-1839.

[12] JetCat RX Turbines with V10 ECU[OL], http://www.jetcat.de/downloads/rx--engines_manual_eng.pdf

IOP Conf. Series: Materials Science and Engineering **157** (2016) 012023 doi:10.1088/1757-899X/157/1/012023

The Quantum Energy Saver design and Fuel-saving application

Xiong Fang[1], Wenwu Mao[2], Xisheng Shen[1], Jianyu LI[1], Wenchao Huang[1] and Zhixin Chen[1]

[1] Fishery Machinery and Instrument Research Institute, Chinese Academy of Fishery Sciences Research Laboratory of marine fishing vessels and equipment, Shanghai, China

[2] School of Engineering, Shanghai Ocean University, Shanghai, China

Abstract: In order to reduce the high fuel consumption of the shipping industry, a new type of quantum energy saver device is studied and developed. According to a period of time to use the energy saving device and the users' feedback, by recording the fuel consumption of diesel engine usage, and comparing the changes in fuel consumption before and after the installation of quantum economizer in the same ship, it can reflected the ability of the fuel consumption. After analyzing the data, it shows that the installation of quantum economizer can significantly reduce the fuel consumption of a diesel engine ship. The analysis and application of this paper can play an important role in saving energy and reducing consumption, and provide a reference for other related research.

1. Introduction

Now shipbuilding industry is very important, it has a direct connection to the 97 industries in the national economy, which provides important equipment for our country's energy transportation, international shipping and ocean development; it is also a very important industry chain in the national economy. The marine diesel engine is the heart of the ship, if it has characteristics of reliable quality and low energy consumption, it will provide a huge driving force to make China's shipbuilding industry develop more quickly, healthy and continuously.

In modern shipbuilding industry, fuel and oil fees accounted for about 50%-60% of the whole cost, even more. With the reducing of the oil production of the whole world and oil price rising, it makes high cost of enterprise operation. In order to save energy and reduce the cost, the actions have been taken by most of the shipbuilding factories. Such as using low price oil and improving featuring cost. [1, 2] In this paper, by the research and application of a new type of quantum economizer, to achiel injection system to reduce fuel consumption. But the methods above are unreliable and it also increase the diesel engine manufacve the purpose of saving energy and reducing consumption.

IOP Conf. Series: Materials Science and Engineering **157** (2016) 012023 doi:10.1088/1757-899X/157/1/012023

2. Quantum energy

In 1990, the German physicist Max Planck found in the research of electromagnetic radiation energy distribution: The emission and absorption of electromagnetic waves are not continuous, but one by one. Everyone has energy; the energy could be a multiple of an elementary unit:

$$\varepsilon_0 = h\nu \tag{1}$$

where ε_0 is energy, h is Planck's constant, and v is the frequency of the radiation. He called the energy unit which is the minimum and cannot be divided as energy quantum. When explain the distribution of the energy of the electromagnetic radiation, Planck first creatively put forward the concept of "quantum" and introduce the Planck's constant; There are more than 100 years when "quantum" was firstly introduced. Further research shows, in the micro field, some physical quantities are also jumping in the smallest units, not continuous, and the smallest units are called the quantum of the physical quantity. Such as: energy quantum, light quantum, flux quantum, sound quantum and etc [3].

In the composition of the universe, there are different levels of the smallest units, such as: all kinds of molecules, atoms, neutrons, protons, electrons, etc. In the range of $10^{-10}m$ and smaller scale, the material movement has obviously shown "quantized" characteristics. And in which, the Planck constant always plays an important role.

At the end of the 19th Century, scientists tried to understand the micro world classical mechanics and macro view, but they failed, which reflected the contradiction between people's subjective understanding and objective reality. In order to solve this contradiction, the only way is to change the view of microscopic particles and get knowledge from practice. From the classical physics, we know that all microscopic particles have the property of the particle and the property of the wave. In 1905, the German physicist Albert Einstein's, 1879 - 1955) introduced the concept of quantum and photons and also put forward the "wave-particle duality" of the light. The concept of particle are mutually exclusive, but in a unified entity and it is shown by De Broglie equation

$$\lambda = h / p \tag{2}$$

The De Broglie wavelength is the wavelength, λ, associated with a massive particle and is related to its momentum, p, through the Planck constant, h [4].

Entire quantum mechanics is based on the concept of wave-particle duality and uncertainty relationship is very important. When it is out of the accuracy which is ruled by uncertainty relationship, we can follow the classical laws, otherwise, follow the quantum laws. Therefore, the uncertainty relation is the boundary of quantum mechanics and classical mechanics; it reflects the process from quantitative change to qualitative change.

3. Quantum technologies

Quantum theory has been perfected and developed from theoretical research to practical application. This kind of wave can be stored in different carrier by the laser sintering machine, which makes it produce energy value in different applications. Materials which have been conversed by quantum energy create billion/sec supermicro wave in the effect of light, force and heat. When the liquid flows

IOP Conf. Series: Materials Science and Engineering **157** (2016) 012023 doi:10.1088/1757-899X/157/1/012023

through the quantum pipe, hydrocarbon chains are rearranged into a long molecular chain, which greatly improves the combustion and explosion, so it can reduce the fuel consumption.

4. Designs and application

4.1. Structure of quantum economizer

The new quantum economizer is a cylinder which is combined with two symmetrical quantum tile pipes. There are two grooves on both sides of the cylinder and it can be fastened in the middle. Grooves and fasten are used to fix the cylinder on the oil pipe, so fluid can flow through the pipe and affected by the cylinder. The tile tubes are used to enhance the physical and chemical properties of the fluid in the pipe and reach the aim of high efficiency and optimization. Structure is shown in Figure 1.

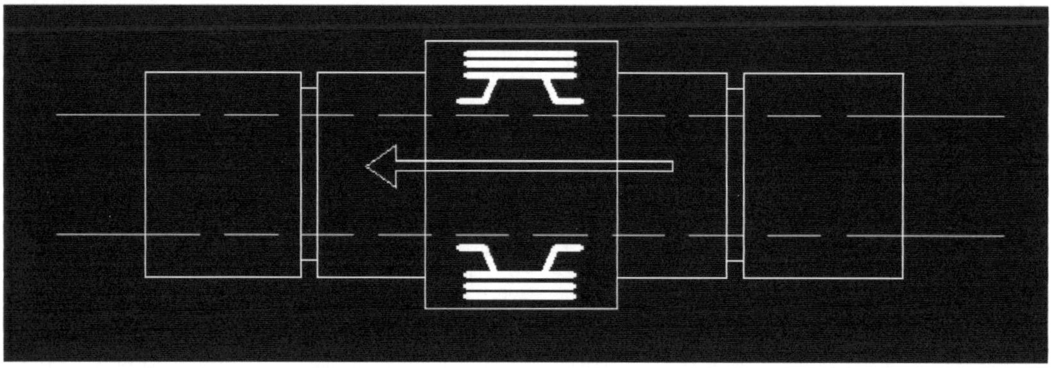

Figure 1. Structure of quantum economizer

4.2. Application on fishing vessel diesel engine

4.2.1. Using objects and operating conditions. In ship industry, fishing vessels has great influence on the ecological environment and highly depend on resources and energy. Fishing vessel is the most traditional and popular fishing platform, both sea farming and sea fishing, and it is driven by diesel engine. According to incomplete statistics, the number of existing fishing vessels in the country is about 1.06 million, accounting or 1/4 of the total number of fishing vessels in the world. [5] So it is a hot topic in the field of fisheries environmental protection for the diesel engine energy saving and consumption reduction. All relevant research institutions have focused on it and hope to have breakthrough. So we also put the fishing vessel diesel engine as the test object to verify the energy saving effect of quantum economizer.

The test fishing vessel is belong to Guangdong Lufeng Kundu ecological Co., Ltd. Ship No. "Yuelufeng 51103", Length: 18.3m, Width: 3.8m. The operation mode is single towing. The parameters of the engine "Jina": Power: 111.8KW, 1500rpm, single engine, single propeller, reduction ratio: 4:1.

Sailing speed is 7-7.2nmi/h, engine speed 700rpm, it depends on wind, flow and wave. When trawling, ship speed is 2.9-3.1nmi/h and it depends on wind and wave in fishing area. Shown in figure 2 and 3.

Figure2. Fishing vessel "Luyuefeng 51103"

Figure3. Installation of quantum economizer

4.2.2. Verification.. The main part of the fuel consumption of the fishing vessel is used for navigation and production, accounting for more than 95%, other operations such as nets out and in, account for little time, fuel consumption can be ignore, so the test use the data of whole fuel consumption and voyage time, calculate the average fuel consumption per hour. Every normal voyage operating time does not exceed 24 hours; other abnormal voyage is not included in the calculation.

4.3. Application result

Before installing the economizer, the fishing vessel has been in operation for 36 months, in normal voyage, the average fuel consumption is 7.6kg/h. After installing the fuel economizer, fuel consumption dropped to 7kg/h, average oil saving is 0.6kg/h, fuel saving effect is about 9.1%. Data is shown in table 1.

In the test of the ship vessel engine, quantum economizer has obvious effect on fuel saving, There are no negative impacts and accidents in the test, the engine operated in good condition and the quantum economizer ran stably.

Table 1. test data

Time	Total fuel consumption（kg）	Time（hour）	Average fuel consumption（kg/h）
1	108.7	15	7.25
2	128.2	18	7.12
3	149.8	22	6.81
4	139.6	20	6.98
5	140.8	20	7.04
6	122.4	17	7.2
7	149.2	22	6.78
8	131.5	19	6.92

5. Conclusion

According to the application above, use of new quantum economizer can obviously save energy and reduce fuel consumption. Conclusion can be drawn by analyzing the data: the time on the sea is longer, the average fuel consumption is lower.

Focus on the problem of high consumption of vessel diesel engine; design the new quantum economizer by the research of quantum theory. There are many advantages such as simple structure, convenient to install and there is no negative impacts.

If the new quantum economizer can be promoted to the shipbuilding industry and all diesel engines install energy economizer, it can save nearly million tons of fuel and billions of cost. Also make the industry develop more healthy and sustainably.

References:

[1] J.Z.Wang and J.Cheng 2010 *ship & ocean engineering* **39** 74-75.
[2] J.J.Cao and B.Guo 2015 *Fishery Modernization* 42 (5) 54-56
[3] R.X.Yu and H.T.Chen 2013 *Electronic Design Engineering* 21(14) 15-16
[4] J.H.Gong and J.F.Gao 2006 *Journal of Teaching and Management* **12** 99-100
[5] J.H.Zhang, J.L.Ding and H.Liu 2015 *Fishery Modernization* 42(5) 69-72

Multi-Attribute Selection of Coal Center Location: A Case Study in Thailand

T Kuakunrittiwong, S Ratanakuakangwan

Department of Industrial Engineering, Faculty of Engineering, Chulaloangkorn University, Bangkok 10330, Thailand.

E-mail: thanyavit.k@student.chula.ac.th.

Abstract. Under Power Development Plan 2015, Thailand has to diversify its heavily gas-fired electricity generation. The main owner of electricity transmission grids is responsible to implement several coal-fired power plants with clean coal technology. To environmentally handle and economically transport unprecedented quantities of sub-bituminous and bituminous coal, a coal center is required. The location of such facility is an important strategic decision and a paramount to the success of the energy plan. As site selection involves many criteria, Fuzzy Analytical Hierarchy Process or Fuzzy-AHP is applied to select the most suitable location among three candidates. Having analyzed relevant criteria and the potential alternatives, the result reveals that engineering and socioeconomic are important criteria and Map Ta Phut is the most suitable site for the coal center.

1. Introduction

The decades of continuously economic growth in Thailand have fueled its GDP and its electricity consumption per capita from 2016 USD and 1454 kWh in 2000 to 6229 USD and 2471 kWh in 2013 according to World Bank. To ensure sustainable economic development and satisfy growing consumption, Thailand Power Development Plan 2015-2036 (PDP2015) is developed. Under this plan, Thailand has committed to diversify its heavily gas-fired electricity generation by increasing other sources of energy, including coal. Despite its controversy, coal has a long history in Thailand and remains the reliable and economical source of electricity.

At the time of writing, coal-fired electricity generation accounts for 21% of total electricity production in Thailand [1]. As the main owner of electricity transmission grids, Electricity Generation Authority of Thailand (EGAT) is recently committed by PDP2015 to implement several power plants with clean coal technology in the southern and central regions, which require approximately ten and seven million tons per annum of coal respectively. To environmentally handle and economically transport unprecedented quantities of sub-bituminous and bituminous coals, a coal center is required. Perhaps, the most important challenge for establishing a coal center is its location since it affects a long-term planning and touches on many aspects. As a result, the goals of this article are to identify important criteria for developing and operating a coal center as well as to select the most suitable location of coal center for the EGAT clean coal projects.

2. Literature Review

2.1. Site Selection

Site selection is considered as one of the most well-known Multi Attribute Decision Making (MADM) problems, which usually refer to problems regarding a discrete number of predetermined alternatives. The main purpose of MADM is to provide the decision makers with the most appropriate alternative according to the levels of preference importance of each criterion. Consequently, certain analyses of criteria importance are performed prior to evaluation of an individual alternative. In high preliminary investment project, where decision makers must prudently choose potential locations [2]; otherwise, localization error could induce bankruptcy [3]. Furthermore, consensus decision is difficult to achieve because of conflicting/competing criteria, hence proper methodologies for site selection should be selected thoroughly.

Among effective techniques applied to MADM problems, Analytical Hierarchy Process (AHP) is outstanding because it can decompose a complex problem into a hierarchical structure, imitating human perception, and can analyze qualitative and quantitative criteria simultaneously. Nonetheless, researchers tend to emphasize more on quantitative than qualitative preferences because of imprecision and uncertainty of qualitative criteria's assessment originated from incomplete information, unobtainable information, and partial ignorance [4]. To enhance precision and certainty in assessments, Fuzzy logic is combined with AHP (Fuzzy-AHP) [5].

2.2. Fuzzy AHP

Fuzzy AHP method was first introduced in van Laarhoven and Pedrycz's work by comparing each fuzzy ratio through triangular membership functions. Buckley substituted triangular membership functions with trapezoidal membership functions, which created a novel version of Fuzzy AHP methods. Chang later proposed the extent analysis method for determining the synthetic extent values of the pairwise comparisons for Fuzzy AHP [6]. However, Wang et al. [7] proved that the extent analysis method may result in a wrong judgment due to a zero weight assigned to useful criteria and attributes; in addition, the weights of criteria or alternatives cannot be derived because this method assigns numerical values, which only represent the priority of a criterion/attribute over the others, to criteria/attributes.

3. Methodology

3.1. Criteria and attributes for coal center site selection

According to Yaron et al. [8], a new approach to bulk unloading in an open sea was proposed. The paper presented site conditions and operating requirements for designing a new terminal. Proper soil condition and geological oceanography were essential for foundation construction. To attain targeted operation performance, physical oceanography was used to determine the specification of facilities. Moreover, environmental concerns were taken into account. Likewise, environmental, health and safety guidelines for port and harbor facilities [9] underlined the importance of minimizing coastal and inland environmental impacts. If any phases of construction or operation undergoing at particular location caused tremendous adverse impacts on environment, a better alternative would be proposed. As a result, a list of criteria and attributes for coal center site selection was developed.

In this study, six criteria are determined as significant criteria, namely socioeconomic (SOC), infrastructure (INF), engineering (ENG), transportation (TRA), environment (ENV), and cost (COS). SOC refers to social perspectives and acknowledgement of the new project and is divided into acceptance of the local community (AOL), approval of the government (AOG), and security of the area (SOA). INF is defined as availability of basic infrastructure at each candidate site, which results in the difficulty levels of construction processes, and is broken down into electricity availability (ELA) and water supply network (WSN). ENG, technical suitability in terms of construction processes and operation conditions required, is characterized by ability to expand capacity (AEC), coastal morphology (COM), meteorology (MET), physical oceanography (POC), and terrain (TER). TRA refers to inland and maritime accessibility as well as modes of transportation from the potential sites to end users and its attributes are proximity to customers (PTC), land routes (LAR), maritime routes and transportation options (MAR), and railways (RAI). ENV concentrates on the degree of environmental changes that might affect a local community and irreversible environmental conditions. Abiotic resources (ABR), biotic resources (BIR), human use value (HUV), and quality of life (QOL) are sub-criteria of ENV. COS is the investments including land, dredging, environmental protection, equipment, and constructing costs. The last criterion consists of construction cost (COC) and damage cost (DMC).

3.2. Evaluation of weights

Buckley's method is utilized as it is comfortable to apply to fuzzy logic and guarantees a unique solution to the reciprocal comparison matrix [10]. The proposed approach is developed with respect to the AHP framework [11]. The analysis steps of such approach are discussed in this subsection.

3.2.1. Construction of the hierarchical structure. To decompose the problem, goal of the problem is firstly defined. Based on the goal, significant criteria are determined. Each criterion is characterized by several essential attributes, which are further used in evaluation of three potential alternatives. The relationship of the four components of hierarchical structure, namely goal, criteria, attributes, and alternatives, are demonstrated as levels and sub-levels shown in figure 1.

3.2.2. Pairwise Comparison. All criteria are compared against all others through the linguistic variables in table 1, which are translated into triangular fuzzy number consisting of lower-bound, most-likely, upper-bound or (l,m,u). Triangular fuzzy numbers is x-y coordinate of (l,m,u) on x-axis and membership function $\mu(x)$ on y-axis. Similarly, all attributes corresponding to their criterion are compared with each other as well as all alternatives are compared with the others pertaining to each attribute. The interval between l and u presents the level of uncertainty; the wider the interval $(l\text{-}u)$, the more uncertainty. Fuzzy comparison matrix (**A**) represents fuzzy relative importance of each pair of criteria i and criteria j where elements of matrix are $a_{ij} = (l_{ij}, m_{ij}, u_{ij})$ and $a_{ji} = (a_{ij})^{-1} = (u_{ij}^{-1}, m_{ij}^{-1}, l_{ij}^{-1})$ as depicted in table 4.

Table 1. Definition of linguistic variables for fuzzy importance scale.

Linguistic Variable		Triangular Fuzzy Numbers (l,m,u)
Not important	(NI)	3/8,3/7,1/2
Very less Important	(VLI)	3/7,1/2,3/5
Less Important	(LI)	1/2,3/5,3/4
Slightly less Important	(SLI)	3/5,3/4,1
Equally Important	(EI)	1,1,4/3
Slightly more Important	(SMI)	1,4/3,5/3
More Important	(MI)	4/3,5/3,2
Very Important	(VI)	5/3,2,7/3
Very strongly important	(VSI)	2,7/3,8/3

3.2.3. Consistency tests for fuzzy weights. According to Buckley [12], if the crisp value of triangular fuzzy number matrix comparison is consistent, the fuzzy comparison matrix is consistent as well. In order to verify the acceptance of the judgment, the consistency evaluations of "most-likely" value are performed. A Consistency Index (CI) examines the consistency of the pairwise comparison matrix through $CI = (\lambda_{max}-n)/(n-1)$, where n is the number of criteria, attributes, or alternatives and λ_{max} is the eigenvalue of the comparison matrix. A Consistent Ratio (CR) or CI/RI measures the judgement consistency, where Consistency Random Index (RI) are generated by numerous simulations of pairwise matrix comparison demonstrated in table 2. Therefore, CR represents the errors caused by decision makers and should be less than 10% of RI or CR<0.1 [13].

Table 2. Consistency Random Index (RI) for n = 1, 2,..., 9 [11]

n	1	2	3	4	5	6	7	8	9	10
RI	0.00	0.00	0.52	0.89	1.11	1.25	1.35	1.40	1.45	1.49

3.2.4 Calculation of the fuzzy weights. To obtain fuzzy weights, the calculations of geometric mean are performed for all criteria using equation (2), an example on lower-bound geometric mean of the *i*th criterion (l_i). Geometric mean of m_i and u_i can also be calculated likewise.

$$l_i = \left[\prod_{j=1}^{J} l_{ij}\right]^{1/J} \text{ and } L = \sum_{i=1}^{I} l_i \tag{2}$$

Where l_{ij} is the lower-bound of the *i*th criterion compared with the *j*th criterion and L is the total of lower-bound geometric mean. The fuzzy weight of *i*th criterion ($\mathbf{W}_i$) is defined as equation (3).

$$\mathbf{w}_i = \left(l_i \cdot U^{-1}, m_i \cdot M^{-1}, u_i \cdot L^{1}\right) \quad \forall i \tag{3}$$

3.2.5 Defuzzification. This study selects the center of gravity approach to compute the crisp number of fuzzy weights from all experts because of its simplicity and efficiency. Consequently, the local weight of each criterion is calculated by equation (4) and equation (5) in accordance with Pan's proposed technique [14].

$$\mu_i(x) = \max\left\{\min\left[\left(\mu_i(x)\right)_{\text{1th expert}}, \left(\mu_i(x)\right)_{\text{2th expert}}, \ldots, \left(\mu_i(x)\right)_{n\text{th expert}}\right]\right\} \tag{4}$$

Where $\mu_i(x)$ is the membership function of the *i*th criterion and $(\mu_i(x))_{n\text{th expert}}$ is $\mathbf{W}_i$ from the first to the *n*th expert's evaluation regarding the *i*th criterion.

$$x_i^* = \int \frac{\mu_i(x) \cdot x dx}{\mu_i(x) dx} \tag{5}$$

Where x_i^* is the local weight of the *i*th criterion. Repeating equation (2) through (5) for all criteria, attributes, and alternatives yields the local weights of all criteria, attributes, and alternatives. The global weight of the *k*th attribute (AG_k) can be computed by equation (6)

$$AG_k = x_i^* \cdot A_{ik} \quad \forall i \forall k \tag{6}$$

Wherer A_{ik} is the local weight of *k*th attribute subordinate to *i*th criterion. Similarly, the final weights of each alternative (D_m) are computed by equation (7)

$$D_m = \sum_{m=1}^{M} AG_k \cdot D_{km} \quad \forall k \tag{7}$$

Where D_{km} is the overall weight of *m*th alternative pertaining to *k*th attribute.

3.2.6 Sensitivity analysis. Prior to the conclusion of the final results, the α - cut concept is conducted to justify the degrees of uncertainty pertaining to the experts' judgments. The triangular fuzzy numbers in table 1 can be denoted in Eq. (8) for SMI, MI, VI, and VSI. The others are the inverses of those variables, except the equally important variable, whose value is (1,1,1).

$$a_{ij}(\alpha) = \left(l + \frac{1}{3}\alpha, m, u - \frac{1}{3}\alpha\right)_{ij} \tag{8}$$

4. Case Study

This study takes the approaching clean coal power plant projects in the central region of Thailand with coal demand of seven million tons per annum as the clients of the coal center. Map Ta Phut, Si Racha, and Tab Sakae are screened and selected as the potential alternatives. Group decision-makers consisting of five experts who have technical expertise and extensive experiences are formed. To further simplify the problem, the hierarchical structure in figure 1 is created, which considers the ultimate goal of this study– selecting the most appropriate site for a coal center – as the highest level and chosen criteria and attributes as sub-levels of the structure. Based on the hierarchical structure, experts' evaluations are employed to calculate weights of criteria, attributes, and alternatives. Pairwise comparisons are made through linguistic variables depicting in table 2. For example, SOC vs. INF is interpreted as the significance level of SOC compared to INF. Pairwise comparisons of criteria in table 3 exemplify the conversion of linguistic variables to triangular fuzzy numbers of the first expert's assessment in table 4.

Table 3. Pairwise comparisons of criteria.

Pairwise criteria			Result				
Reference			1st Expert	2nd Expert	3rd Expert	4th Expert	5th Expert
SOC	vs.	INF	MI	MI	VI	VI	MI
SOC	vs.	ENG	SLI	LI	SLI	EI	LI
SOC	vs.	TRA	EI	SLI	EI	MI	SLI
SOC	vs.	ENV	MI	MI	MI	MI	EI
SOC	vs.	COS	VI	VI	VSI	VSI	MI
INF	vs.	ENG	NI	NI	NI	VLI	NI
INF	vs.	TRA	VLI	VLI	LI	LI	VLI
INF	vs.	ENV	EI	EI	EI	SLI	SLI
INF	vs.	COS	MI	MI	MI	EI	EI
ENG	vs.	TRA	MI	EI	MI	MI	MI
ENG	vs.	ENV	VI	MI	VI	VI	VI
ENG	vs.	COS	VSI	VSI	VSI	VSI	VSI
TRA	vs.	ENV	MI	MI	MI	MI	MI
TRA	vs.	COS	VI	VSI	VI	MI	VI
ENV	vs.	COS	MI	MI	MI	MI	MI

Table 4. Pairwise comparisons of criteria.

	SOC	INF	ENG	TRA	ENV	COS
SOC	(1,1,1)	(2,7/3,8/3)	(3/5,3/4,1)	(1,1,4/3)	(4/3,5/3,2)	(2,7/3,8/3)
INF	(3/8,3/7,1/2)	(1,1,1)	(3/8,3/7,1/2)	(3/8,3/7,1/2)	(3/8,3/7,1/2)	(1,1,4/3)
ENG	(1,4/3,5/3)	(2,7/3,8/3)	(1,1,1)	(4/3,5/3,2)	(5/3,2,7/3)	(2,7/3,8/3)
TRA	(1,1,4/3)	(2,7/3,8/3)	(1/2,3/5,3/4)	(1,1,1)	(1/2,3/5,3/4)	(5/3,2,7/3)
ENV	(1/2,3/5,3/4)	(2,7/3,8/3)	(3/7,1/2,3/5)	(4/3,5/3,2)	(1,1,1)	(2,7/3,8/3)
COS	(3/8,3/7,1/2)	(1,1,4/3)	(3/8,3/7,1/2)	(3/7,1/2,3/5)	(3/8,3/7,1/2)	(1,1,1)
					Consistency Ratio 0.02	

5. Result and Discussions

The results of analyses of criteria importance indicate that engineering is the most important criteria and socioeconomic, transportation, environment, infrastructure, and cost are ranked respectively. Engineering is considered important most as facility design depends highly on the suitability of an area for construction and operation. Specification and Number of facilities– berth, breakwater, jetty, dredging, stockyard area, and heavy equipment – vary considerably with COM, POC, and TER and have a large effect on cost. Since the presence of the coal center is a tool to advance the prosperity and growth of the nation, acknowledgement and well-being of the local community must not be overpassed. Consequently, socioeconomic is ranked in second. The third ranking is transportation. PTC is the highest attribute importance because the closer end users, the lower delivery cost and delivery time. Additionally, marine transportation is the best transportation mode due to capacity and cost per transshipment. The weight of environment is only a little lower than transportation's because of modern construction techniques, the state-of-the-art facilities, and preventive measures, which have abilities to lessen the adverse environmental impacts significantly. The second lowest and lowest ranks are infrastructure and cost since transmission grid and water supply usually exist although the improvement is required and the coal center construction should not be limited by budget.

In terms of criteria weights, attribute weights and alternative weights, the alternative ranking is Map Ta Phut, Si Racha, and Tab Sakae presented in table 5. Map Ta Phut has the remarkable advantages on engineering and socioeconomic because it is situated in an industrial area with permission granted by the government. Its location is distant from the local community area and is well-equipped with proper engineering conditions for construction and operations the coal center. However, it has a constraint on limited expandability. On the other hand, Si Racha is selected as the contingency alternative due to its high ability to expand capacity even though there is a tradeoff between expandability and conditions for construction and operations.

To justify the reliability of the results, the α-cut of one is employed. The result shows that the most appropriate alternative is still unchanged, Map Ta Phut (0.385), Si Racha (0.330), and Tab Sakae (0.285). The experts also ensure the rationality and correctness of the results. This strengthens that the approach is applicable and efficient.

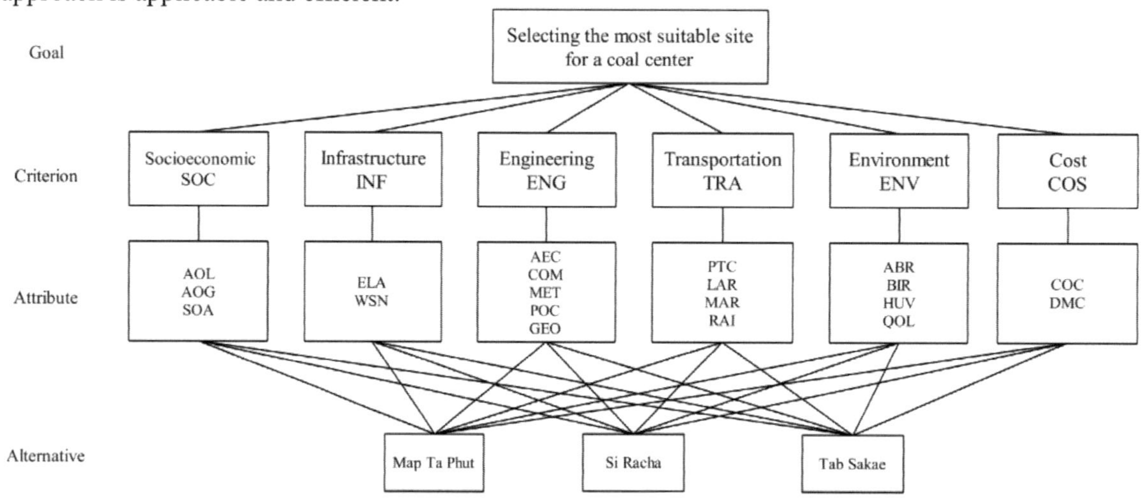

Figure 1. Hierarchical structure for coal center site selection.

Table 5. The results of the alternatives.

Criterion		**Attribute**			**Alternative**			**Final weight**		
	Local weight		Local weight	Global weight	Map Ta Phut	Si Racha	Tab Sakae	Map Ta Phut	Si Racha	Tab Sakae
SOC	0.22	AOL	0.40	0.08	0.44	0.28	0.30	0.037	0.024	0.025
		AOG	0.34	0.07	0.47	0.28	0.26	0.034	0.020	0.019
		SOA	0.27	0.06	0.40	0.29	0.32	0.023	0.016	0.018
INF	0.09	ELA	0.60	0.05	0.40	0.20	0.40	0.022	0.011	0.022
		WSN	0.40	0.04	0.33	0.33	0.33	0.012	0.012	0.012
ENG	0.26	AEC	0.16	0.04	0.10	0.30	0.60	0.004	0.012	0.024
		COM	0.25	0.06	0.53	0.33	0.14	0.033	0.021	0.009
		MET	0.18	0.05	0.33	0.33	0.33	0.015	0.015	0.015
		POC	0.24	0.06	0.50	0.25	0.25	0.030	0.015	0.015
		TER	0.19	0.05	0.41	0.29	0.31	0.019	0.014	0.015
TRA	0.19	PTC	0.34	0.06	0.30	0.60	0.10	0.019	0.039	0.006
		LAR	0.19	0.04	0.37	0.42	0.21	0.013	0.015	0.008
		MAR	0.28	0.05	0.36	0.41	0.23	0.019	0.022	0.012
		RAI	0.21	0.04	0.42	0.15	0.43	0.017	0.006	0.017
ENV	0.18	ABR	0.28	0.05	0.42	0.28	0.30	0.021	0.014	0.015
		BIR	0.19	0.03	0.44	0.31	0.25	0.015	0.011	0.009
		HUV	0.33	0.06	0.36	0.34	0.32	0.021	0.020	0.019
		QOL	0.23	0.04	0.36	0.30	0.36	0.015	0.012	0.015
COS	0.10	COC	0.50	0.05	0.50	0.30	0.20	0.024	0.015	0.010
		DMC	0.50	0.05	0.45	0.27	0.28	0.022	0.013	0.014
							Total	0.415	0.326	0.297

6. Conclusion

Precise and accurate selection of the most appropriate site literally generate the success of the coal center. This paper proposes the Fuzzy-AHP approach to resolve the difficulty of dealing with qualitative and quantitative concurrently and to attain the result with certainty. Based on criteria importance computed by Fuzzy-AHP, high suitability of construction and operational conditions with low adverse effects on the local community are the main reason that justify Map Ta Phut as the most appropriate site for the coal center.

Acknowledgments
The authors acknowledge the assessments received from the project experts and the support provided by Center for Engineering Service (CES), Chulalongkorn University.

References

[1] Energy Policy and Planning Office (EPPO), Ministry of Energy 2016 *Energy Statistics of Thailand 2015* (Bangkok: Amarin Printing and Publishing)
[2] Tabari M, Kaboli A, Aryanezhad M B, Shahanaghi K and Siadat A 2008 A new method for location selection: A hybrid analysis *Applied Mathematics and Computation* **206** 598-606
[3] García J L, Alvarado A, Blanco J, Jiménez E, Maldonado A A and Cortés G 2014 Multi-attribute evaluation and selection of sites for agricultural product warehouses based on an Analytic Hierarchy Process *Computers and Electronics in Agriculture* **100** 60-9
[4] Özdağoğlu A and Özdağoğlu G 2007 Comparison of AHP and Fuzzy-AHP for Multi-Criteria Decision Making Processes with Linguistic Evaluations *İstanbul Ticaret Üniversitesi Fen Bilimleri Dergisi Yıl* 65-85
[5] Lima Junior F R, Osiro L and Carpinetti L C R 2014 A comparison between Fuzzy AHP and Fuzzy TOPSIS methods to supplier selection *Applied Soft Computing* **21** 194-209
[6] Ertuğrul İ and Karakaşoğlu N 2008 Comparison of fuzzy AHP and fuzzy TOPSIS methods for facility location selection *The International Journal of Advanced Manufacturing Technology* **39** 783-95
[7] Wang Y, Jung K-A, Yeo G-T and Chou C-C 2014 Selecting a cruise port of call location using the fuzzy-AHP method: A case study in East Asia *Tourism Management* **42** 262-70
[8] Yaron S L, Jacob S, Tzachar C and Zwemmer D 1982 The Hadera Offshore Coal Unloading Terminal: A New Approach to Unloading of Bulk in the Open Sea *the 14th Annual OTC in Houston*
[9] International Finance Corporation (IFC) 2007 *Environmental, Health, and Safety Guidelines for Ports, Harbors, and Terminals*
[10] Ayağ Z and Özdemir R G 2006 A Fuzzy AHP Approach to Evaluating Machine Tool Alternatives *Journal of Intelligent Manufacturing* **17** 179-90
[11] Saaty T L 1984 *Quantitative Assessment in Arms Control: Mathematical Modeling and Simulation in the Analysis of Arms Control Problems,* ed R Avenhaus and R K Huber (Boston, MA: Springer US) pp 285-308
[12] Buckley J J 1985 Fuzzy hierarchical analysis *Fuzzy Sets and Systems* **17** 233-47
[13] García J L, Alvarado A, Blanco J, Jiménez E, Maldonado A A and Cortés G 2014 Multi-attribute evaluation and selection of sites for agricultural product warehouses based on an Analytic Hierarchy Process *Computers and Electronics in Agriculture* **100** 60-9
[14] Pan N-F 2008 Fuzzy AHP approach for selecting the suitable bridge construction method *Automation in Construction* **17** 958-65

IOP Conf. Series: Materials Science and Engineering **157** (2016) 012025 doi:10.1088/1757-899X/157/1/012025

The mathematics of Soft logic

Moshe Klein[1,2,a] **and Oded Maimon**[3,b]

[1]Department of Industrial Engineering, Tel Aviv University, 69978 Israel
[2] Department of mathematics, Ohalo College Katzrin , 1290000 Israel
[3]Department of Industrial Engineering Tel Aviv University 69978 Israel

[a]gan_adam@netvision.net.il; [b]maimon@tau.ac.il

Abstract. We strive to understand, by using mathematical tools, the phenomena where the observer can interpret simultaneously opposite situations, such as the distinct interpretations of the Necker cube. In this study we present a new coordinate system that draws a distinction between -0 and +0. The central theorem presented here postulates that when divisions and multiplications of 0 are combined with real numbers, the Mobius strip provides a model for non-standard analysis. We also suggest that the axis 0 is the fifth dimension extension of the quaternions. We also propose to add the soft logic capability to humanoid type robots, as a way to overcome the Turing test.

1. Introduction

The mathematician Gödel proved in 1931 that there are sentences in number theory, for which it is impossible to determine whether they are true or false, such as the so-called "liar paradox" (when a person says he is a liar). This surprising result presented by Gödel is true in all mathematical theories that contain number theory. It breaks with the prevalent deductive view, which has been developing since Euclid's *Elements.*

Following this result of Gödel, Wittgenstein delivered a seminar in 1939, at Cambridge University, about the foundation of mathematics. In this seminar, he suggested re-examining the basic assumptions of mathematicians. According to Wittgenstein, the mathematician is an inventor rather than a discoverer. Therefore, there is always a subjective dimension to the process of creating mathematics.

Wittgenstein hinted the idea that mathematical statements always have a dual meaning, both empirical and mathematical. Accordingly, there are two ways of examining them, an empirical way and a mathematical way. For example, there are two ways to validate that $\boldsymbol{625 = 25 \times 25}$. The first way is to test if the weight of 25 objects, each of which weighs 25, is in fact 625. Any other result would suggest that the statement is false. The second way is to obtain the result by the derivation of certain mathematical rules. An empirical meaning involves experiments and calculations, and a mathematical meaning involves mathematical logic and proofs.

In 1969, George Spencer-Brown published his book *"Laws of Form"*. In the book, he developed a new mathematical operator, which he called a “distinction”. Distinction is the operation of pointing to an object, and by doing so, separating it from its environment. This operation have a new symbol ⏋, which enables us to develop a new mathematical language with only one symbol.

This language is a primal language – since it contains the most fundamental that we can do - observe an object. Spencer-Brown produced this "primitive language" from the most abstract building blocks of language- that can be the basis for any language. Through the sign, the observer and the phenomena are united. In other words, the sign marks the active attention, which is a creative activity.

IOP Conf. Series: Materials Science and Engineering **157** (2016) 012025 doi:10.1088/1757-899X/157/1/012025

The distinction is a single operation, which agrees with the fact that we can really pay attention to only one thing. The act of separating a figure from its background always creates a single image (even if this figure is multiple).

The Boolean logic (the logic of the truth-values 0 and 1) is a higher level of logic than the one set by the Spencer-Brown language, which is a finer and more primitive one. Zero, unlike the empty place, is already creating. The operation of distinction is an act of marking a circle, with the intention of saying "it". According to Spencer-Brown it does not matter what you are looking at - what matters is the act of observation.

The basic axioms of Spencer-Brown are:

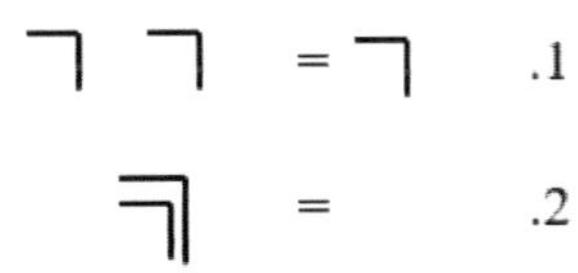

Figure 1. The laws of form

The first law says that if you go back to that which you have already noticed, and notice it again, it will not change anything. The second law interprets as follows: the creation of something from nothing. To distinguish inside the distinction is like nothing. Once more, we must remember that Spencer-Brown is looking for abstraction and for the most basic extension – in this primitive world there is only one object and one observer who is drawing a distinction.

According to studies of Marcelo Dascal, the mathematician and philosopher Leibniz, as a young researcher, aspired to develop a universal language with a single signal. As suggested by Dascal, Leibniz converted this first vision into a new one: to discover and develop a mathematical language that will demonstrate a softer logic than that of truth and falsehood. Leibniz had an ambitious plan to construct a universal language, which will prevent misunderstandings between people as well as serve as a scientific language that reflects thought. According to Leibniz, language is also a tool for thinking and it influences thinking. Therefore, precise formal language, such as expression and precise thinking, is necessary to reduce errors and increase certainty, thus allowing for the resolution of disputes. Dascal argues that Leibniz knew that no rational thinking, not even a "soft rationality", is included in the formal computational model of rationality on computational language. In fact, Leibniz wrote in many occasions that the logic of two states is insufficient.

In this study, we propose to see the Necker cube phenomenon as a basis for the development of a mathematical language in accordance with Leibniz's vision of soft logic. By using the new coordinate system, that was developed by Yale Landsberg, we make a distinction between -0 and +0. By observing the Necker cube phenomenon with the Quaternions number system as was done by Martin Hay, we discover the existence of a fifth dimension which is the 0 axis. All this allows us to prove that the Mobius strip can be a model for non-standard analysis.

2. The Necker cube and ± 0

If we draw a hexagon **Figure 2** and its three principal diagonals, we obtain a flat figure that can be interpreted as a three-dimensional cube.

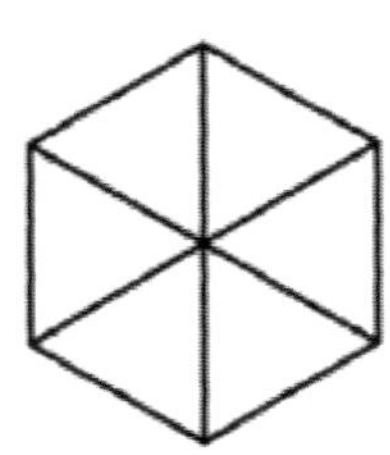

Figure 2. An hexagon with 3 main diagonal

This form is called a Necker cube and It can be interpreted in two different ways, as follows:

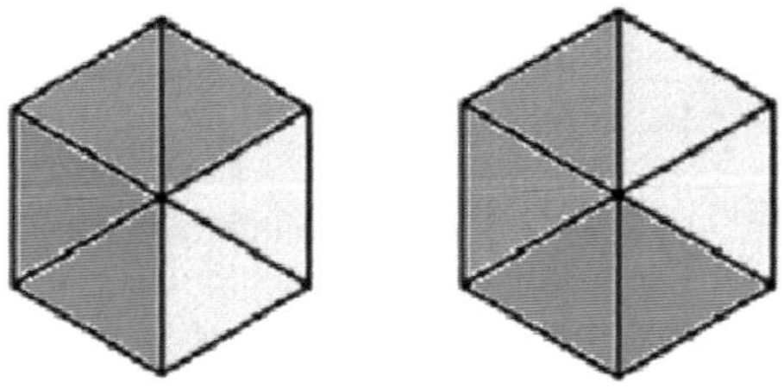

Figure 3. Two interpretations of the Necker cube

A cube has 6 faces, 8 vertices and 12 edges. The flat figure has 6 vertices and one point,at the center of the hexagon, that can appear as either of two more vertices.

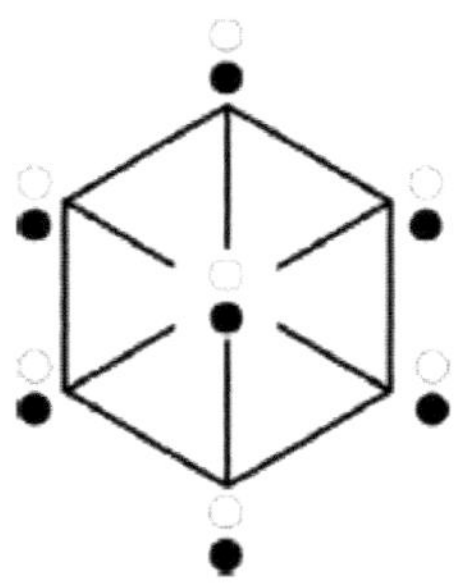

Figure 4. Polarity in hexagon

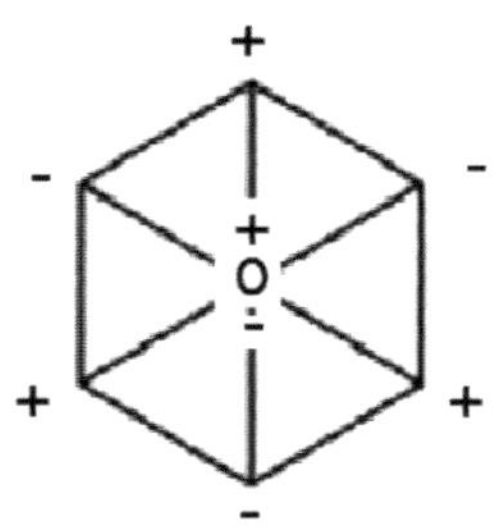

Figure 5. A bipartite graph

We shall use the signs of - and + to create a Bipartite graph **Figure 5**. with the edges of the hexagon, by denoting the vertex of each diagonal with one of these signes. The center of the hexagona shall be marked by the letter O.

Each interpretation of the Necker cube brings to mind opposite signs ± of O, either + or -. Therefore, we can conclude that: -O=+O . The center depends on the point of view of the observer. If the vertices are to attain some numerical values, the only conclusion will be as follows:

$$O = \emptyset$$

The Hexagon's vertices will be $\pm\emptyset$ Thus, we discover a distinction between - $\emptyset$ and + $\emptyset$. A distinction between $\pm\emptyset$ also exists in the theory of calculus. The two marks of $\emptyset$ appear while looking at the limit of a function's derivative from left or right. In the following section, we shall extend the concept of $\pm\mathbf{0}$ by building a new coordinate system.

3. The new coordinate system

The ability to interpret a drawing in two different ways also exists in **Figure 6.** The diagonal can be interpreted as moving upwards or as moving away from us, sideways. This property may be more easily observed in a simple corner **Figure 7.** Now, we will create a coordinate system. The first number that was invented was 1 so we shall locate it in place of B **Figure 8.** Then, the number 2 was invented, so we shall locate it in place of C: **Figure 9.**

IOP Conf. Series: Materials Science and Engineering **157** (2016) 012025 doi:10.1088/1757-899X/157/1/012025

Figure 6.

Figure 7.

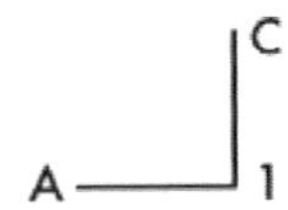

Figure 8.

Figure 9.

The numbers 3,4,5,6, etc., which were invented later, shall be placed along the vertical axis, as follows:

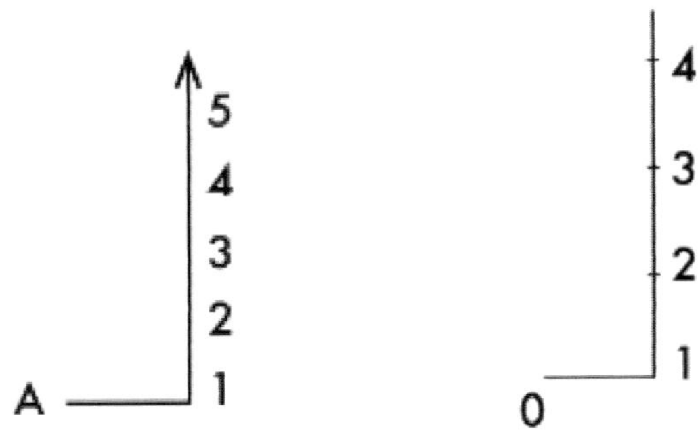

Figure 10. New location for 0

The invention of the number 0 revolutionized mathematics. Thus, as it is befitting, we propose to position the number 0 perpendicularly, in place of A. Let us keep in mind the following statements:

- The left end of the number's coordinate is marked with 0.
- It is important to note: We measure from 0, but count from 1.
- It is possible to build a basis that will be a different length than 1.

Having established the foundation for this number coordinate, we can begin the construction of our argument.

1st lemma: The horizontal segment (0, 1] is dual to the vertical segment $(1, \infty]$

Proof:

Let us consider the following function **Figure 11.**: $f(x) = \frac{1}{x}$

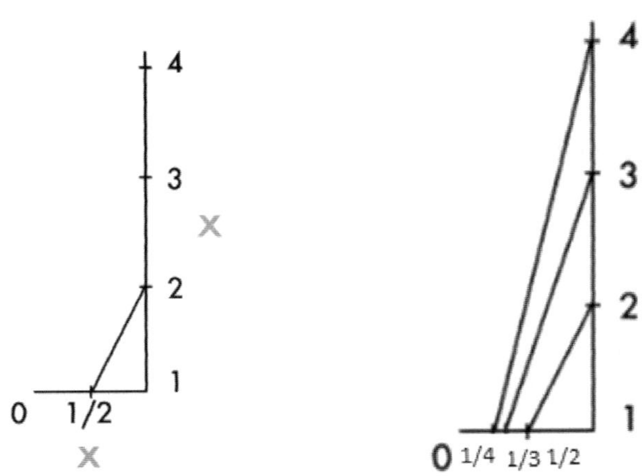

Figure 11. The connection between x to 1/x

This function creates a one-to-one correspondence between the segment (0, 1] and the segment$(\mathbf{1}, \infty]$.
Q.E.D

We would like to stress that the line we drew connects two different points on the same coordinate.

2nd lemma: The new number's coordinate connects 0 to infinity.

Proof:
Let us consider again the function $f(x) = \frac{1}{x}$
When x tends to 0, 1/x tends to infinity and vice versa.
Q.E.D

Thus, we hope, and as it is demonstrated below, the link between 0 and infinity becomes more intuitive.

3rd Lemma: All lines between opposite points intersect at one point.

Proof:
Let us observe the following drawing **Figure 12.** connecting x to 1/x:

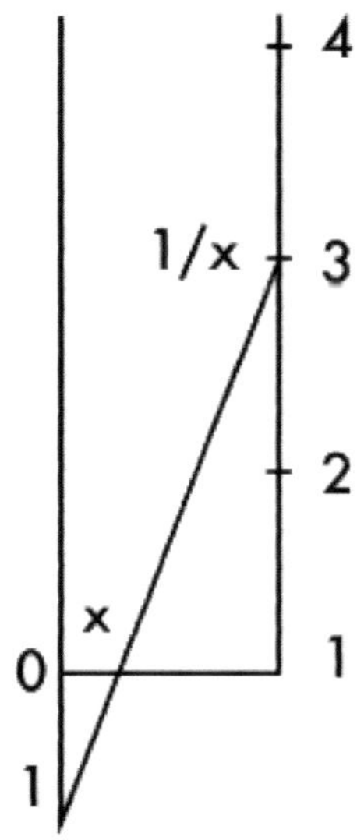

Figure 12. A proof of lemma 3

Let us apply the properties of similar triangles:

$$\frac{x}{1} = \frac{1 - x}{\frac{1}{x} - 1}$$

All the lines intersect at one point, which is located one unit below the 0.
This same point is the beginning of the Cartesian coordinate.
Conclusion:

4th Lemma: The height of 0 from the beginning of the Cartesian coordinate is 1.

5th lemma: A distinction between -0 and +0 can be made by applying the new coordinate system.

Proof:
As we have already seen, the height of 0 is 1, in relation to the Cartesian coordinate's origin point. We would also like to suggest extending this new coordinate system to the negative numbers. Let us consider the next drawing:

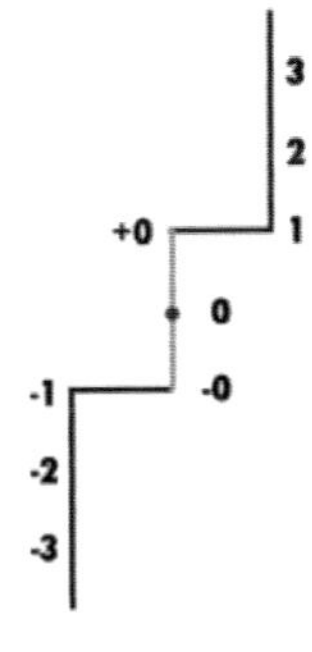

Figure 13. A distinction between -0 and +0

The number 0 is also opposite to the number -1, but it is not identical to the 0 opposite to the number 1. Hence, we suggest to denote the two different "zeros" as -0 and +0.

Q.E.D

6th lemma: There is a line of 0.

Proof:
As we have seen, the new coordinate system draws a distinction between -0 and +0.

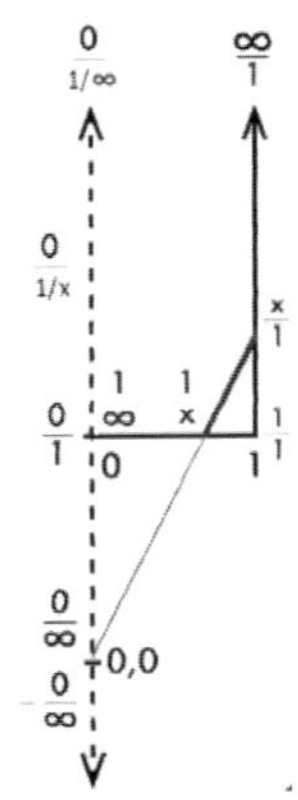

Figure 14. The line of 0

As seen in the drawing above, a line can be drawn to pass through them. This line will be comprised of all of 0's multiplications and divisions.

Q.E.D

Non-standard analysis is a mathematical theory that extends the real number line by adding infinite and infinitesimal numbers (for more details see Appendix A).
Conclusion:
7th lemma: A model for nonstandard analysis can be constructed.

4. The Mobius strip model

The Möbius strip, having only one side, can serve as a model for the understanding of the relationship between mathematics and physics. This is due to its ability to represent the fact that the viewer is not merely a watcher, but that he is also able to see himself as an observer in the world.

Theorem 1: The Mobius strip is a model of a non-standard analysis

Figure 15. A Mobius strip

Proof:

As we have observed, while presented on a plane, the new numbers coordinate creates a non-standard analysis model. To demonstrate that the Mobius strip includes the new coordinate system in it, let us observe the next drawing:

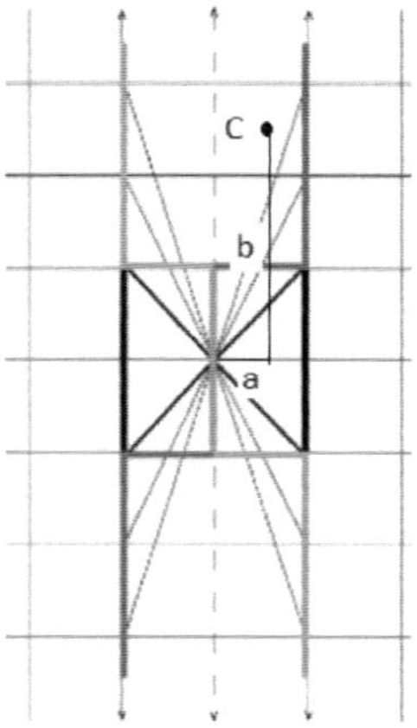

Figure 16. A model for the Mobius strip

Let us take a straight piece of paper and on it, draw the new coordinate system in the plane. Draw a square in the middle and fold the piece of paper twice, along the diagonal line. The resulting form will be a Mobius strip. The coordination of the point C on the Mobius strip are:

$$\boldsymbol{a + b\bar{0}}$$

Q.E.D

5. Quaternions and the fifth dimension

Hamilton invented the field of quaternions in 1843. It is a four-dimensional extension of the complex numbers. Each element in the field written by using the following expression:

$$a + bi + cj + dk$$

For more details, see appendix B.

Theorem 2: We can extend the quaternions to the 5th dimension by the 0 axis.

Proof:

Chirality is a fundamental symmetry in nature. It occurs at all scales, from the chiral molecules in the living body and up to spiral galaxies. As demonstrated below, chirality can be manifested as in these four different objects (red, yellow, green, blue), arranged at the corners of a tetrahedron.
Such a tetrahedron has two formations which are related, as (+) to (–) or as clockwise to anticlockwise. (Note the order of green, blue and yellow in the photographs below).

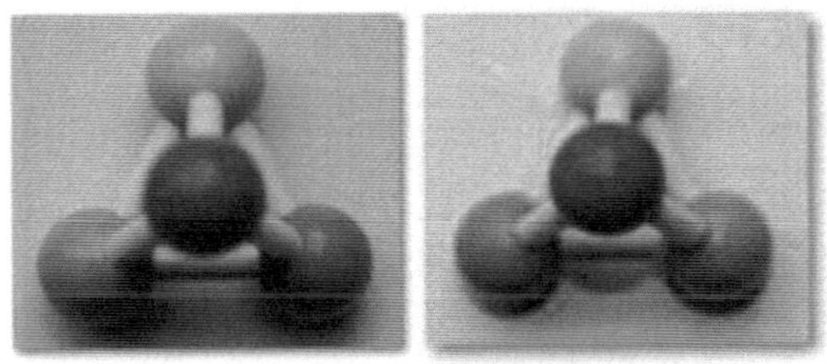

Figure 17. Chirality of the tetrahedron

In one formation, all the corners can be denoted (+) (or "is"), while in the other formation all the corners can be denoted (-) (or "not").

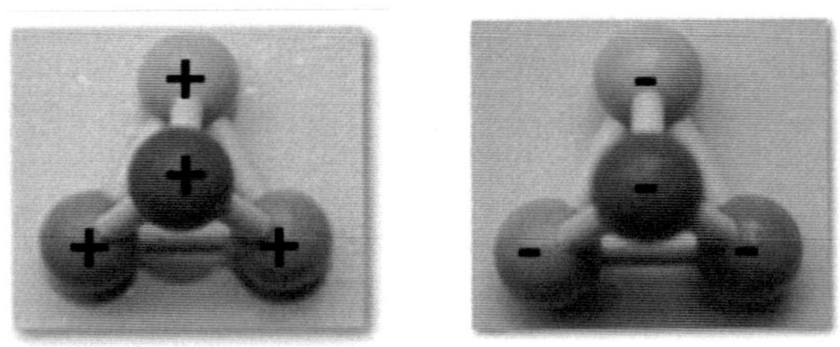

Figure 18. Two opposite chirality's

The two formations can be nested in a cube, so that opposite corners of the cube are labelled as (+) and (–) opposites. Each face of the cube then has two (+) corners and two (-) corners.

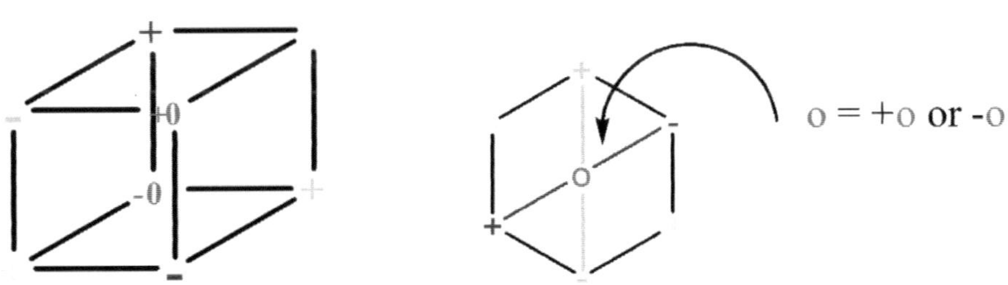

Figure 19. The Necker cube and ± 0

Now the Necker cube effect can be viewed in terms of the nested tetrahedrons. The brain exploits the difference between what a chiral tetrahedron is (+) and what it is not (-) to draw distinctions. It is a chiral system. The universe's chirality provides the basis for consciousness. It is possible to relate the red, yellow, green and blue vertexes to quaternions, conventionally represented as ± 1, $\pm i$, $\pm j$ and $\pm k$.

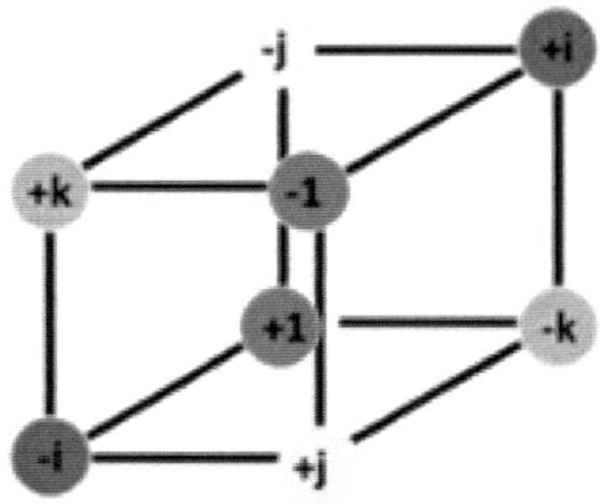

×	1	*i*	*j*	*k*
1	1	*i*	*j*	*k*
i	*i*	−1	*k*	−*j*
j	*j*	−*k*	−1	*i*
k	*k*	*j*	−*i*	−1

Figure 20. The Necker cube and quaternions

Each face of the cube can be represented by 4 units of the vertices of the face. We use the multiplication Table and check which unit multiplication will make all the results become negative.

face	i	J	k	-i	-j	-k
[(-1)(-i)(j)(k)]	[(1)(-i)((j)(-k)]	**[(-1)(-i)(-j)(-k)]**	[(-1)(i)(j)(-k)]	[(-1)(i)((- j)(k)]	[(1)(i)(j)(k)]	[(1)(-i)(-j)(k)]
[(-1)(i)(j)(-k)]	**[(-1)(-i)(-k)(-j)]**	[(-1)(-i)(-j)(k)]	[(1)(i)(-j)(-k)]	[(1)(i)(k)(j)]	[(1)(i)(j)(-k)]	[(-1)(-i)(j)(k)]
[(-1)(i)(-j)(k)]	[(-1)(-i)(j)(k)]	[(1)(-i)(-j)(k)]	**[(-1)(-i)(-j)(-k)]**	[(1)(i)(-j)(-k)]	[(-1)(i)(j)(-k)]	[(1)(i)(j)(k)]
[(1)(-i)(j)(-k)	[(1)(i)(-j)(-k)]	[(-1)(i)(j)(-k)]	[(1)(i)(j)(k)]	[(-1)(-i)(j)(k)]	[(1)(-i)(-j)(k)]	**[(-1)(-i)(-j)(-k)]**
[(1)(-i)(-j)(k)]	[(1)(i)(j)(k)]	[(1)(-i)(-j)(-k)]	[(-1)(-i)(j)(k)]	**[(-1)(-i)(-j)(-k)]**	[(-1)(i)(j)(k)]	{(1)(i)(-j)(-k)]
[(1)((i)(-j)(-k)]	[(-1)(i)(-j)(k)]	[(1)(i)((j)(k)]	[(1)(-i)(-j)(k)]	[(1)(-i)(j)(-k)]	**[(-1)(-i)((-j)(-k)]**	[(-1)(i)(j)(-k)]

For every face there is only one such unit and this is the way we assign to each face of the cube a unit element. The 4 numbers coding a face behave under multiplication as a unitary quaternion. The two interpretations of the Necker cube are like jumping from the positive faces to the negative faces.

We can add a fifth dimension:

$$\boldsymbol{a + bi + cj + dk + e\overline{0}}$$

Q.E.D

6. Conclusion and Application

The possibility to see the Necker cube in two different ways is analogous to the measurement process in quantum theory and the influence the observer has on the measurement's result. We also note here a possible application in the study of quantum mechanics - more specifically, on the zero point energy.

The physics world today holds a conflict between the idea of locality (the speed of light limit) and the idea of nonlocality. In our study, we propose a new coordinate system which may unify these two concepts under one framework.

In the development of special relativity, Einstein developed the idea of measuring the length of a body by measuring the time movement of light. Later, Minkowsky developed the geometry of the fourth dimension by using the Lorentz transformation. The new coordinate system, presented here, is based on a method, developed by Minkowsky, of measuring a distance by Manhattan metrics (Taxicab geometry).

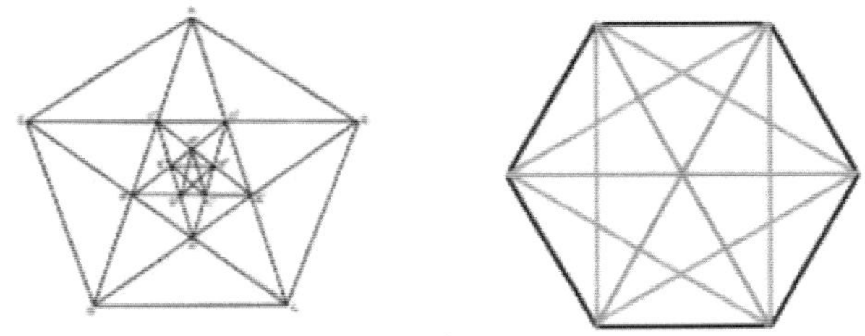

Figure 21. Two polygons and Paradigm shifts

In Pythagoras's time, the discovery of irrational numbers such as the golden ratio in the pentagon gave rise to a paradigm shift. We believe that a similar paradigm shift is possible today, through the discovery of non-local numbers, which appear on the 0 axis. This axis arises from a deep observation of the hexagon and its diagonals.

7. Acknowledgment

The writers of this article would like to thank Yale Landsberg, who developed the Beth-El Non-Standard Geometry of the Real Number Line as well as Martin Hay, the inventor of Chiralkine systems http://chiralkine.com .

8. References

[1] Hilbert D. 1900 Mathematical Problems, *Bulletin of the American mathematical society* (new series), Vol. **37,** No. 4, 407-436
[2] Marks T, Hay, M. and Klitzner, H. (October 2015). Quaternions, chirality, exchange interactions: A new tool for neuroscience Society for Chaos Theory in Psychology & Life *Sciences newsletter*, Vol.**23**, No.1, 8-14.
[3] Dascal M. (Ed.) (2008) Leibniz: *What Kind of Rationalist*? Logic, Epistemology, and the Unity of Science, vol. **13**, Springer Netherlands.
[4] Spencer-Brown, G. (1969/1979) *Laws of Form*. New York: E. P. Dutton.
[5] Robinson, A. (1974) *Non-standard Analysis*. Princeton: Princeton University Press.
[6] Rapoport, D, 2013 Klein bottle logophysics: a unified principle for non-linear systems, cosmology, geophysics, biology, biomechanics and perception *Journal of Physics: Conference* Series **437** 012024

9. Appendix

A. Non-standard analysis

The earliest development of the infinitesimal calculus by Newton and Leibniz was formulated using expressions such as infinitesimal number. These formulations were criticized by George Berkley and others. There was a challenge to develop a consistent theory on analysis using infinitesimal and the first one who did that was Abraham Robinson.

Non-standard analysis is a mathematical theory that extends the real line number by adding to them infinite numbers and infinitesimal numbers. This new set of numbers called hyperreal number. Is marked as R *.

Robinson wrote: "The idea of infinitely small or *infinitesimal* quantities seems to appeal naturally to our intuition. At any rate, the use of infinitesimals was widespread during the formative stages of the Differential and Integral Calculus. As for the objection [...] that the distance between two distinct real numbers cannot be infinitely small, Gottfried Wilhelm Leibniz argued that the theory of infinitesimals implies the introduction of ideal numbers which might be infinitely small or infinitely large compared with the real numbers but which were *to possess the same properties as the latter*."

IOP Conf. Series: Materials Science and Engineering **157** (2016) 012025 doi:10.1088/1757-899X/157/1/012025

By looking on the set of all infinite sequences of real numbers named M, we define addition and multiplication operations between the sequences by action on each component. All fixed sequences (c,c,c,...) represent the real number c. The infinitesimal numbers are those sequences that are 0 from a certain place.

By using the axioms of set theory (Zorn's Lemma) we show the existence of a maximum set D while group R * is defined by the quotient of their division.

$$R^* = M / D.$$

It is important to note here the non-constructive character of this building.

B. Quaternion

Quaternions are a number system that extends the complex numbers This system invented by William Hamilton in 1843, and applied to mechanics in three-dimensional space. Hamilton knew that the complex numbers could be interpreted as a point in a plane. He was looking for a way to do the same for points in three-dimensional space. One day he discovered that he can solve the problem in 4-dimeson space.

The Quaternions have 4 unit elements 1,I,j,k with the following rules of multiplication.

×	1	*i*	*j*	*k*
1	1	*i*	*j*	*k*
i	*i*	−1	*k*	−*j*
j	*j*	−*k*	−1	*i*
k	*k*	*j*	−*i*	−1

A feature of quaternions is that the multiplication of two quaternions is noncommutative Hamilton defined a quaternion as the quotient of two directed lines in a three-dimensional space or equivalently as the quotient of two vectors.
Each element in the field can be written by using the following expression:
a + bi + cj + dk

Quaternions have many applications in theoretical and applied mathematics. For example, calculation involving three-dimensional rotation. Is also has many application in in physics and computer graphics.

Structure Topology Optimization of Brake Pad in Large-megawatt Wind Turbine Brake Considering Thermal-structural Coupling

S F Zhang, J Yin, Y Liu*, Z H Sha, and F J Ma

School of Mechanical Engineering, Dalian Jiaotong University, Dalian China 116028

Email: liuyu@djtu.edu.cn

Abstract. There always exists severe non-uniform wear of brake pad in large-megawatt wind turbine brake during the braking process, which has the brake pad worn out in advance and even threats the safety production of wind turbine. The root cause of this phenomenon is the non-uniform deformation caused by thermal-structural coupling effect between brake pad and disc while braking under the conditions of both high speed and heavy load. For this problem, mathematical model of thermal-structural coupling analysis is built. Based on the topology optimization method of Solid Isotropic Microstructures with Penalization, SIMP, structure topology optimization of brake pad is developed considering the deformation caused by thermal-structural coupling effect. The objective function is the minimum flexibility, and the structure topology optimization model of brake pad is established after indirect thermal-structural coupling analysis. Compared with the optimization result considering non-thermal-structural coupling, the conspicuous influence of thermal effect on brake pad wear and deformation is proven as well as the rationality of taking thermal-structural coupling effect as optimization condition. Reconstructed model is built according to the result, meanwhile analysis for verification is carried out with the same working condition. This study provides theoretical foundation for the design of high-speed and heavy-load brake pad. The new structure may provide design reference for improving the stress condition between brake pad and disc, enhancing the use ratio of friction material and increasing the working performance of large-megawatt wind turbine brake.

1. Introduction

With the gradual consumption of traditional energy, wind energy and other renewable energy resources have become vigorously developed energy resources all over the world. Wind turbines are developing to larger capacity level. Large-megawatt wind turbine has high rotational speed and heavy load. It has been a key point to study on braking system of the wind turbine. Wind turbine brake works without people to ensure the safety production during a long time [1]. Brake pads are not rigid body and thermal insulation, meanwhile a large amount of heat is generated by the effect of friction, which reduces the coefficient of sliding friction in friction pair, and has an influence on the efficiency of braking, ultimately results in the abnormal operation. With the combined action of the brake force, the friction resistance and the heat, the brake pad comes to deformation. The actual contact area is reduced, which causes serious non uniform wear of brake pads and further reduces the braking performance and friction material utilization. The working condition of the brake pad is analyzed in this paper. After the simplification of brake pad model, the thermal-structural coupling analysis is been finished, whose result is seen as a target to continue with structure topology optimization of the brake pad using SIMP method. According to the topological result and the machinability of the structure, the reliability of rebuilt model is verified in the thermal-structural coupling analysis. Compared with former structure, advantages of new structure are shown.

2. Brake working condition analysis and its modeling

2.1. Brake condition analysis

In the braking process, the brake force makes the brake pad contact with the brake disc. Therefore, there is a brake force with certain size and direction perpendicular to the contact surface in central position on the top of brake pad. In this process, there is a friction resistant with direction opposite to the rotation speed. Besides, in order to limit the relative motion of the brake pad, constraints should be imposed on the model. In order to ensure the normal operation of the brake pad, the movement of the contact surface between the brake pad and brake disc needs to be constrained. The brake pad has been designed groove on both left and right sides. In addition, it is assumed that the contact style between the brake pad and brake disc is surface to surface, so the free degree of the bottom surface on the brake pad is required constraints.

2.2. Establishment of brake pad model

After the model is simplified, there are 6780 elements and 8646 nodes after meshing. Material of large-megawatt wind turbine brake is copper based powder metallurgy, and its material properties are shown in Table 1. The circular area of the model is applied with brake force, and based on the coulomb friction law, the lower surface is applies with friction force. Constrain X, Y directions on both sides in the groove, which can limit the traverse between the brake pad and the brake disc. Constrain Z direction at the bottom surface of the brake pad can restrain the contact form. between the brake pad and the brake disc contact surface. The established model is shown in Figure 1.

Table 1. The properties of copper-base powder metallurgy material.[2-4]

Density (kg/m^3)	Elasticity Modulus (MPa)	Thermal Conductivity W/(m℃)	Specific Heat Capacity J/(kg℃)	Poisson's Ratio	Linear Expansion Coefficient /℃	Coefficient of Sliding Friction
5250	5200	30	550	0.3	1.5×10^{-11}	0.3

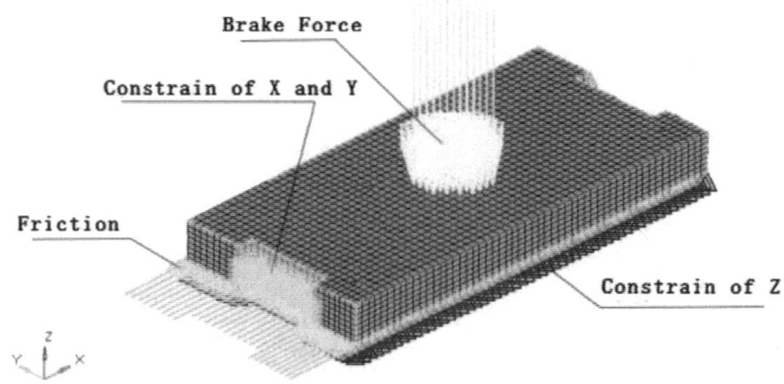

Figure 1. The model of wind turbine brake pad.

3. Thermal-structural Coupling Analysis

Through the heat transfer calculation, relevant thermal conditions are developed, and the thermal-structural coupling analysis of brake pad is carried out with ANSYS finite element analysis software, which obtains the strain and stress contours as shown in Figure 2 and Figure 3 respectively. From the strain contour in Figure 2, the maximum deformation occurs in the top surface of the brake pad, in other words, the maximum deformation locates in the brake under the positive pressure. This is because large-megawatt wind turbine brake needs larger brake force to meet the requirement of braking, but the action area of brake force on the top surface is very small, which makes the stress concentration, and further causes large strain in this area, as shown in Figure 3. In addition, in Figure 2, there is large strain in the bottom left area of the brake pad, which corresponds to the inlet side of the friction resistance in Figure 1. Due to the direction of the friction and braking pressure, the effect of resultant makes that the brake has the trend of extrusion to the export side, which means the brake pad becomes a more flexible deforming cantilever beam. The deformation makes the brake pad easier to separate from the brake disc, which decreases the braking performance directly. In the stress contour, as shown in Figure 3, it is easily to find the extrusion trend. There is obvious stress concentration on the export side. In order to improve the issues of the existed deformation and stress concentration of the brake pad with actual working condition, decrease the non-uniform wear and improve the braking performance as well as the use ratio of friction material, structure topology optimization considering thermal-structural coupling of the brake pad is developed to optimize the distribution of friction blocks on the brake pad.

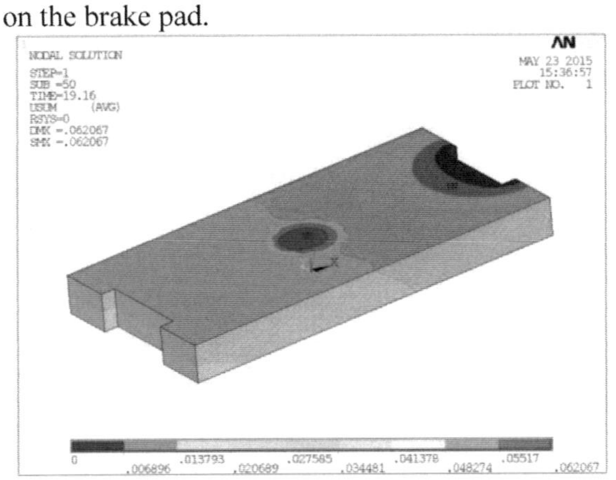

Figure 2. Strain contour of heat structural coupling analysis

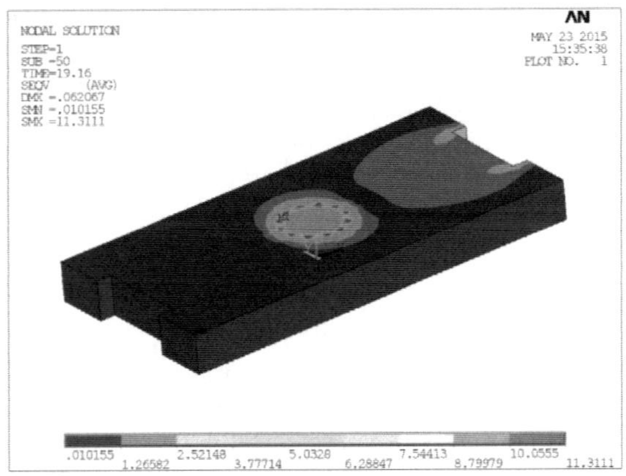

Figure 3. Stress contour of heat structural coupling analysis

4. Structure Topology Optimization

Topology optimization is a mathematical method to optimize the distribution of materials in a certain region with given loads, constraints and performance indexes [5]. At present, the continuum topology optimization method includes variable density method, homogenization method, etc. [6-8]. Relationship between the material property and density is defined in the variable density method, which means that the density of element is the design variable.[9-11] With the advantages of easy program and high efficiency, this method is widely used in the continuum topology optimization which chooses stress and strain as responses.[12] The design variable of topology optimization method based on Solid Isotropic Microstructures with Penalization is the density of every element in the design area, which changes between 0 to 1 continuously.

Based on Solid Isotropic Microstructures with Penalization method and combined with the result of thermal-structural coupling analysis, structure of brake pad is optimized in this paper. Firstly, the actual working condition is determined, and the finite element model is established with the condition of constraints and loads. Secondly, the thermal-structural coupling analysis is developed. The performance of brake pad is analyzed according to the analysis result, which is also seen as the boundary conditions of the optimization. Thirdly, the mathematic model is established to carry out the topology optimization. Considering the processing technology of the brake pad and the machinability of the structure, the result of topology optimization is reconstructed. To ensure the new structure meeting the requirement of the optimization, the thermal-structural coupling analysis is developed again. If it meets the requirement, the analysis is over, if not, it should be returned to modify the model.

4.1. Result of Structure Topology Optimization

Based on the result of the thermal-structural coupling analysis, the value of strain on every node is taken out, which is imported to the model of structure topology optimization as imposed displacement to simulate the actual working condition of the brake pad. The design variable is the density of every element. The boundary conditions are (1) the strain of area with braking force on the top surface of the brake pad should be no more than 0.05mm, (2) the volume fraction of the brake pad should be between 50% to 80% to increase the use ratio of the friction material. Considering the effect of thermal-structural coupling, the structure topology optimization of wind turbine brake pad is shown in Figure 4. Because of the thermal effect, the density of most elements is about 0.35 in Figure 4(a). Choose the threshold value as 0.35 to control the density with [0.35,1], which is shown in Figure 4(b).

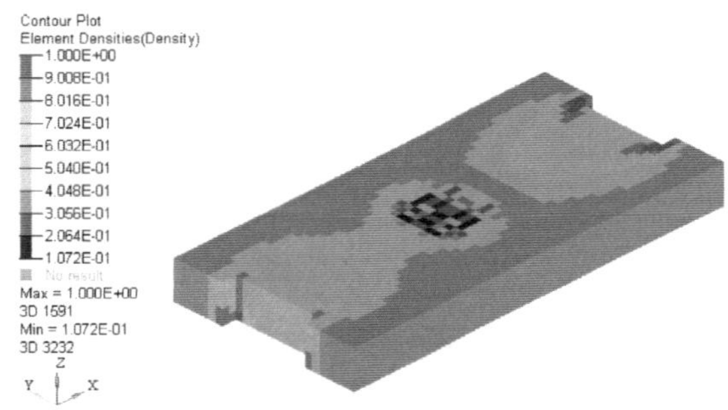

(a)Density form 0 to 1

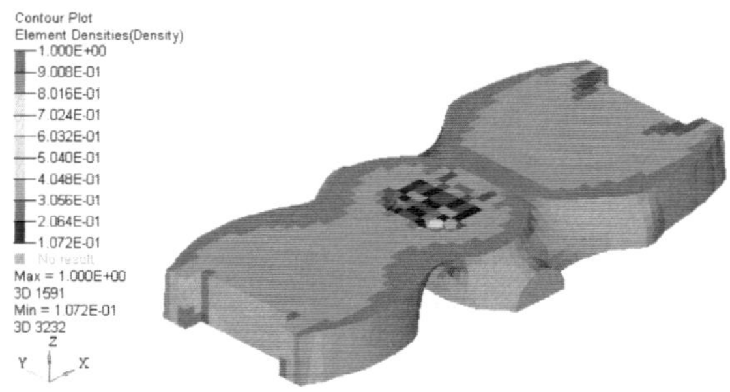

(b) Density form 0.35 to 1

Figure 4. Structure topology optimization result density contour

4.2. Structure Topology Optimization Result Analysis

Topology model of brake pad can be seen in Figure 4(b) obviously. Materials on two sides of direction Y are deleted, and most of the materials are stored around the two grooves. Materials in the middle of the brake pad near the entry of the friction are deleted, but there is an obvious bulge in the center of the brake pad. Corresponding to the position of maximum strain in Figure 3, few materials are deleted in the center on the top surface of the brake pad in Figure 4(b). The maximum contact stress does not appear on the surface, but the subsurface below near the surface. To meet the requirement of manufacturing and the contact condition of braking process, materials here should be retained. To remit the stress concentration caused by braking pressure and friction during braking, pad width of the entry of friction is smaller than that of the exit. Compared with the result of structure topology optimization only considering static condition in Reference 13, there is an obvious change that is the interspace on the bottom of the brake pad. Considering the thermal effect, only the uniform change of temperature, can the brake pad deform and wear uniformly. The interspace can make the heat transfer more fluently and decrease the materials store in the exit of friction. To meet the stiffness requirement of the brake pad in the center, two bulges are designed on each side of the brake pad on the bottom where contacts with the brake disc.

4.3. Reconstruction of the Brake Pad and Result Verification

4.3.1. Reconstruction of the Brake Pad. The final structure of wind turbine brake pad can be reconstructed based on the topological model. The machinability of the reconstructed structure should be concerned. As the large-megawatt wind turbine brake pad is made in copper based powder metallurgy material through sintering and pressing with high temperature. and the corresponding moulds will be designed according to the structure of brake pad, a principle taking manufacturing of moulds conveniently should be considered, the final structure of brake pad is shown in Figure 5.

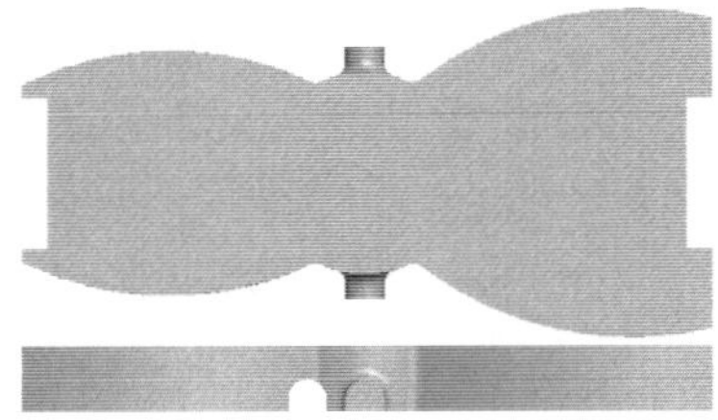

Figure 5. Structure of reconstruction brake pad

4.3.2. Verification of the Reconstructed Structure. The stress and strain contours shown in Figure 6 are the results of verification finite element analysis of the reconstructed structure. The maximum strain of the brake pad is 0.045 mm, which is less than 0.05 mm and meets the requirement of the optimization object.

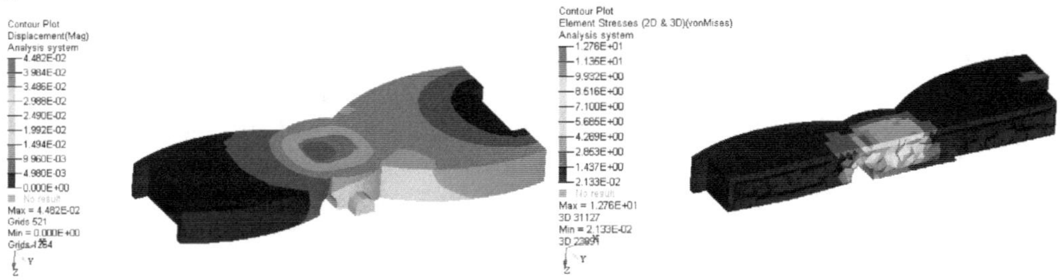

(a) Strain Contour (b) Stress Contour

Figure 6. Reconstruction model validation result contour

Compared with the result of thermal-structural coupling analysis of the brake pad, the degree of deformation of the brake pad decreases a little. In the stain contour of verification analysis shown in Figure 6(a), there is only a little or no deformation on both entry and exit sides of the friction, especially the entry side, which means the topology optimization result is good to the stability of the braking process and has inhibiting effect on the non-uniform wear of the brake pad. In the stress contour shown in Figure 6(b), the maximum stress appears in the middle of the brake pad, the value of stress of most materials on both entry and exit sides is about 0.02 MPa, which distributes evenly. Compared with the stress contour shown in Figure 4, because of the effect of thermal-structural coupling, stress in the area of brake force and the exit side of the friction is higher than other areas, which means the optimized stress distribution is much better than the origin one. Besides, both the mass and volume of the brake pad decrease, which increases the use ratio of the friction materials indirectly.

5. Conclusions

For the issues of severe non-uniform wear of brake pad in large-megawatt wind turbine brake during the braking process, the indirect thermal-structural coupling method is used in this paper to develop the stiffness condition of the brake pad considering thermal-structural coupling. The analysis result is imposed as equivalent displacement in the optimization model, with the topology optimization method of Solid Isotropic Material with Penalization, the maximum stiffness of the brake pad is set as the optimization object to meet the requirement of the thermal-structural coupling working condition, and the distribution of the friction block on the brake pad is designed. The volume and mass of the new structure is less than the origin one, which remits the non-uniform wear of the brake pad and increases the use ratio of the copper based powder metallurgy friction material. The verification finite element analysis of the optimized structure under the same working condition shows the advantages of working performance of optimized structure in the actual working condition.

Acknowledgments

The financial support from the National Natural Science Foundation of China under Grant No. 51475066, Natural Science Foundation of Liaoning Province under Grant No. 2015020114 and

Outstanding Talent Project of Education Department of Liaoning Province under Grant No. LR2015012 are acknowledged.

References

[1] Xiong Weijun, Yan Hongwen. Calculation of Mechanical Brake Torque for Mw Level Wind Generating Unit. *Machine Building & Automaton*, 2011. 5: 137-138.

[2] LIU Yu, ZHANG Shengfang, SONG Xueping, et al. Topological optimization design on frame structure of power platform trailers. *Chinese Journal of Construction Machinery*, 2014.2(12):34-37.(in Chinese)

[3] Chang Junkang, Duan Baoyan. An improved variable density method and application for topology optimization of continuum structures. *Chinese Journal of Computational Mechanics*, 2009, 26(2): 188-192. (in Chinese)

[4] Li Jing, Geng Kuihua. Topological optimization and research on the fixed platen of injection molding machine by variable density method. *Machine Design & Research*, 2008, 24(4): 107-109. (in Chinese)

[5] Chen Yongdang, Zhang Jianhua, Tong Hongjie, et al. Variable Density and Multi-load Topology Optimization Method for Rocket Sled Chassis Design. *Journal of Computer-Aided Design & Computer Graphics*. 2012.8: 1108-1112. (in Chinese)

[6] Yu Xiao, Guo Zhimeng, Hao Junjie, et al. Preparation and properties of copper-based powder metallurgy brake pad for wind turbine. *Materials Science and Engineering of Powder Metallurgy*, 2014.19(1): 66-70. (in Chinese)

[7] Li Jisha, Jiao Biaoqiang, Gu Leilei, et al. Working Theory and Measuring Methods of Leveling Valve. *Railway Locomotive & Car*, 2011.5: 93-96. (in Chinese)

[8] Sun Zhigang, Zhao Yong. Analysis of Heat-Structure Coupling of Screed of 4.2m Pavers. *Road Construction and Machinery*. 2015,5:48-51. (in Chinese)

[9] Liang Tang,Jie Wu,Jinhao Liu,et al. Topology Optimization and Performance Calculation for Control Arms of a Suspension. *Advances in Mechanical Engineering*. 2014: 1-10.

[10] Markus J B, Bernhard B, Gordon G P. Topology optimization of smart structures using a homogenization approach. *Journal of Intelligent Material Systems and Structures*, 2004, 15(8): 655-667.

[11] Hassani B, Hinton E. A review of homogenization topology optimization I – homogenization theory for media with periodic structure. *Computers and Structures*, 1998, 69(6): 707-717.

[12] Kolpakov A G. Effect of influation of initial stresses on the homogenized characteristics. *Mechanics of Materials*, 2005,37(8): 840-854.

[13] Zhihua Sha, Jian Yin, Shengfang Zhang, et al. Structure Topology Optimization Design of Brake Pad in Large Megawatt Wind Turbine Disc Brake. *Advances in Intelligent Systems Research*,2015.5:1451-1455.

Study of seismic response and vibration control of High voltage electrical equipment damper based on TMD

Chuncheng Liu[1], Chongyang Wang[1], Long Mao[1], Chuanming Zha[2]

[1]Northeast Dianli University Institute of Civil Engineering, Jilin, China
[2]State grid xinjiang economic research institute, Xinjiang, China

E-mail: wangchongyang0411@163.com

Abstract. Substation high voltage electrical equipment such as mutual inductor, circuit interrupter, disconnecting switch, etc., has played a key role in maintaining the normal operation of the power system. When the earthquake disaster, the electrical equipment of the porcelain in the transformer substation is the most easily to damage, causing great economic losses. In this paper, using the method of numerical analysis, the establishment of a typical high voltage electrical equipment of three dimensional finite element model, to study the seismic response of a typical SF_6 circuit breaker, at the same time, analysis and contrast the installation ring tuned mass damper (TMD damper for short), by changing the damper damping coefficient and the mass block, install annular TMD vibration control effect is studied. The results of the study for guiding the seismic design of high voltage electrical equipment to provide valuable reference.

1. Instruction

China belongs to one of earthquake-prone countries. When the earthquake disaster, there are prone to damage electrical equipment in the composition of ceramic [1, 2].Install the damping shock absorber is an important way to improve the seismic performance of electrical equipment [3]. TMD damper is mounted on the structure of a device used to reduce the dynamic response of the structure, which consists of two parts, mass and damping blocks. TMD damper full name of the tuned mass damper, as early as 1909 proposed by Frahm [4]. With the development of the industry, the TMD damper also obtained the further research. In 2006, Chenglin Zhang proposes two broadband: Girder Model Tuned Mass Damper – GTMD and Ring - Model Tuned Mass Damper – RTMD [5], these two kinds of Damper can in a wider frequency range effectively suppress the vibration of the structure, which has been widely used. In 2014, Bai Wen, who use multiple loop TMD damper towering cylindrical porcelain electrical equipment for the damping research conclusions show tuned mass damper can be used on electrical substation equipment [6].

Based on previous studies and the seismic design specifications at home and abroad [7, 8], using numerical analysis method, selection of best TMD damper installation position, choosing typical electrical equipment, refining damper parameters, study the structure of the different parameters on the vibration reduction effect.

2. Seismic response analysis of typical SF_6 circuit breaker

This paper chooses LW36-126A/T3150-40 high-pressure sulfur hexafluoride AC breaker, as shown in figure 1. Using ANSYS software simulation circuit breakers, aluminum flange, ceramic column, stent using SOLID45 unit, a control box with SHELL281 unit. The finite element model of the circuit breaker shown in figure 2. Modal analysis performed first.

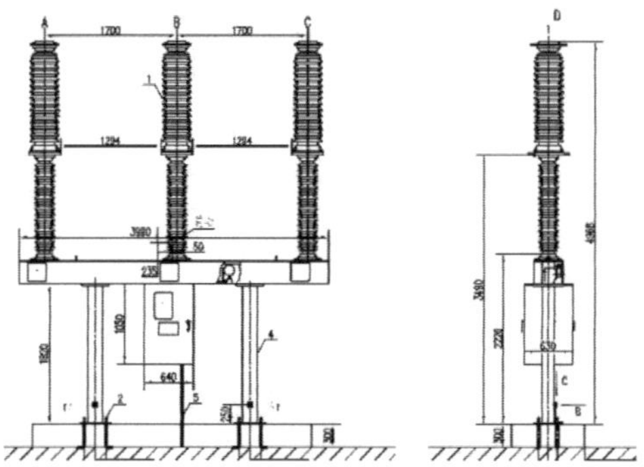

Figure 1. SF_6 circuit breaker structure size figure

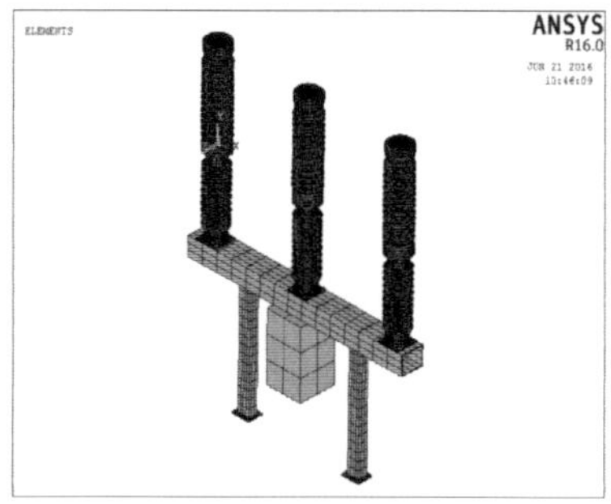

Figure 2. Finite element model of SF_6 circuit interrupter

In modal analysis, pre-select the circuit breaker 10 modal vibration mode. Figure 3 shows the front of the breaker modes Figure 4 order. Table 1 shows the 10 modal frequency of each order.

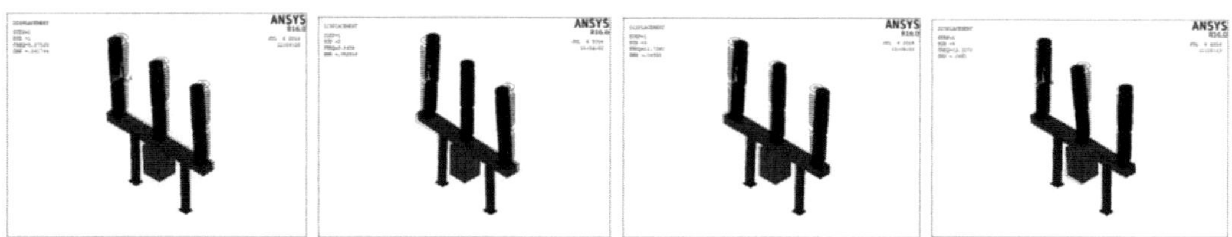

(a) First order (b) Second order (c) Third order (d) Fourth order

Figure 3. Modal figure of SF6 circuit interrupter

Table 1. Circuit breaker 1-10 order modal frequencies

Modal	1	2	3	4	5	6	7	8	9	10
Frequency	3.8753	8.5434	11.726	13.587	15.295	15.857	17.907	23.851	25.144	47.313

2.1. Select seismic waves

This paper selects the natural El Eentro and Taft waves recorded to load the input circuit breaker. Figure 4 for EL Centro wave and Taft wave acceleration time history curve.

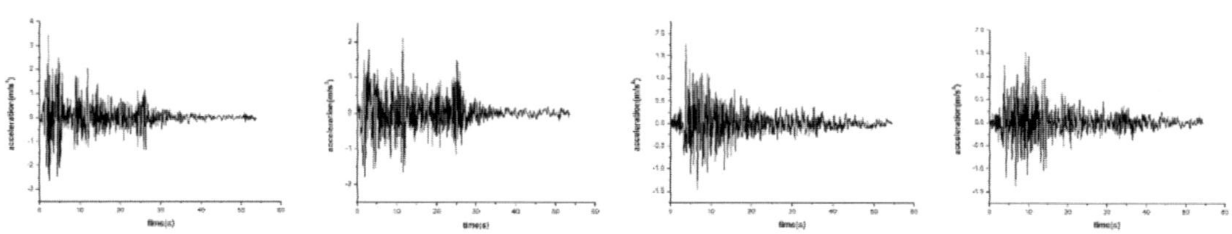

(a) El Eentro-ew (b) El Eentro-ns (c) Taft-ew (d) Taft-ns

Figure 4. EL Centro wave and Taft wave acceleration time history curve.

2.2. Breaker pole Stress analysis

Circuit breaker pole column is mainly composed of ceramic material, its brittleness is big and bending shear capacity is weak. When the earthquake loads, its root cause greater stress, causing brittle fracture occurs. Table 2 shows the circuit breaker three-phase column under seismic load stress. To stress analysis, a column suffered the biggest, the most prone to brittle failure.

Table 2. The stress of circuit breaker three-phase polar column under the seismic action/Pa

Column	Direction	EL Centro		Taft	
		Minimum stress	Maximum stress	Minimum stress	Maximum stress
A	X	-816985	757606	-314145	286336
	Z	-655480	613028	-252771	225969
B	X	-229794	205720	-108192	82843.2
	Z	-378321	312740	-163768	125569
C	X	-147303	161028	-115583	86528.2
	Z	-224396	195077	-88207.8	70657.5

2.3. Breaker top displacement analysis
Circuit breaker top-heavy, the maximum displacement occurs at the top. From table 3 can be concluded that, under seismic load, the three-phase is most prone to column damage fracture was born in column A, C, and A column under strong earthquake are more likely to happen.

Table 3. The top displacement of circuit interrupter three phase column/mm

Seismic wave	Direction	Displacement (mm)		
		A	B	C
EL Centro	X	0.696	0.587	0.505
	Z	17.84	15.368	12.979
Taft	X	0.69	0.43	0.7
	Z	11.4	11.2	11.7

3. Vibration Control Analysis

3.1. Breaker - ring TMD system
First, install the circular TMD damper. Then, through the damper for vibration control of circuit breaker, reduce the seismic load to the destruction of the circuit breaker. Formula for equivalent damping system as shown in (1), with ξ_e for equivalent damping system, $\overline{m}$ for TMD damper than with the quality of the circuit breaker, ξ for TMD damping coefficient of dampers, ξ_d as the main body structure damping coefficient [6].

$$\xi_e = \frac{\overline{m}}{2}\sqrt{1 + \left(\frac{2\xi_d}{\overline{m}} + \frac{1}{2\xi}\right)^2} \tag{1}$$

3.2. Breaker - ring TMD system finite element model
Choose suitable ring TMD damper, install it in very top three phase. As shown in figure 5 is the damper structure diagrams, figure 6 is the damper installation location.

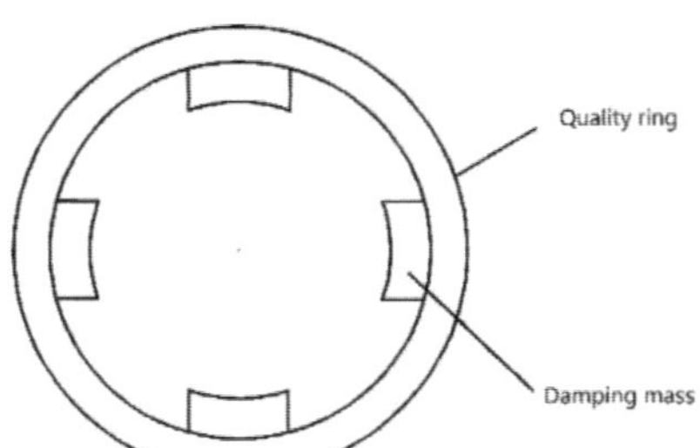

Figure 5. The structure diagram of circular TMD

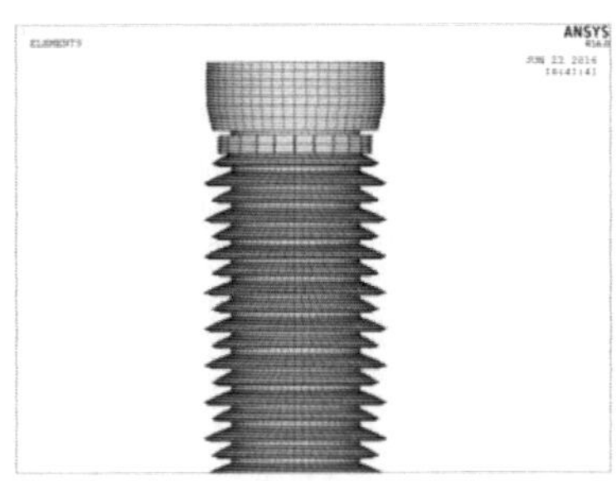

Figure 6. The installation drawing of circular TMD

3.3. Vibration damping control impact
From the above analysis, the vibration control effect depends on factors both TMD mass and damping coefficient. This paper set the TMD damper mass block is the main structure of the 7.5%, by changing the damping coefficient of the TMD damper damping effect.

As shown in figure 7, not installed TMD damper, the displacement of structure is very big, after the installation of TMD damper, the maximal displacement obviously reduced, and after the change of damping coefficient, damping effect is also corresponding change, when factor take 0.04, damping effect is better. As shown in figure 8, for the installation of TMD displacement before and after contrast, it can be seen that the damping effect is obvious.

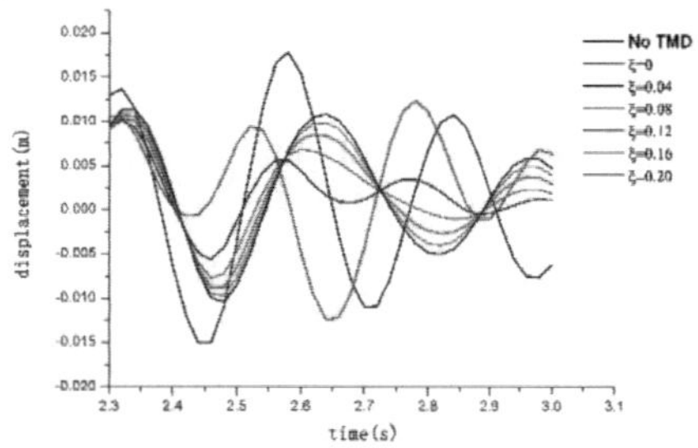

Figure 7. The displacement time history diagram of without TMD and different damping coefficient of TMD

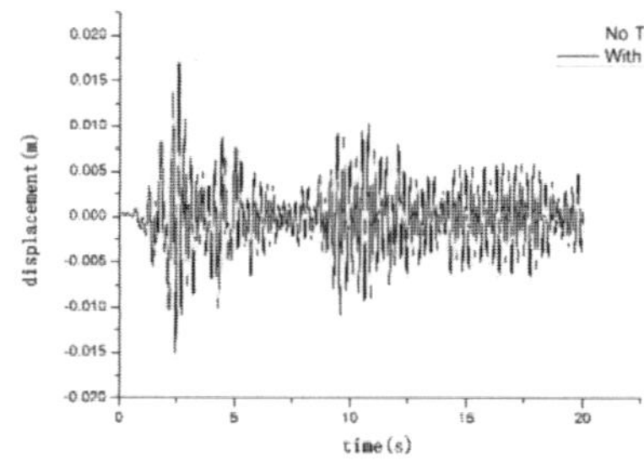

Figure 8. Installation of TMD compared with not installed TMD displacement diagram

Damping ratio is calculated according to the simulated data. Formula is as follows.

$$\text{Damping ratio} = \frac{\text{No TMD response} - \text{With TMD response}}{\text{No TMD response}} \tag{2}$$

Extracted on the basis of above analysis, the damping ratio under different damping coefficient, as shown in figure 9, from the tendency of the line can be concluded that the displacement of the shock absorption effect of stronger before they are weakened with the increase of damping coefficient, damping effect and stress first changes obviously with the increase of damping coefficient, and then leveled off, acceleration shock absorption effect of the minimum acceleration shock absorption effect increases gently lower, after the first maximum acceleration shock absorption effect increases after the first down. Comprehensive analysis of TMD damping coefficient at around 0.05 vibration control effect is best.

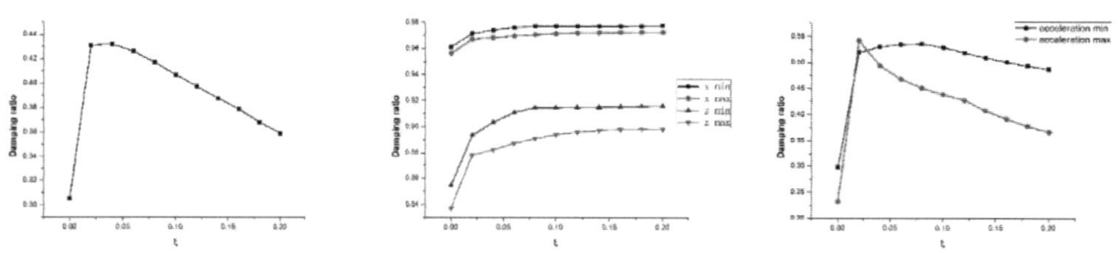

(a) displacement (b) stress (c) acceleration

Figure 9. The damping effect of each response

3.4. *Effect of the mass ratio of vibration control*

In addition to the damping coefficient and damper quality will also have influence on TMD damper. Different mass block, vibration effect is also different. This paper set up 7.5% of the mass of the main structure units of quality, TMD damping coefficient is 0.04.Mass ratio calculation formula as shown in (3).Ratio from 25% to 225% in each 0.25 in a group, by changing the parameters of quality than the TMD dampers for structural vibration control effect.

$$\text{quality ratio} = \frac{\text{quality of the mass}}{\text{unit mass}} \times 100\% \tag{3}$$

Calculate the acceleration damping ratio. As shown in figure 10, analysis of the data found when quality than 1.25(the main structure TMD mass block quality 9.375%), the structure vibration control effect is better.

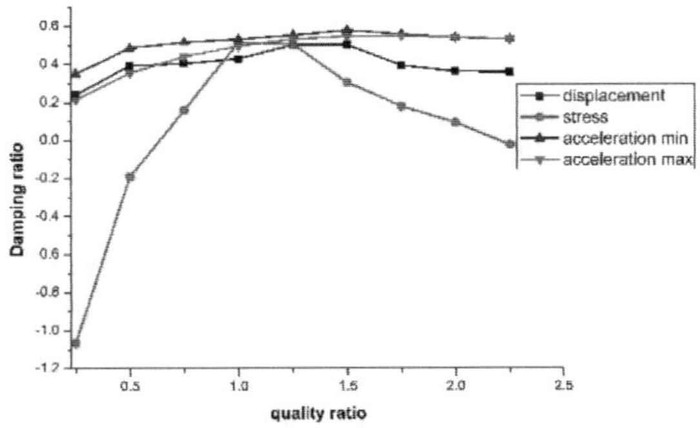

Figure 10. The damping effect of different quality ratio

4. Conclusion

(1) Through the study of the seismic response analysis of a typical circuit breaker, the circuit breaker in three-phase extremely column in the direction perpendicular to the plane (Z) displacement is larger, and with flange binding site in the biggest stress in the root, easy to brittle fracture.

(2) In large part, installing annular TMD damper can reduce the moving of the circuit breaker pole pillars.

(3) Ring TMD damper for vibration control effect of the circuit breaker with the change of the ratio of damping coefficient and quality change, the circuit breaker at the top of the displacement, acceleration and polar column root stress as the parameter change into regular change, and when the damping coefficient of 0.05, quality than 9.375%, the quality of structure of the TMD damper vibration control effect is best.

References

[1] Li Xiyu, Dang Zhenping. Bending strength reliability analysis of the post porcelain insulator for UHV[J]. Insulators and Surge Arresters, 2013(2):17-23.

[2] Liu Zhenlin, Dai Zebing, Lu Zhicheng. Seismic damage probability analysis of procelain power equipment based on Weibull distribution[J]. Power System Technology, 2014, 38(4): 1076-1081.

[3] Lu Zhicheng, Dai Zebing, Cao Meigen,Zhu zhubing. Study of method for aseismic design of electrical equipments with vibration absorbing system [J]. Engineering Journal of Wuhan Univeristy, 2010, 43(Supplement): 187-190.

[4] Zhao Ling, Tuned mass damper is used in the study of large span floor vibration control [D], Beijing: Beijing university of technology master's degree thesis, 2007.

[5] Zhang Chenglin Zhang jianrun,Sun Qinghong,Chen nan,Peng Wen,Yao Shujian. Two broadband tuned mass damper simulation and application [J]. Journal of manufacturing automation, 2006, 28 (8) :7-9.

[6] BaiWen. Porcelain column type high multiple tuned mass damper damping ring type electrical equipment technology research [D]. Harbin: institute of engineering mechanics, China seismological bureau. 2014.

[7] IEEE Std 693—2005. IEEE recommended practice for seismic design of substations[S].IEEE Power Engineering Society, 2005.

[8] GB 50260—2013 Code for seismic design of electrical installations[S].Beijing, China: China Planning Press, 2013.

IOP Conf. Series: Materials Science and Engineering **157** (2016) 012028 doi:10.1088/1757-899X/157/1/012028

Seismic Performance Research of Transmission Tower in Consideration of the Pile-soil-structure System

Chuncheng Liu[1], Long Mao[1], Chongyang Wang[1], Chuanming Zha[2]

[1]Northeast Dianli University Institute of Civil Engineering, Jilin, China
[2]State Grid of Xinjiang Electric Power Company Institute of Economic and Technology, Xinjiang, China

E-mail:maolongvip@163.com

Abstract. The seismic performance of transmission tower in consideration of pile-soil-structure dynamic interaction is researched through numerical simulation. Based on a transmission tower of a specific project, pile-soil-transmission tower coupled system is established. By using the method of time history, the pile-soil- transmission tower system dynamic response under seismic load were calculated, and comparing with the results without considering interaction system. Results show that, after considering interaction of the system, the period of the structure have extended and the mode of the structure lagged. On soft sites, compare with no considering the interaction, the results have a big difference, the relative increment of the maximum displacement at the top of the tower is 39.82%, respectively. Therefore it is suggested that the pile-soil-structure dynamic interaction should be fully considered in aseismic design of transmission tower on soft sites and medium soft sites.

1. Introduction

Transmission tower is an important part of power transmission, also is the foundation of safe and reliable operation of the electric power network, in the event of damage will cause big loss to national economy, seriously affected the quality of people's lives. When making the design of the transmission tower, the main control load generally took the wind load, ice load, bolt load, etc., rarely consider the effect of seismic load. And with the rapid development of China's economy, the demand for electricity also presents the fast growth tendency, so in recent years, more and more large span, long distance transmission lines are occurred. The condition of the transmission tower lines which have broken down under seismic load is constantly occurred. Such as, the earthquake of Kobe in Japan in 1995, according to statistics, there are different degrees of damage for 20 transmission towers, most of the damage form is at the basic part of the structure, the earthquake caused widespread users in this region losing power [1]. Chi-chi earthquake in Taiwan in 1999, result a large area of tower have collapsed and deformed of the structures [2], which caused huge losses. On May 12, 2008 Wenchuan earthquake, according to the statistics [3], more than 20 110 KV lines have occurred seriously damage, and also the same condition for 8 500 KV lines, which paralyzed the whole region transmission network, caused great obstacles to the rescue work. Surveys show that most damages of the transmission tower occurs at the foundation of the structure. And also seriously damage to ground soil. So to explore the dynamic characteristics of ground, soil, foundation and the transmission tower is an urgent job.

2. The whole finite element model of pile-soil-transmission tower system

2.1. Model of Transmission Tower

In this paper, beam elements were used to establish a three-dimensional finite element model of a transmission tower. Literature [4] pointed out that, the result of the transmission tower was calculated by the beam element is more in agreement with the results of the experimental. If all nodes of the transmission tower are regarded as a hinge point. For space model, outside plane of the single truss will be variable geometry system. So when calculate the seismic response of transmission towers that in consideration the dynamic interaction of pile-soil-transmission tower system, all connecting node of the transmission tower is regarded as rigid node.

2.2. Pile-Soil Contact Surface Model

In this paper, ABAQUS was used to simulate the pile-soil contact surface, using master-slave contact algorithm, first of all we need to establish surface separately on the pile and the soil, and we should to establish the corresponding contact pairs. The principle we choose the master surface and the slave surface is that the slave surface must be meshed more precise. If the mesh density of the master surface and the slave surface is very similar, we should choose a stiffer surface as the master surface. Here choose the surface of the pile as the master surface. Since it has a big difference between pile radius and soil radius which we need to calculate, this article take the 40~60 times pile radius as the soil radius [8]. The beam element is used to establish the transmission tower in this article.

2.3. Model of Soil

2.3.1. The Boundary of the Soil. Deeks, Jingbo Liu [5] [6] propose the viscous-spring artificial boundary. In seismic response analysis, we should intercept limited soil area, in which we establish a viscous-spring artificial boundary, and its equivalent to continuous spring-damping system that distribute on artificial boundary.

2.3.2. The Size of Soil Element. When using time history analysis method to calculate a model, the soil element size have a great influence on the calculation result and the convergence. If the grid size of the soil element is too small, it will increase the number of degrees of freedom, computing time growth, the results may be difficult to convergence; In contrast, soil element size is too big, it will have a significant impact on the high-frequency seismic waves. Finally, it will make a great error of results. Generally, for the S-wave, element height should satisfy the following relation [7]:

$$H_{\max} = (\frac{1}{5} - \frac{1}{8})\lambda_{\min} \tag{1}$$

$$\lambda_{\min} = (\frac{1}{5} - \frac{1}{8})V_s / f_{\max} \tag{2}$$

Where, λ_{min} is called minimum wavelength; V_s is called shear wave velocity of the soil; f_{max} is called the highest frequency.

3. A numerical example

3.1. Project Overview

In this section, the example is a straight tower of drum shape, the height of the tower is 48.4m, the size of the transmission tower at the bottom is 11.6m* 11.6m, one tower legs with one pile, the length of the pile is 10m, radius of the pile is 0.3m, Angle steel of transmission tower used in the model are Q235 and Q345. Plan measure of the transmission tower was shown in Figure 1.

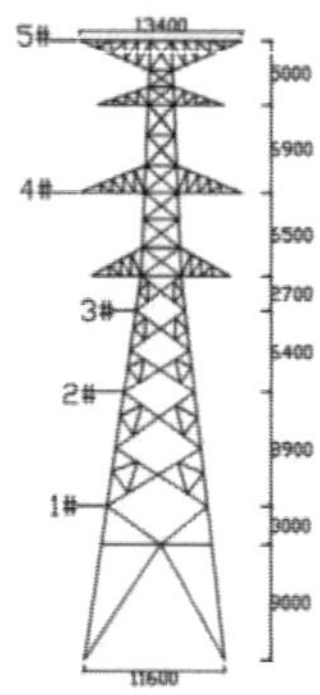

Figure 1. The measure of transmission tower/mm

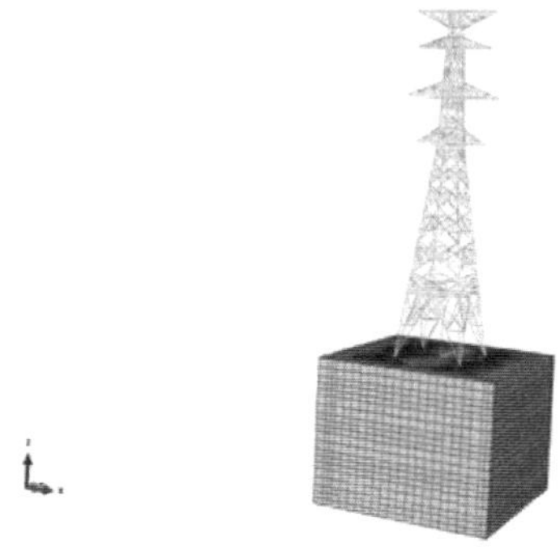

Figure 2. Model of pile-soil-transmission tower coupling

3.2. The Establishment of The Finite Element Model

This article is based on pre-treatment procedures of ABAQUS, use which to create one three-dimensional model of transmission tower (referred to as TW), and two pile-soil-transmission tower dynamic interaction model (referred to as PST1, PST2). It is correspond to the two types of site classification in Table 1.

Table 1 Parameters of four kinds of site soil

Serial number	Thickness （m）	Shear wave velocity （m/s）	Density （kg/m^3）	Elastic modulus (MPa)	Cohesion （Pa）	Internal friction angle
	5	560	2000	1505.3	25000	15
1	5	580	2100	1695.5	35000	20
	10	590	2200	1838	50000	25
	5	100	2000	48	25000	15
2	5	120	2100	72.6	35000	20
	10	130	2200	89.2	50000	25

According to the literature [8] the length, width and height of soil area which we need to calculate is respectively 30m, 30m and 20m. According to formula (1), (2) we can calculate the height of soil element. Take 10HZ as f_{max}, equivalent shear wave velocity of two categories of site soil are 118.6m/s and 579.7m/s. The article is selected 2m as the height of soil element. Finite element model of pile-soil-structure is shown in Figure 2.

3.3. Modal Analysis

To study the dynamic response of the whole structure of the transmission tower, firstly we should analysis the dynamic characteristic. Table 2 shows five natural cycle of the two kinds site conditions. As can be seen from the table, after considering pile-soil-structure interaction, natural period increased compared with the structure that no considering the interaction. With the softening of the site conditions, the rate of increase of natural period gradually increases. This shows that after taking into account the interaction, the nature period becomes long compared with no considering the interaction. The more soft ground conditions is, the effect of natural period is more significant.

Table 2 The nature period

	T1/s	T2/s	T3/s	T4/s	T5/s
TW	0.383	0.379	0.246	0.184	0.183
PST1	0.499	0.496	0.246	0.191	0.189
PST2	0.687	0.645	0.642	0.635	0.634

Figure 3 (a), (b), (c) is the 1, 3, and 6 mode of TW model. As can be seen from the figure, The first mode is lateral displacement deformation, the first mode is local bending deformation, and the third mode is Local bending deformation and torsional deformation.

Figure 4 (d), (e), (f) is the 1, 5, and 7 mode of PST model. Respectively corresponding to 1, 3, 6 mode of TW model. We can see from the figure that after considering soil-pile-structure dynamic interaction, the mode of the structure lagged when compared with no considering the interaction.

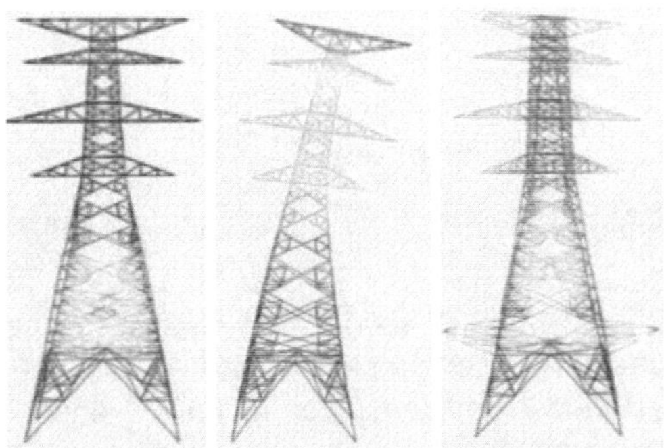

(a)First mode (b) Third mode (c) Sixth mode
Figure 3. The modes of TW model

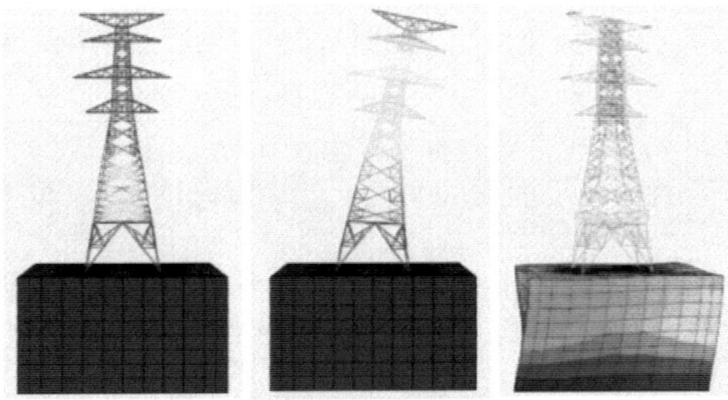

(d) First mode (e) Fifth mode (f) Seventh mode
Figure 4. The modes of PST model

3.4. Time history analysis under seismic load

Seismic Response Analysis commonly used response spectrum method and time history analysis. Response spectrum method and time history analysis method are used to analysis the seismic response of structures. The response spectrum method can only solve the maximum response of the structure, and the time history analysis can get the response of the structure in the time of the whole seismic load. So the time history analysis method is adopted in this paper.

According to the current seismic code of our country, the principle of choosing the seismic wave is that the main period of the seismic wave must be similar with the remarkable period of the construction site, so that the characteristic of the input earthquake wave is consistent with the condition of the construction site soil. According to the above principles we select two natural earthquake records as shown in table 3, Corresponding to the two sites in Table 1. The duration that we selected is 20s. Two kinds combination of working condition was calculated in this paper as shown in Table 4.

Table 3 Two natural seismic records

Serial number	Earthquake	Magnitude	Recording point	Peak value	Site class
1	Tangshan aftershock	6.3	Qian'an	118.9	Hard soil site
2	Tangshan aftershock	7.1	Ninghe	200	Soft soil site

Table 4 Case combination

Case combination	Work condition 1	Work condition 2
Computational model	PST1	PST2
Seismic record	1	2

Figure 5 is the time history of displacement curve at the top of tower under the two work conditions, and they are in sharp contrast. It can be seen from the figure that, when considering the pile-soil-structure interaction, the maximum value of the top displacement of the tower under the two conditions is larger, and the peak time of displacement appeared in advance compared with no considering the interaction.

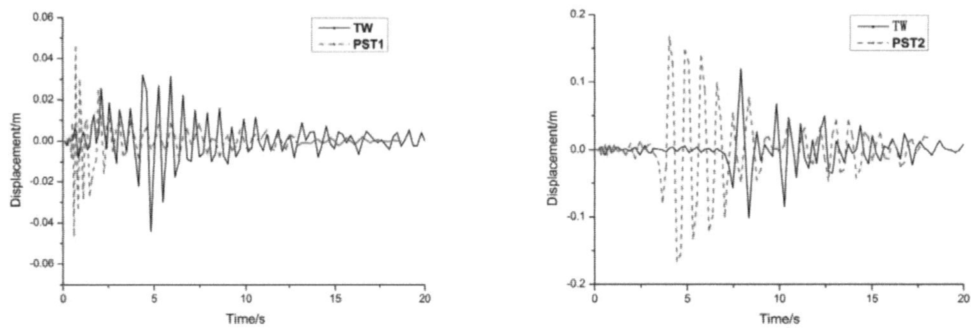

Figure 5. Time history response of displacement at the top of the tower under two kinds of work conditions.

Table 5 lists the maxima of nodes displacement of TW and PST model two conditions, Data observed in Table 5, we can obtain the following conclusions: after considering pile-soil-structure interaction, the maximum value of displacement at the top of the tower has been increased, and with the increase of the height, the relative increment of the nodal displacement becomes smaller, and the absolute increment of the nodal displacement is increasing.

Table 5 the maximum value of nodal displacement under each work condition

		The maximum value of nodal displacement /cm				
Node		1	2	3	4	5
	TW	0.275	0.704	1.202	2.31	4.396
Condition 1	PST1	0.632	1.275	1.846	2.675	4.575
	H%	129.82	81.11	53.58	15.80	4.07
	TW	0.806	2.049	3.413	6.412	11.967
Condition 2	PST4	2.8	5.299	7.385	11.021	16.732
	H%	247.39	158.61	116.38	71.88	39.82

4. Conclusions

In this paper, the seismic response of the transmission tower considering pile-soil-superstructure interaction is studied by using the time history analysis method, and it is compared with the seismic response of transmission tower under the condition of rigid foundation, Draw the following conclusions:

(1)After considering the pile-soil-structure dynamic interaction, the nature period of the structure becomes longer, and the mode lag.

(2)In the soft soil site, considering the dynamic interaction of the pile-soil -transmission tower, the absolute increment and the relative increment of the displacement and velocity at the top of the tower are obviously increased; the stress at the root of the transmission tower is increased. Thus, in soft soil site, pile-soil-structure dynamic interaction has great influence on the seismic performance of transmission towers. Full consideration should be given to the seismic design of transmission towers in this area.

(3)In the hard soil site, the dynamic interaction between pile-soil-structure is very small and can be neglected in the seismic response of transmission towers

5. References

[1] Shinozuka M. The Hanshin-Awaji earthquake of January 17, 1994: performance of lifelines[R]. State University of New York at Buffalo, NCEER report: 95—0015.

[2] Yin R H, Li D L, Liu G L, et al. Seismic damage and analysis of power transmission towers [J]. World Earthquake Engineering, 2005, 21（1）: 51—54.

[3] Ziyin Zhang, Biao Zhao, Weiwei Cao. Investigation and preliminary analysis of damages on the power grid in the Wenchuan earthquake of M8.0. [J].Electric Power Technologic Economics, 2008, 20（4）:1—4.

[4] Prasad Rao N, Kalyanaraman V. Nonlinear behavior of lattice panel of angle towers [J]. Journal of Constructional Steel Research, 2001, 57(12): 1337—1357.

[5] Deeks A J, Randolph M F. Axisymmetric time-domain transmitting boundaries [J]. Journal of Engineering Mechanics, 1994, 120(1): 25—42..

[6] Jingbo Liu, Zhenyu Wang, Xiuli Du, Yixin Du. Three dimensional viscous-spring artificial boundaries in time domain for wave motion problems[J]. Engineering Mechanics, 2005, 22(6): 46—51.

[7] Lysmer J, Udaka T, Tsai C F. FLUSH a computer program for approximate 3-D analysis of soil-structure systems [R]. Report No. EERC75-30, Earthquake Engineering Research Centre, University of California, Berkeley, 1975.

[8] Jing Chen. Numerical simulation of friction pile which consider consolidation by ABAQUS[J]. Henan Science，2010，25（11）: 1—3.

IOP Conf. Series: Materials Science and Engineering **157** (2016) 012029 doi:10.1088/1757-899X/157/1/012029

Multisensor system for toxic gases detection generated on indoor environments

C M Durán[1], P A G Monsalve[1], C J Mosquera[2]

[1]Multisensor System and Pattern Recognition Research Group, Electronic Engineering Program, Engineering and Architecture Faculty, Universidad de Pamplona, Km 1 Vía Bucaramanga, Pamplona (N.S) Colombia
[2] Grilex Research Croup, Faculty of education, Universidad de Pamplona, Km 1 Vía Bucaramanga, Pamplona (N.S) Colombia

cmduran@unipamplona.edu.co

Abstract. This work describes a wireless multisensory system for different toxic gases detection generated on indoor environments (i.e., Underground coal mines, etc.). The artificial multisensory system proposed in this study was developed through a set of six chemical gas sensors (MQ) of low cost with overlapping sensitivities to detect hazardous gases in the air. A statistical parameter was implemented to the data set and two pattern recognition methods such as Principal Component Analysis (PCA) and Discriminant Function Analysis (DFA) were used for feature selection. The toxic gases categories were classified with a Probabilistic Neural Network (PNN) in order to validate the results previously obtained. The tests were carried out to verify feasibility of the application through a wireless communication model which allowed to monitor and store the information of the sensor signals for the appropriate analysis. The success rate in the measures discrimination was 100 %, using an artificial neural network where leave-one-out was used as cross validation method.

1. Introduction

Nowadays wireless communication systems for different industrial applications have important advantages (compared to the conventional wiring), since they tend to increase the efficiency and safety in each of the processes. One of the main advantages that provide these types of systems is low cost of installation and maintenance [1]. Most of the Industrial facilities are rigorous and expensive because they are demanding in terms of quality of cabling requirements. Therefore, less cables means to use fewer elements in the system, especially when it comes to modernizing or upgrading components within the industry. For this reason, a wireless communication system coupled with an electronic nose can be a solution to be implemented in environments which required to install large amount of wiring lines and can detect dangerous gases as well [2].

1.1. Underground coal mines

At present, mining is a task that annually takes many lives because the vast majority of companies that do this work do not have detection methods to provide safety for workers inside an underground coal mine. Extreme security actions would avoid fatal accidents that every year leave great human losses, material, social damage and a host of problems due to diggings and emissions of toxic gases. In most cases, 95% of underground coal mines in Colombia do not have a system for detecting toxic gases in real time, as evidenced by experimental work [3].

1.2. Gas monitoring
Inside the underground coal mines is necessary to implement a permanent and continuous monitoring of methane (CH_4) and oxygen gases in transport and ventilation [4]. Also, where have active fire outbreaks must be implemented a system of permanent and continuous monitoring of carbon monoxide (CO) and oxygen (O_2). These gases should be monitors continuously and permanently in places where vehicles are used with internal combustion engine. Before each use of the measuring equipment, it must make a clear verification of the range (±10% standard value of the pattern). In this study, Xbee wireless modules for communication between nodes were implemented in gas sensor arrays, which are important part to develop an electronic nose in order to detect toxic gases usually generated in coal mines [5].

2. Materials and methods
The following section shows the components used for the design and development of multisensor prototype. This consists basically of a sampling and measurement system than comprises of a set of six metal oxide gas sensors (Type MQ) and a data processing stage used to classify a set of acquired samples by data acquisition card wirelessly. The wireless technology used in this study was ZigBee, which is a communication protocol that can use a single coordinator, one or more routers and as well as several mobile devices (i.e., e-noses).

2.1. Communication model
The topology of the wireless communication system is a tree-like structure, which can be used in confined spaces. For this reason, it is possible to implement the distribution of an electronic nose within a location with limited space, such as a coal mine or indoor environments [6]. In turn, it is possible to obtain better communication between nodes of the same type. The model was implemented with a standard configuration API (Application Programming Interface), which allows more information node and thus know the status of it. The data or frame is transmitted in serial mode to the coordinator, which is interconnected to a Xbee-USB-PC adapter to send requests to the present node and handles this request depending on the case (i.e., routers or several e-noses. If an end-device were selected, this will indicate a default router which will serve as a repeater for that coordinator receives data from the remote point.

2.2. Data processing methods
A total number of 41 measurements were acquired with the multisensor system from a set of toxic gas cylinders for calibration of the system (Table 1). A brief description of these gases and measurements (#M) are given below: Gas mixture (9M): CH_4 (2.5 %), H_2S (25 ppm), O_2 (18%), CO (100 ppm), Carbon monoxide (12M): CO (100 ppm), Liquefied gas (10M): LG (98% purity), Carbon dioxide (9M): CO_2 (2.5 %).

All measures in voltage values were acquired and transformed in conductance values. We used the static parameters to extract information from sensor array where the most relevant information from the sensor response was obtained to reduce the amount of information of data set. The main static parameters such as: maximum conductance (Gmax), minimum conductance (Gmax), initial conductance (Gi) and final conductance (Gf) were extracted to calculate the conductance increment (ΔGmax= (Gmax-Gmin)), normalized conductance increment or also called fractional difference maximum (ΔGnmax=ΔGmax/Gmax)), absolute differences logarithm (ΔGlog=log (|Gmax-Gmin|) and absolute fractional differences maximum logarithm (ΔGlog=log (|Gmax-Gmin|/Gmax)). Thus, more combinations could be calculated and applied using Gi and Gf. The data matrix size was 41x6 that corresponded to 41 measurements and 6 variables, which was used in order to determine the success rate of data classification from the sensor responses.

Linear multivariate methods are often used for discrimination, reduction, identification and classification of different Volatile Organic Compounds (VOC's). In this study two linear transformation PARC methods and a Neural Network were applied to classify a set of toxic gases.

The data set were pre-processed using a statistical parameter such as *increased maximum*, see equation (1), to obtain the discrimination of the data. Finally, the normalized data were processed applying pattern recognition methods such as Principal Components Analysis (PCA) and Discriminant Function Analysis (DFA).

$$\Delta max = Gmax - Gmin \quad (1)$$

where, Gmax is the response or the maximum conductance (resistance inverse) of the sensors in presence of odour, and Gmin is the steady-state of minimum conductance or baseline signal.
Once trained the system, a new measure was acquired randomly through the cylinders in order to discriminate which class belongs.

Table 1. Identification of categories

# Gas containers	Compound	PCA Labels	DFA Labels	Categories Colour
1	Gas mixture $CH4$ (2.5 %), H_2S (25 ppm), O_2 (18%), CO (100 ppm)	A	Δ	Yellow
2	Methane CH_4 (2.5 %)	B	◊	Red
3	Carbon monoxide CO (100 ppm)	C	□	Green
4	LG(Liquefied gas) butane (98% purity)	D	*	Blue
5	Carbon dioxide (2.5 %)	E	●	Purple

3. Results and discussion

Figure 1 illustrates the electronic nose prototype which was developed in order to be applied in indoor environments but with alternative to cover a wide range of applications, in this work we developed an Electronic Nose to detect toxic gases frequently generated in underground coal mines. Although the multisensory system has not been used to tests on-site (inside of mine), the wireless communications between coordinator and routers was successful, reaching a distance of 110 meters indoors (obstacles in classrooms).

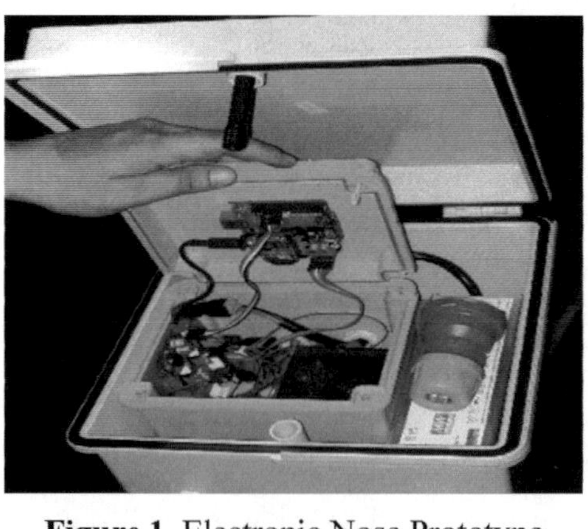

Figure 1. Electronic Nose Prototype

IOP Conf. Series: Materials Science and Engineering **157** (2016) 012029 doi:10.1088/1757-899X/157/1/012029

The data pre-processing and processing algorithms were done with tools provided by Matlab having added a Graphical User Interface (GUI). It is important to clarify that both PCA and DFA were used by being linear supervised and unsupervised methods to project data set from different sensors to a two-dimensional plane [7]. Figure 2 depicts the classification realized with the PCA and DFA methods for discrimination of the data set. PCA was selected in this application because is an effective unsupervised linear method to project data from several sensors to a two-dimensional plane and it is possible to discriminate easily the data set as well. Figure 2(a) illustrate the clusters of gas mixture (A, yellow colour) and CH_4 (B, red colour) compounds which are very close, so this could have some overlaps with samples near these two groups.

Figure shows the first two principal components represented in 99.74% of the variance in the data set. DFA technique was applied to discriminate the groups or clusters, to define the better way to separate groups and remove variables which are little related to group distinctions. Therefore a number of two latent variable (factors) of a linear combination of independent variables were chosen. Figure 2(b) shows the results of DFA that were obtained from 41 measurements mentioned above, where a new measure carried out with 2.5% of CH_4 was classified correctly with the cluster of gas mixture (i.e., included 2.5 % of CH_4). It was observed that the distance among the different groups showed good discrimination in both cases, achieving an adequate selectivity and repeatability of the system and demonstrating a high performance of the equipment. In addition, this analysis showed that DFA is a good supervised method to distinguish different varieties of toxic gases present in underground coal mines. The two first factors achieved up 99% of variance in the data set.

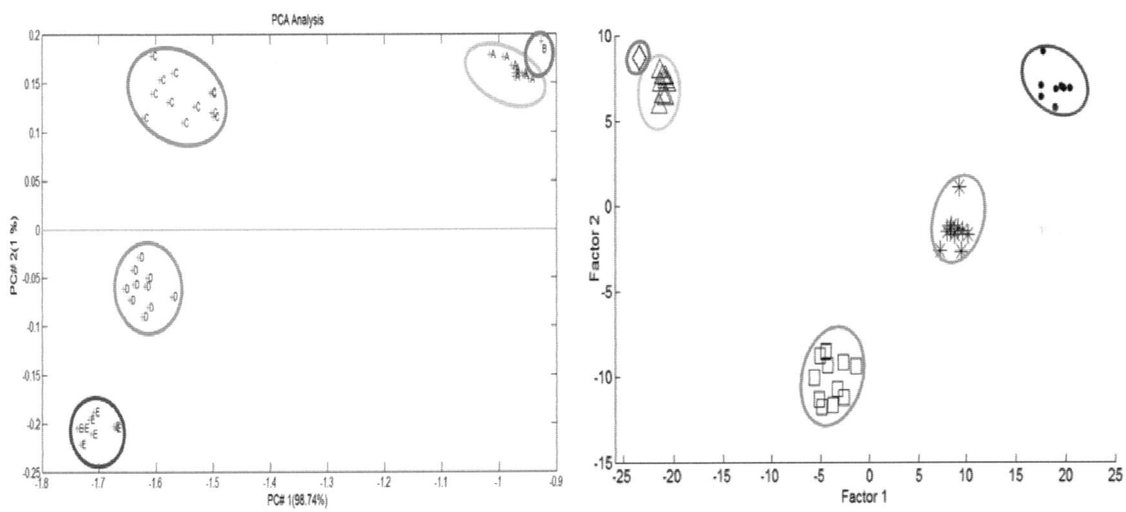

Figure 2. a) PCA graph with different types of cluster (Toxic gases) and new measurement projection, b) DFA performed on the mean-centred response matrix with statistical parameter Δmax

In this study a supervised method such as probabilistic neural network was applied to classify the data set, it provides a solution to measures classification problems. PNN was performed by Specht [8]. Figure 3 illustrates the PNN response using normalized data and targets, in order to estimate the classification where a success rate was obtained with 41 measures and an average of 97.6 %. Clearly the new measure of methane gas was grouped within the gas mixture category, for obvious reasons the general classification was 100%. The cross validation method generates N evaluation procedures (1 for each measurement), for each iteration a different measurement was left out, while the remaining measurements were used for network training (the one not used for training). This is repeated N time (one for each measurement) so that the final result is the average success of entire iterative process. This method did not present over-fitting because the evaluation measurement was not used in training data.

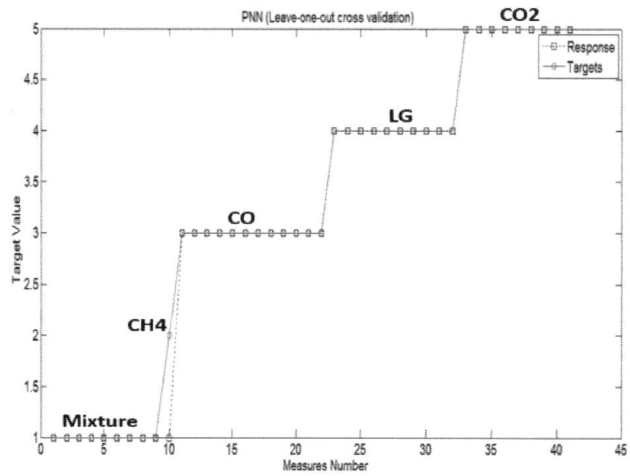

Figure 3. Toxic gases classification using the supervised PNN

4. Conclusions

The sensitivity and selectivity obtained by the WEN system was good and the repeatability shown by the equipment was satisfactory. The e-nose developed on a ZigBee wireless network shows the viability to implement a measurement instrument to detect VOCs in indoor environments (Underground Coal Mines) due to low-power consumption, reliability and high portability. Through the results obtained with the WEN is possible to develop real-time applications for monitoring and control of various processes, requiring a multisensory system of such characteristics to be implemented in different sectors (e.g, industrial sector, environmental, health, etc). The pattern recognition methods (DFA and PCA) were applied on the normalized array and two first discriminant factors or scores reached more than 99% of variance. The classification of data set was successful achieving 100% with the PNN supervised learning.

Acknowledgments

We would like to give our most sincere thanks to the Pamplona University by funding an important part of this project

References

[1] Chen S, Yao J and Wu Y, 2012 Analysis of the Power Consumption for Wireless Sensor Network Node Based on Zigbee, *Procedia, Engineering*. **29** 1994-98.

[2] Mahmood A, Nadeem J, Razzaq S, 2015 A review of wireless communications for smart grid, *Rene and Sust Ener Revi*, **41** 248-60.

[3] Restrepo E, Ríos C and Jiménez B, 2012 Detección y control de atmósferas explosivas en minas subterráneas de carbón usando programación estructurada, *Revi Educ en Inge*, **7** 10–21.

[4] Fakundu B, Andrews G, Phylaktou H, 2014 Gas Explosion Venting: Comparison of Square and Circular Vents, *Chem Engi Tran*, **36** 163-8

[5] Durán A.C.M and Rodríguez J.C, 2008 Sistema de Olfato Electrónico para la Detección de compuestos Volátiles. *Revi colo de tecno de Avan*, **12** 20-26

[6] Baronti P, Pillai P, Chook W. V, Chessa S, Gotta A, Hu Y, 2007 Wireless sensor networks: A survey on the state of the art and the 802.15.4 and zigBee standards, *Comp Comm*, **30** 1655-95

[7] Martinez A.M, Kak A, 2001 PCA versus LDA, *IEEE transactions on pattern analysis and machine intelligence*, **23**, p 228

[8] Amari A, El Barbri N, Llobet E, El Bari N, C Xavier and Bouchikhi B, 2006 Monitoring the Freshness of Moroccan Sardines with a Neural-Network Based Electronic Nose, *Sensors*, **6** 1209-23.

Position controlled Knee Rehabilitation Orthotic Device for Patients after Total Knee Replacement Arthroplasty

Patsiri Wannaphan and Teeranoot Chanthasopeephan*

Mechanical Engineering Department, King Mongkut's University of Technology Thonburi, 126 Pracha-uthit Road, Bang Mod, Tung-kru, Bangkok 10140, Thailand

Abstract. Knee rehabilitation after total knee replacement arthroplasty is essential for patients during their post-surgery recovery period. This study is about designing one degree of freedom knee rehabilitation equipment to assist patients for their post-surgery exercise. The equipment is designed to be used in sitting position with flexion/extension of knee in sagittal plane. The range of knee joint motion is starting from 0 to90 degrees angle for knee rehabilitation motion. The feature includes adjustable link for different human proportions and the torque feedback control at knee joint during rehabilitation and the control of flexion/extension speed. The motion of the rehabilitation equipment was set to move at low speed (18 degrees/sec) for knee rehabilitation. The rehabilitation link without additional load took one second to move from vertical hanging up to 90° while the corresponding torque increased from 0 Nm to 2 Nm at 90°. When extra load is added, the link took 1.5 seconds to move to 90°. The torque is then increased from 0 Nm to 4 Nm. After a period of time, the speed of the motion can be varied. User can adjust the motion to 40 degrees/sec during recovery activity of the knee and users can increase the level of exercise or motion up to 60 degrees/sec to strengthen the muscles during throughout their rehabilitation program depends on each patient. Torque control is included to prevent injury. Patients can use the equipment for home exercise to help reduce the number of hospital visit while the patients can receive an appropriate therapy for their knee recovery program.

1. Introduction

Knee osteoarthritis is the most common disease in orthopedic. The osteoarthritis of knee is a chronic condition that affects the lives of the elderly. After total knee arthroplasty, patients can return to their daily activities with an improvement as compared to the preoperative condition. Rehabilitation after total knee replacement arthroplasty aims to decrease pain, recovery the range of motion, and help patients improve knee function for their daily activities. After the surgery, patients are usually evaluated with Risk Assessment and Prediction Tool (RAPT)[1] by doctors and specialist to lead to rehabilitation program for each specific patient. The numbers of patients who need to go through rehabilitation process are increased. The increased number of patient who requires knee rehabilitation results in the increasing demand of physical therapy and time to assist all patients. Therefore hospitals world-wide create their rehabilitation equipment and initiated rehabilitation programs to help assist patients' knee performance.

In the past, researchers designed and developed control techniques for rehabilitation equipment for patients. Many designed rehabilitation equipment are developed for one degree of freedom, considering the knee's range of motion and the amount of torque handle by the knee during its movement[2]. Biomechanical analysis of lower limb simulation helped researchers gathered information and understood the relationship of the limb motion and the forces which occur during each rehabilitation activity. These information was design requirement of the device for rehabilitation in each program[3]. The Continuous passive motion(CPM) was developed with accurate alignment of the same actual human knee joint using linear actuator to drive the link structure mechanism[4]. An example of the knee and hip rehabilitation for patients after a spinal cord injury[5] was a three degrees of freedom system with Human-Machine Interface which provided connection between user and the designed device during rehabilitation exercise. Nikitczuk et al. applied eletrorheological fluid (ERF) to provide resistive torque during the motion and designed a control law to control the amount of torque exerted atthe knee joint. Control techniques were applied on the orthosis through adjustable actuate torque for the flexion/extension at the knee joint[6]. To obtain a robust system for knee rehabilitation, Mefoued, et al. applied Sliding Mode Control(SMC) to improve the response of the actuator in knee joint by tracking trajectory of the motion and compared it with the one after applied a PID control system [7]. In gait rehabilitation, a compact robot was designed with controllable actuator to assist patients in gait exercises with the knee-ankle-foot robot[8].

In this study, we propose our design of knee rehabilitation equipment for patients after total knee replacement arthroplasty in sitting position. The equipment main focus is on knee flexion and extension of knee joint in rehabilitation. The study includes the design of a control system to control knee joint position to assist user along with torque monitoring to provide user safety and prevent injury during rehabilitation.

2. Hardware design

2.1. Design requirements

The designed knee rehabilitation equipment has one degree of freedom at the knee joint and the device is used in sitting position for flexion and extension. The length of thigh link is adjustable within the range of 38 – 50cm. The calf link is 28-40cm long and can be altered depends on the size of the patient's limb. The rehabilitation equipment can be used for exercise within the range of 0°-80°. The speed of the motion is set at 40 degrees/sec, which is normal speed for knee recovery program. However, the speed can be increased up to 60 degrees/sec to strengthen the muscles after a period of rehabilitation program.

2.2. The device kinematics

The forward kinematics of the device is based on two links which are the thigh link and the calf link in sagital plane, as shown in figure 1. The DH (Denevit-Hertenberg) parameter of the device is shown in table 1.

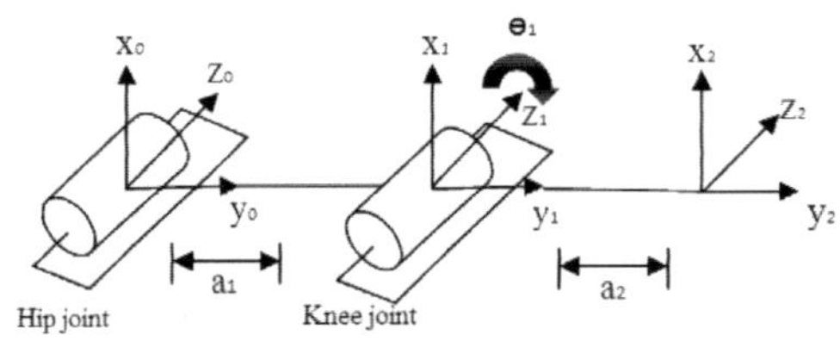

Figure 1. Kinematics diagram of the device.

Table 1. Table of DH (Denevit-heartenberg) parameter.

Link i	a_{i-1}	α_{i-1}	d_i	θ_i
1	0.4	0	0	0
2	0.52	0	0	0

Substitute the parameter in the table 1 in equation (1) to the transformation matrix in equation (1) and workspace of the end effector of the device in figure 2.

$$T_2^0 = A_1 A_2 = \begin{bmatrix} c\theta_1 & -s\theta_1 & 0 & a_1 c\theta_1 + a_2 c\theta_1 \\ s\theta_1 & c\theta_1 & 0 & a_1 s\theta_1 + a_2 s\theta_1 \\ 0 & 0 & 1 & 0 \\ 0 & 0 & 0 & 1 \end{bmatrix} \quad (1)$$

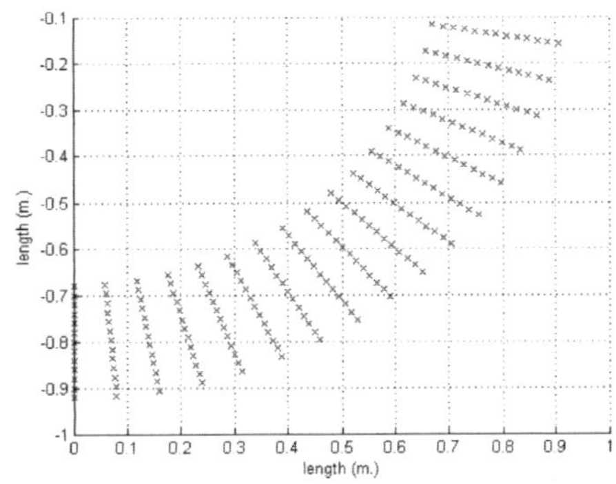

Figure 2. Workspace of the designed knee rehabilitation equipment.

2.3 *Design mechanism*

The knee rehabilitation device consists of a 36 Watt DC motor which was connected to the gear box, the encoder (Roundss RCC38S6) and the torque sensor (Sensor Technology RWT321-DG). The system was connected to the analog output NI9215. The DC motor was connected to the NI9505 full H-bridge brushed dc servo drive module. The NI9215 module and NI9505 module were plugged into the National Instruments (NI) CompactRIO 9104, 8 slot field programmable gate array (FPGA) which is the real-time controller and controlled through LabVIEW. A bevel gear was used to transmit and change the direction of the dc motor drive to the thigh link during the motion. The system together with the lower limb was attached to the set as shown in figure 3. User gets to exercise one leg at a time, while the link is attached to the leg and the support.

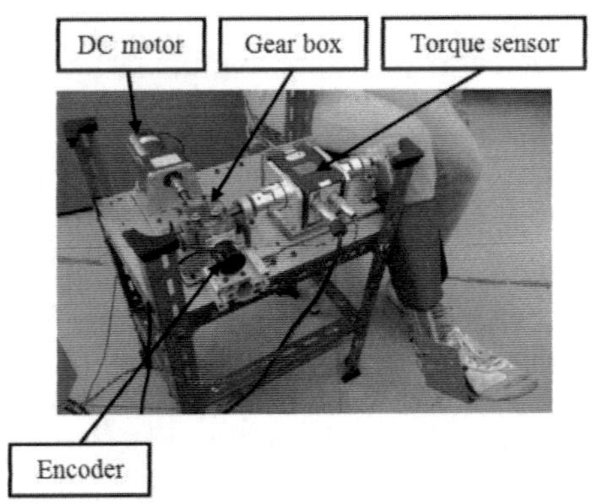

Figure 3. The designed prototype.

3. Control system design

During a physical therapy, the system is designed to prevent injury and at the same time assist user with the provided torque. We simulated the movement of a device model with a human model using Matlab. First, the equipment was modeled through Lagrange-Euler equation as shown in equation (2).

$$I\ddot{\theta} - mg\frac{L}{2}\sin\theta = T \quad (2)$$

Where L is the length of the link, I is the inertia, T is the input torque from the motor, and θ is the angle of the link as measured from the vertical direction. The transfer function (3), which was derived from (2) where input is torque and output is θ, is used in control simulation in Matlab.

$$\frac{\theta(s)}{T(s)} = \frac{1}{Is^2 - mg\frac{L}{2}} \tag{3}$$

3.1. Human model

Human model is represented by the relation between the torque requires (input) to move the lower link to 80° at the knee joint (output) for rehabilitation. Where the inertia of the leg (I_{leg}) is 0.43 kgm^2mass of the patient (m_{human}) is 4.8kg, and length of the leg (L_{human}) is 0.52m.

3.2. Device model

Device model is relation between the torque requires (input) to move the lower link to 80° at the knee joint (output) for rehabilitation. Where the inertia of the device (I_{device}) is 0.07 kgm^2, mass of the device (m_{device}) is 0.8kg, and length of the device (L_{device}) is 0.52m.

The human model and the device model are similar except for the value of the parameters. These two models are used to determine the controller through simulation in Matlab. The position control of the rehabilitation equipment is necessary; the position of the setup should not exceed the range of motion of the legs to prevent injury. The interaction between human legs with the device is controlled through position feedback of the human leg which moves along the desired position input of the system. The system is controlled using PID controller. The system is controlled through Labview program with graphical user interface (GUI) on a computer which is connected to the real-time controller (cRIO-9012) via communication with I/O devices. In this experimental set up, we use NI9505 module by National instrument to control the speed of the DC motor which drives the moving mechanism of the system and used NI9215 to acquire an analog output signal from the torque sensor (Sensor Technology RWT321-DG).

4. Result and Discussion

The designed equipment was tested at 18 degrees/sec, which is considered low speed for the flexion and extension of the knee rehabilitation[5] to ensure that the designed position control is suitable for the system. The motor speed was monitored through NI9505 and NI9215module was used to collect the torque signal from the equipment. We observed that when the link (with no extra load) was set to move at 18 degrees/sec, the calf angle started moving from 0° at vertical position and up to 90°. The link took approximately 1 second to move up to 90° as shown in figure 4a). The torque increased up from 0 Nm to 2 Nm at 90° and then the calf moved down from $90^o$with the torque 2 Nm and moved down to 0° returned to the vertical position with the torque 0 Nm as shown in figure 4b). When extra load is added, the corresponding position and torque profile are shown in figure 5a). However, the link took approximately 1.5 seconds to move to 90°. The torque increased up from 0 Nm to 4 Nm and then the link returned to the vertical position with the torque 0 Nm as shown in figure 5b). The test was performed continuously and the result appeared to be the same through each run. The position of the equipment are shown in figure 4a) and 5a). The actual position (dashed line) is following up the desire position the controller will have to be adjusted with careful consideration of the effect of acceleration on the position of the equipment.

a)

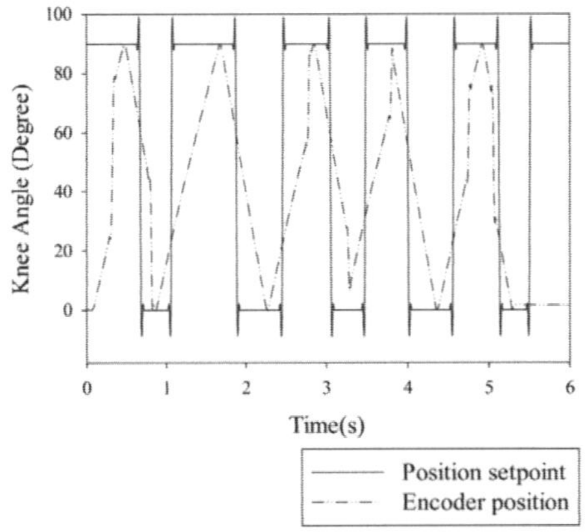

b)

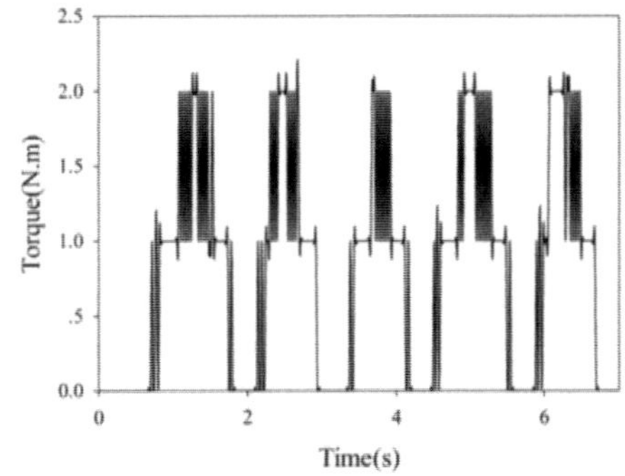

Figure 4. a) Desired position and actual position of the calf link at the velocity 18 degrees/sec. b) Torque at the knee joint in the experiment at the velocity 18 degrees/seconds.

a)

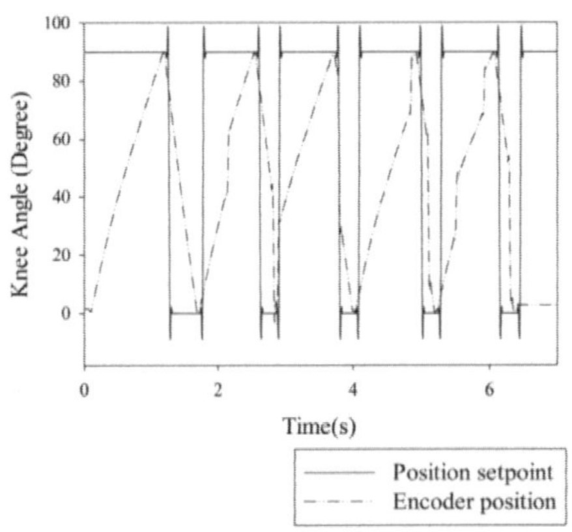

b)

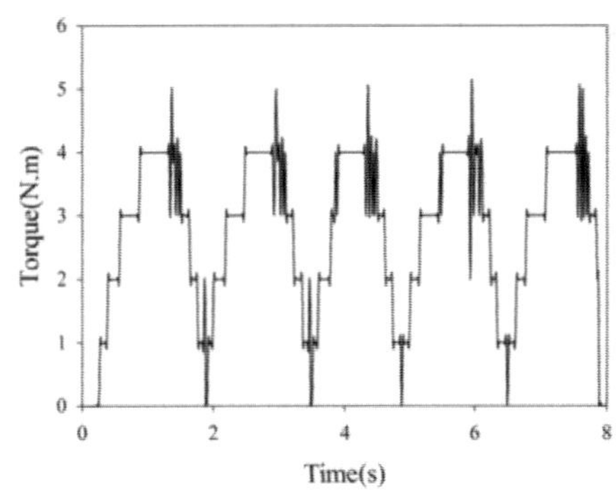

Figure 5. a) Desire position and actual position of the calf link with a 1 kg mass at the velocity 18 degrees/sec. b) Torque at the knee joint in the experiment at the velocity 18 degrees/seconds.

5. Conclusion

In this paper, we presented knee rehabilitation equipment for patients after total knee replacement arthroplasty in sitting position. The position applied by the motor on the patient's knee was controlled in order to assist patients for their knee exercise as well as prevent themselves from injuries. The result show that the link can move to the desire position. The motor torque was monitored in order to confirm device safety while using this equipment. In the future, the measured torque will be used to model a torque feedback system to help assist the patients when they have difficulty moving the knee at the beginning session of the therapy. The level of exercise can be challenged up to 40 degrees/sec for knee recovery and up to 60 degrees/sec to strengthen the muscles. In future work, the impedance control will be imposed the position of the system and control the motor torque when function together with the torque exerted by user during rehabilitation. With this knee rehabilitation equipment, the system helps recovery the range of motion for knee joint while the patients can use the device for exercise at home to decreases the number of hospital visit and receive an appropriate therapy.

Acknowledgments

Research is financially supported by the Higher Education Research Promotion and National Research University Project of Thailand, King Mongkut's University of Technology Thonburi and National Research Council of Thailand.

References

[1] M.G. Benedetti, D.Sarti, S. Bonfiglioli Stagni, and E. Mariani, "Setting, Clinical Pathways, Fast-Track and Rehabilitation Following Primary Knee Arthroplasty: A Literature Review," *The Open Rehabilitation Journalpplied,* vol. 8, pp. 17-24, 2015.
[2] M. Cenciarini and A. M. Dollar, "Biomechanical Considerations in the Design of Lower Limb Exoskeletons," in *2011 IEEE International Conference on Rehabilitation Robotics*, ETH Zurich Science City, Switzerland, 2011, p. 6.
[3] Y. Pei, Y. Kim, G. Obinata, E. Genda, and D. Stefanov, "Robot-Aided Motion Planning for Knee Joint Rehabilitation with Two Robot-Manipulators," presented at the 35th Annual International Conference of the IEEE EMBS, Osaka, Japan., 2013.
[4] K. Kim, M. Kang, and Y. Choi., "Conceptualization of an exoskeleton Continuous Passive Motion(CPM) device using a link structure," presented at the IEEE International Conference on Rehabilitation Robotics 2011, Zurich, Switzerland, 2011.
[5] Erhan Akdog˘an and M. A. Adli, "The design and control of a therapeutic exercise robot for lower limb rehabilitation:Physiotherabot," *Mechatronics,* vol. 21, p. 14, 12 February 2011.
[6] Jason Nikitczuk, Brian Weinberg, Paul K. Canavan, and C. Mavroidis., "Active Knee Rehabilitation Orthotic Device With Variable Damping Characteristics Implemented via an Electrorheological Fluid," *IEEE/ASME TRANSACTIONS ON MECHATRONICS,* vol. 15, DECEMBER 2010.
[7] S. Mefoued, S. Mohammed, and Y. Amirat, "Knee Joint Movement Assistance Through Robust Control of an Actuated Orthosis," presented at the IEEE/RSJ International Conference on Intelligent Robots and Systems., San Francisco, CA, USA, 2011.
[8] H. Yu, M. STA Cruz, G. Chen, S. Huang, C.Zhu, E.Chew, *et al.*, "Mechanical Design of a Portable Knee-Ankle-Foot Robot," presented at the 2013 IEEE International Conference on Robotics and Automation (ICRA), Karlsruhe, Germany, 2013.

Research on the Sensor of Welding Seam Tracking Based on the Same Plane Electrode

Zhendong Shang, Weiguang Wang

School of Mechatronics Engineering, Henan University of Science and Technology, Luoyang, Henan, 471003, China

Email:806122137@qq.com

Abstract. Capacitive sensors have a wide range of applications, but its edge effect is severely restricted by the detection accuracy. In this paper, we put forward a kind of capacitance sensor based on the edge of capacitance. It has a wide range of static and dynamic measurement (mm) and high measurement accuracy. And its mathematical modeling, through the analysis proves that the model is correct and feasible.At the same time also used in the finite element analysis software ANSYS, three-dimensional model of the sensor was established, and the optimization and analysis of the sensor's output characteristic of sensor specific structural parameters, the final determination of the optimal structure parameters of the capacitive sensor. Finally, the experimental results show that the output signal of the sensor is good.

1. Introduction

The urgent problem of modern manufacturing industry is how to improve the production efficiency [1]. As is known to all, the traditional manual welding, not only the working environment of workers actively hostile, and welding of the quality of the products also can't meet the requirements, so the welding automation is the world industry is presented, and the necessary requirement of welding [2].Therefore, this paper presents an edge effect type capacitive seam tracking sensor to realize the automatic tracking of the weld seam. It has the advantages of simple structure, stable performance, strong anti-interference, groove to a wide range, and can adapt to different types of groove tracking, only need a certain structure changes of the sensor according to different forms of groove.

2. The establishment of mathematical models

The space area occupied by the two electrode of the same plane capacitance sensor is Ω_1 and Ω_2.The space occupied by the detected object is Ω_2. The boundary surface of the space Ω_1, Ω_2 and Ω were $\partial\Omega_1$, $\partial\Omega_2$ and $\partial\Omega$. The derivation of the mathematical model of the same plane capacitance sensor is introduced in the following details:

$$\begin{cases} \nabla_2 V = 0,(x, y, z) \in \Omega \cup \left(R_3 - \overline{\Omega}_1 - \overline{\Omega}_2 - \overline{\Omega}\right) \\ V|\Omega_1 = V_1, V|\Omega_2 = V_2 \\ \lim\limits_{r=\sqrt{x^2+y^2+z^2}\to\infty} V = 0 \\ V_{i|\partial\Omega_1} = V_{0|\partial\Omega_1} \\ d\left.\dfrac{\partial V_1}{\partial n}\right|_{\partial\Omega} = \tilde{d}\left.\dfrac{\partial V_0}{\partial n}\right|_{\partial\Omega} \quad \Omega \end{cases} \tag{1}$$

For the same plane capacitance sensor model, the spatial region occupied by the two electrode is:

$$D = \left\{ (x, y, z) \left| \begin{matrix} -b/2 \le x \le b/2 \\ -k/2 \le y \le k/2 \\ 0 \le z \le c \end{matrix} \right. \right\} \tag{2}$$

$$D_2 = \left\{ (x, y, z) \left| \begin{matrix} -b/2 \le x \le b/2 \\ -k/2 \le y \le k/2 \\ 0 \le z \le c \end{matrix} \right. \right\} \tag{3}$$

The space occupied by the measured work-piece is in the range of:

$$D = \left\{ (x, y, z) \left| \begin{matrix} -m/2 \le x \le m/2 \\ -1/2 \le y \le 1/2 \\ c + d \le z \le c + d + n \end{matrix} \right. \right\} \tag{4}$$

Through the above for both electrodes of the capacitive sensor and the work-piece in the boundaries of space constraints, The steps to obtain the relationship between the capacitance of the same plane capacitance sensor and the measured work-piece are as follows: The two electrodes of the sensor and the work-piece are D, D_1 and D_2, When the electrode D_1 and D_2 on the potential to determine the V_1 and V_2, It can be assumed that the distance between the sensor and the work-piece is d. By the following differential equation:

$$\begin{cases} \nabla_2 = 0,(x, y, z) \in D \cup \left(R_3 - \overline{D}_1 - \overline{D}_2 - \overline{D}\right) \\ \lim\limits_{r=\sqrt{x^2+y^2+z^2}\to\infty} V = 0 \\ V|_{\partial D_{in}} = V|_{\partial D_{out}} \\ V|_{D_1} = V_1, V|_{D_1} = V_2 \\ d\left.\dfrac{\partial V_1}{\partial n}\right|_{\partial D_{in}} = \tilde{d}\left.\dfrac{\partial V_0}{\partial n}\right|_{\partial D_{out}} \end{cases} \tag{5}$$

The potential distribution function of the space in the working range of the capacitance sensor can be determined as V (x, y, z).

The electric field intensity distribution is obtained by the gradient of the potential distribution function V (x, y, z):

$$\bar{E} = -grad V \tag{6}$$

Using the change of the spatial electric field intensity, the relationship between the sensing capacitance value C and the distance D of the work-piece can be obtained from the same plane capacitance sensor:

$$c = \frac{E 4\pi \varepsilon s}{d(V_1 - V_2)} \tag{7}$$

In the formula, V_1 and V_2 two electrode potential, S under the working area of the capacitive sensor, ε as a dielectric constant interval; E Electric field strength for electrode excitation. By using the above formula, we can see that when the external factors interfere with the welding process, when the potential difference between the two electrodes is a fixed value, the dielectric constant and electric field intensity are the excitation set to a fixed value. So the distance between the work-piece and the capacitance value is a nonlinear change. The traditional capacitance calculation formula is similar, which further shows the correctness of the model.

3. Modeling and Simulation of the same plane capacitance sensor based on Analysis

According to the mathematical model of the same plane electrode, the mathematical model of the capacitance sensor can be used to deal with the fixed parameters only when the simulation is used [3-4], Some factors which influence the electric field distribution and the electric field size of the sensor are also neglected, Interference factors consider only the quantitative parameters of the sensor itself, welding position of welding gun and welding process in the possible interference [5]. Finite element analysis method is used to simulate the interaction of the electrode with the same plane capacitance sensor.3D Analysis analysis model was established [6], as shown in Figure 1.

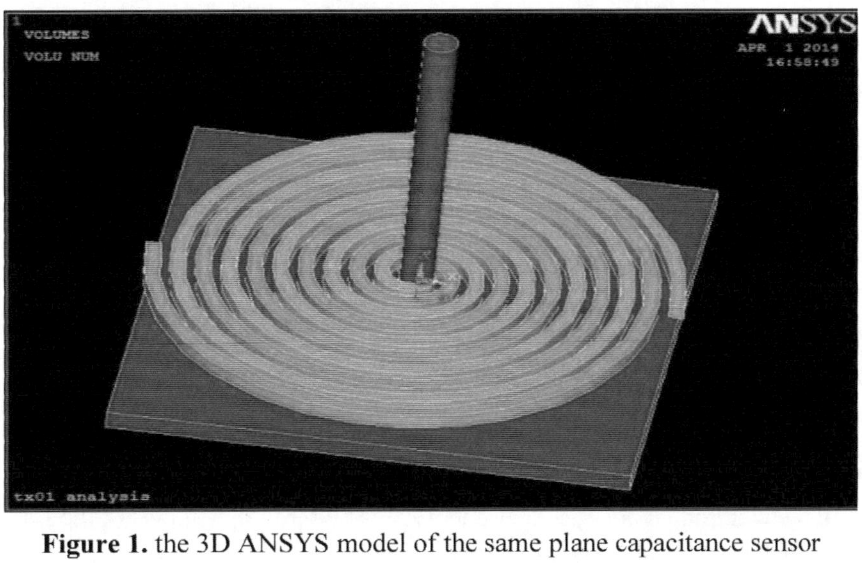

Figure 1. the 3D ANSYS model of the same plane capacitance sensor

In the process of ANSYS modeling, considering the welding environment of the sensor, the whole analysis model is set up with the air bag, which is 4 times the size of the whole analysis area [7] Unit type selection of three dimensional solid element solid122, the degree of freedom is the electric potential. The material attribute is defined as a silicon dioxide material, and the electrode is a copper electrode. The mesh is divided into free grid, and the results are shown in Figure 2.After loading, solving and post-processing results.

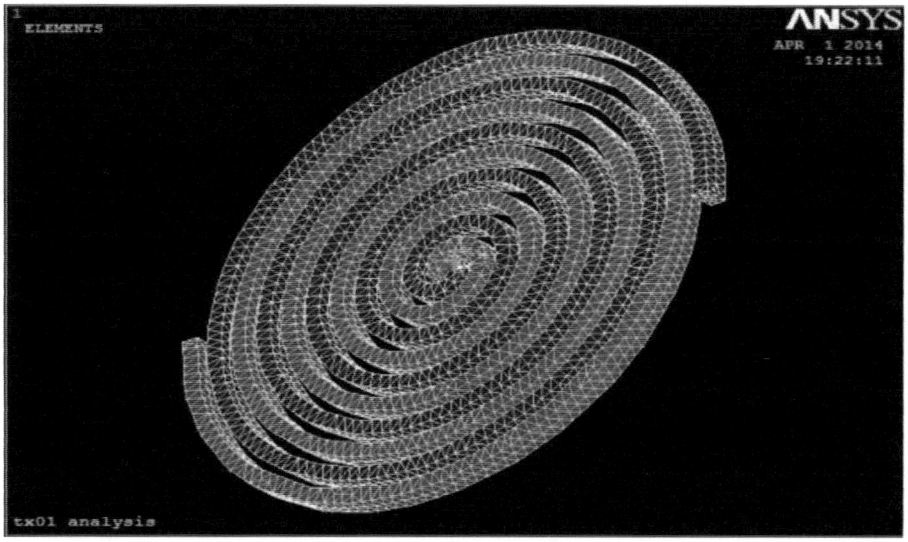

Figure 2. 3D ANSYS mesh of the same plane capacitance sensor

4 . Simulation results and Analysis

4.1. The effect of electrode parameters on the capacitance of the sensor

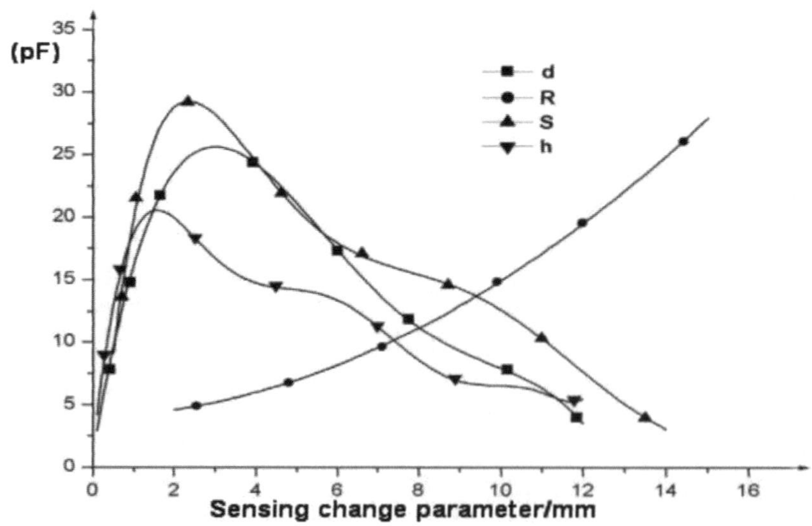

Figure 3. Relationship between the main parameters of the electrode and the capacitance value

In the simulation, the relationship between the electrode thickness D, the electrode radius R, the distance between the electrode h and the electrode width s and the capacitance of the sensor is mainly studied. When the measurement is assumed, the distance between the work-piece and the sensor is

5mm, and the size of the measured object is 50 x 50 x 5mm metal sheet. At the same time, the simulation of a single variable in the study of a single variable Where D, R, h, s and the simulation of the simulation of the relationship between the capacitance as shown in Figure3.

From the Figure can be concluded that D, h, s have a maximum value, more than the maximum value of the sensor capacitance will become smaller. And the electrode radius R will be as the radius of the sensor capacitance becomes larger, taking into account the assembly and practicality, and other factors, the choice of a radius of 10mm as the standard.

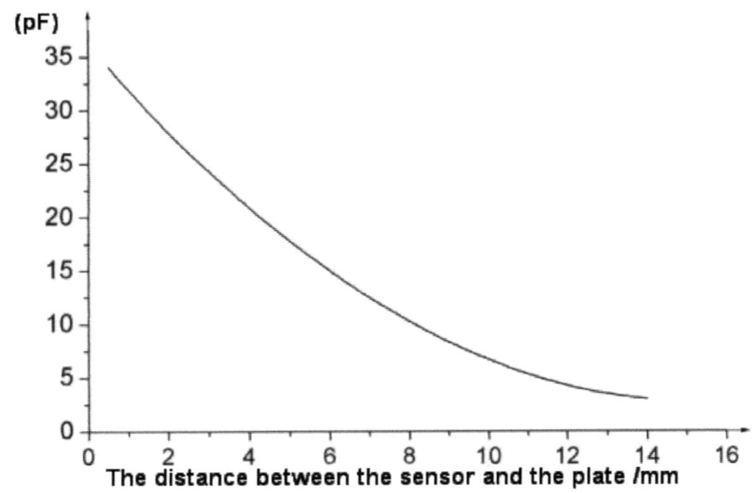

Figure 4. characteristic of capacitance sensor

4.2 . Research on sensing properties

In the study of the effect of the distance between the electrode and the work-piece on the capacitance of the sensor, the size of the electrode is selected as the optimum size of the electrode, and the distance from the 2mm to the 14mm is changed from the distance between the work-piece and the work-piece. As shown in Figure 4, the output capacitance value and the variation law of the distance between the work-piece is approximately proportional to the inverse proportion, and the formula (7) is consistent with the change trend.

5. Experimental verification and Analysis

Based on Analysis finite element simulation software for electrode wound with each structure design of planar capacitance sensors and analyzing the influential factors of the output results. A simple sensor probe is made in this experiment, Specific electrode structure parameters with reference to the conclusions drawn from the above simulation: electrode thickness 2.5mm, electrode radius 10mm, the distance between the electrode and the 2mm, the electrode width 1.5mm.Then by using the waveform generator input frequency 50Hz sine wave oscillation excitation, through the power amplification circuit input to two sensor electrodes and two electrodes produce changes in the capacitance of the output by capacitance voltage conversion circuit through digital oscilloscope view. The sensor itself is not fixed together with the torch in front of the torch 1cm and 4Hz frequency swing. The sensor overall probe and welding of steel plate work-piece for V groove welding, torch and work-piece weld distance 7mm, sensor and the work-piece weld distance of 1cm, plate 5mm thick. By constantly adjusting the welding torch with respect to the weld position, check the voltage waveform at the output of the digital oscilloscope, specific experimental data as shown in Figure 5.

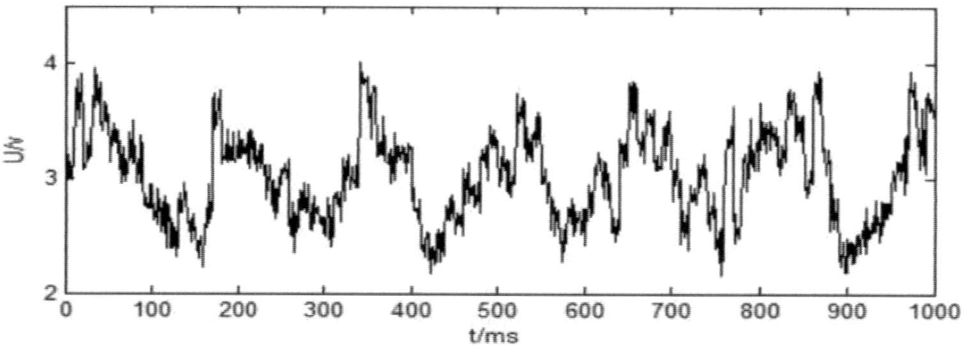

Figure 5. the output curve of the capacitive seam tracking sensor

By the diagram shows the capacitance sensor each swing scanning through the weld, the output voltage waveform presents certain regularity, by waveform can clearly analyze sensor at the moment the position, so you can see changes in capacitance sensor to sense the position of welding seam in good condition, can meet the requirements of welding conditions are required for the signal.

6. Conclusion

- A new type of electrode cross wound with the same plane capacitance welding seam tracking sensor based on edge effect is presented.
- Analysis finite element analysis software was used to build the three - dimensional model of the sensor and the sensor electrode structure of the various parameters were optimized.
- The output characteristic of the sensor is analyzed and the structure of the capacitive sensor suitable for seam tracking is determined. Finally, the conclusion is validated by experiments.

Reference

[1] Jiang, Huanzhong.Welding method and equipment (First Edition) [M]. Beijing: Mechanical Industry Press, 1981

[2] Huang, Shisheng etc.Research and Prospect of welding seam tracking technology [J]. electric welding machine,1995（5）:1-5.

[3] Puers. Robea. Capacitive sensors:when and how to use them, Sensors and Actuators, 1993:93-105.

[4] Hong, Bo. Mathematical modeling and Simulation of arc sensor moving welding robot [D]. Xiangtan: Doctoral Dissertation of Xiangtan University, 2005

[5] Yan, Jun-Guang.Based on [D]. detection and processing of weld seam tracking signal of capacitance sensor for Hu-nan: Master Dissertation of Xiang-tan University, 2013

[6] Dong,Yonggui;Xiang,Li. The electrode structure and sensing properties of the same surface scattered field capacitance sensor [J]. Journal of Qing-hua University (natural science edition), 2004 44(11):9-37.

[7] Noltingk B E, A Novel Proximity Gauge, Journal of Physics E: Scientific Instruments, 1969,2(2):356-360.

Design of a Micro Cable Tunnel Inspection Robot

Wei Song[1], Lei Liu[2], Xiaolong Zhou[1], Chengjiang Wang[3]

[1]State Grid Chongqing Electric Power CO. Electric Power Research Institute, Chongqing;
[2]Harbin Institute of Technology Shenzhen Graduate School, Shenzhen;
[3]Shenzhen Institute of Advanced Smart Technology, Shenzhen.

Abstract: As the ventilation system in cable tunnel is not perfect and the environment is closed, it is easy to accumulate toxic and harmful gas. It is a serious threat to the life safety of inspection staff. Therefore, a micro cable tunnel inspection robot is designed. The whole design plan mainly includes two parts: mechanical structure design and control system design. According to the functional requirements of the tunnel inspection robot, a wheel arm structure with crawler type is proposed. Some sensors are used to collect temperature, gas and image and transmit the information to the host computer in real time. The result shows the robot with crawler wheel arm structure has the advantages of small volume, quick action and high performance-price ratio. Besides, it has high obstacle crossing and avoidance ability and can adapt to a variety of complex cable tunnel environment.

1. Introduction

With the rapid development of robot technology, the application of the robot in the power industry is more and more common. The inspection robot for cable tunnel will greatly save manpower, material and financial resources. More importantly, it can greatly reduce the damage to the inspection personnel in the harsh environment of the tunnel.

In foreign countries, the United States was the first to develop a cable tunnel inspection robot. Bing Jiang and Alanson P. Sample, University of Washington, developed a set of underground cable inspection robot named Rangers in 2005[1]. The robot is equipped with two 12V DC gear motors in the ends and designed with a single-platform multi-section modular. So it has a stable steering capability. In addition, a multi segmented structure is designed to allow the robot to carry more sensors. The robot platform is divided into two sections, which are used for the control of the robot and the data processing of the sensor. The robot, 1.2 meters in length, can walk along the cable line of the diameter of 4~8 cm, cross the obstacles encountered along the way and enter the ground to crawl along the cable to identify the location of the fault.

In China, due to more and more attention to the research and development of robot, some research institutions have made some achievements in the tunnel robot. The articulated tracked robot for cable duct was developed by Shanghai Jiao Tong University. The robot is attached to an infrared camera with PTZ and a infrared temperature sensor, in addition to various types of gas sensors. Through collecting and processing the data of these devices, the robot can understand toxic gas concentration of the tunnel in real time and test whether the cable is damaged[2][3].

In 2012, a cable tunnel patrol robot was developed by Zhejiang University. The robot has the functions of autonomous inspection, emergency handling and remote monitoring, which could sustainable work 4 hours. In addition, it is equipped with infrared cameras and temperature, gas and other sensors. It can monitor the deformation, temperature and harmful gas in the tunnel and make the case of the tunnel upload to the monitoring center in real time. Therefore, the robot can replace manual to inspect independently, effectively reduce the casualties, and improve the efficiency of cable inspection.

2. Overview of tunnel inspection robot

Based on the analysis on the characteristics of the tunnel environment and combined with the existing technical conditions, a cable tunnel inspection robot is designed, as shown in figure 1. The features and main functions are as follows:

(1) The robot body platform is 26cm length, 27cm width, 17cm height, and about 5 kg weight, which can easily enter narrow occasions.
(2) The robot can move flexibly and has strong climbing and obstacle avoidance ability. It can carry out reconnaissance and patrol in the high temperature, toxic, flammable and other threats to detection personnel security occasions.
(3) It has a broad visual angle.
(4) It can collect temperature and harmful gas concentration, measure distance through ultrasonic wave, and send the information back to the console in real time.

Figure 1. Cable tunnel inspection robot

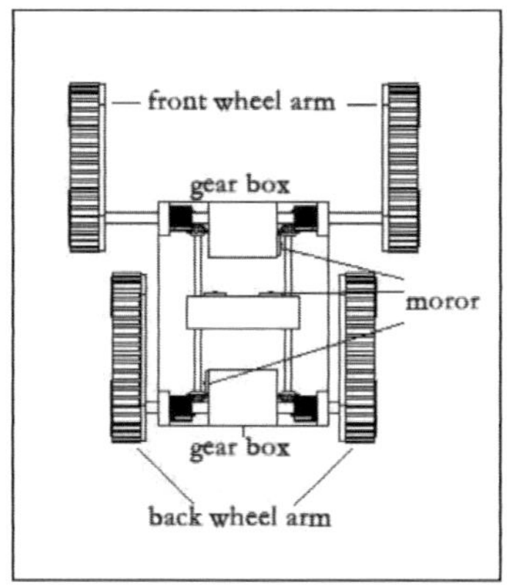

Figure 2. Structure of cable tunnel inspection robot

3. Mechanical structure design

3.1 Motion structure of the body

The robot body adopts the track wheel arm structure. The crawler type wheel arm structure is a new way to combine the wheel motion mode, walking mode and track mode. The robot with wheel arm type structure has the advantages of both wheel, walking and the crawler robot, which can achieve climbing stair, crossing ditch, obstacle, steering and other functions. So it has strong maneuverability in complex environment.

The wheel arm adopts structural control. Four wheel arms were divided into two groups. The first two wheel arms were the first group and the later two arms were the second groups. Each arm movements are consistent, and the two groups do not interfere with each other. Four tracks are divided into two groups. The left two are the first group, the two on the right are the second group. Each set of tracks is consistent, and the two groups of tracks do not interfere with each other. This control method makes the robot more stable and flexible. The wheel arm is driven by a servo motor with a speed reducer, which is convenient to measure and control, and it is advantageous to drive and control in higher precision.

The application of track increased the ability of robot obstacle. The robot can walk over a small volume of obstacles. Crossing large obstacles can be achieved by controlling the operation of personnel. In addition, the track can increase friction, so that the robot has a greater advance power and thus adapts to the more complex environment [4]-[10]. The structure design of the cable tunnel inspection robot is shown in figure 2.

3.2 Vehicle body gait analysis

(1) Up the stairs process:

The motion process of robot upstairs is as shown in figure 3. It shows that the maximum height of climbing stair mainly depends on the difference between the wheel arm and the big radius. The robot is designed with the wheel arm length of 200mm and the big wheel arm radius of 30mm. The maximum stair climbing height is 170mm. The actual height of the stairs is basically 160mm, so the robot have the ability to cross the floor crossing.

(2) Down the stairs process:

The motion process of robot down stairs is as shown in figure 4.

(3) Crossing ditch process:

The process of robot crossing groove is as shown in figure 5. The width of crossing groove of the robot depends on the position of its gravity center and the overall length. Robot body is 350mm length and the total length is 600mm after unfolding front and back wheel arm. The distance between the center of gravity and the front of the wheel arm is 250mm. In normal driving conditions, the maximum crossing channel width is 240mm. In the case of a larger initial speed, the robot can across a wider trench. In addition, by adjusting the expansion angle of the front wheel arm, making the robot be in a certain angle with the horizontal direction so that the robot can cross the ditch with a certain gap with the aid of inertia in case of a larger initial velocity[11][12][13].

(4) Turning process:

The robot turning process is as shown in figure 6.

1) Turn left: close the left drive (the left track does not move), open the right drive (the right track moves forward).
2) Turn right: turn off right drive (the right track does not move), open the left drive (the left track moves forward).
3) Turn left in place: drive the left track forward and drive the right track reverse.
4) Turn right in place: drive the right track forward and drive the left track reverse.

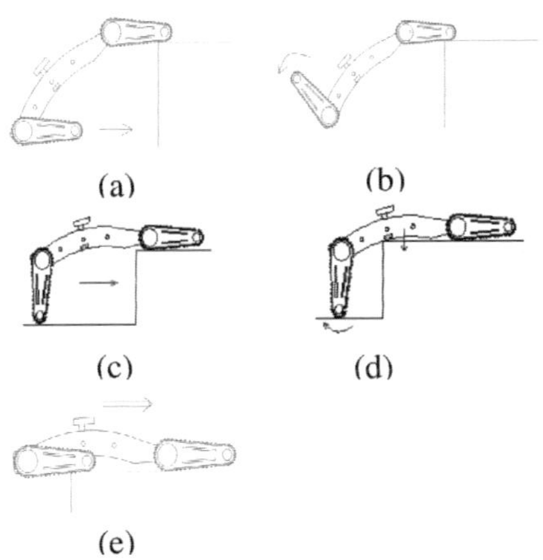

Figure 3. Motion process of robot on stairs

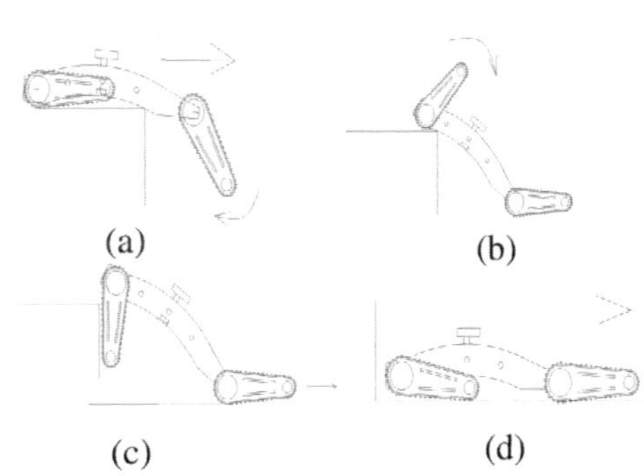

Figure 4. Motion process of robot down stairs

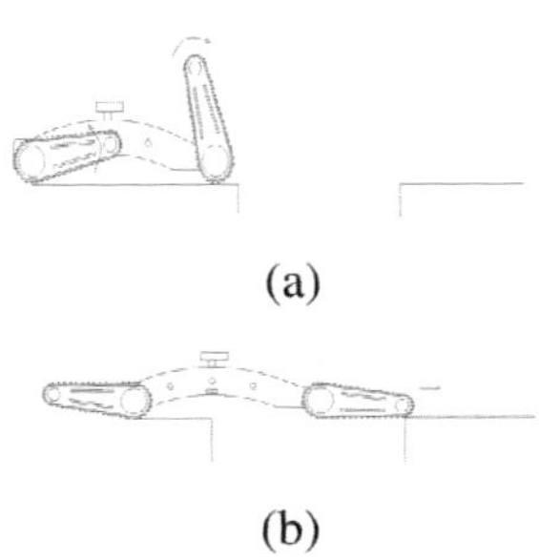

Figure 5. The process of robot crossing groove

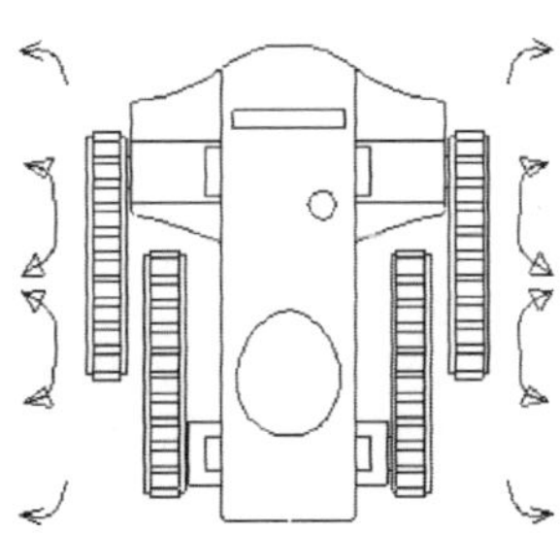

Figure 6. Robot turning process

3.3 Calculate the obstacle height

The crossing obstacle process of crawler wheel arm structure robot is as shown in figure 7.

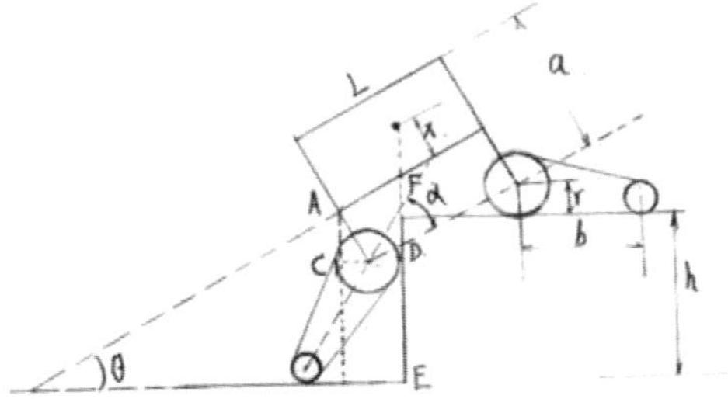

Figure 7. Schematic diagram of the obstacle

L-body length, a-leg length, b-center distance between two track tension wheel, h-barrier height, r-tension wheel radius, x-distance between the body's center of gravity and the chassis, α-Track maximum deflection angle, θ-robot body inclination angle.

Analyze the length relation of each part in the diagram and get:

$$L\sin\theta + a\cos\theta + b/2 * \sin(\alpha+\theta) + r = h + r + a\cos\theta \tag{1}$$

That is $h = L * \sin\theta + b/2 * \sin(\alpha+\theta)$

Because $CD = a * \sin\theta + b/2 * \cos(\alpha+\theta) + r$ and $CD = L/2 * \cos\theta - x * \sin\theta$, merge two formula above and get:

$$L/2 * \cos\theta - (a+x)\sin\theta - b/2\cos(\alpha+\theta) - r = 0 \tag{2}$$

In addition, the bottom surface of the robot platform can not be touched with the steps of the endpoint (h<EF), that is:

$$h < b/2 * \sin(\alpha+\theta) + a * \cos\theta + r + (L/2 - x * \tan\theta) * \sin\theta \tag{3}$$

In fact, when the track is full contacted with the step plane, just make α+θ=90° and the situation of α+θ> 90° does not occur. So the range of α+θ is 0°-90°, and take 90° when it is larger than 90°. In general, in order to make the robot cross obstacle better and more stably, the track and the step plane need to complete contact as far as possible. Generally the value of α is greater than 60 degrees and θ is also greater than 30 degrees when meeting the obstacle limits, which would make α+θ> 90°. So according to the above principles, take α+θ= 90° in the above three types.

In this way, the formulas (1), (2), (3) can be written as:

$$\begin{cases} h = L\sin\theta + b/2 \\ L/2 * \cos\theta - (a+x) * \sin\theta - r = 0 \\ h < b/2 + a\cos\theta + r + (L/2 + x\tan\theta) * \sin\theta \end{cases} \tag{4}$$

When calculating the maximum crossing obstacle height of the robot, it is also necessary to ensure that the bottom surface of the robot platform can not be touched with the step end points (i.e. no floor interference) when the step height is h. Suppose the body inclination angle is θ′, there is

$$\begin{cases} L\sin\theta' = h \\ a/\cos\theta' + r > h \end{cases} \tag{5}$$

After eliminating θ′, the inequality can be obtained:

$$a/\sqrt{1-h^2/L^2} + r > h \tag{6}$$

The maximum crossing obstacle height of the robot can be obtained by formula (4) and (5). θ, L, a, b and rare independent variables. The method for solving is as follows.

Objective function: $h = -(L\sin\theta + b/2)$

Equality constraints: $L/2 * \cos\theta - (a+x) * \sin\theta - r = 0$

Inequality constraints:

$$-b + a\cos\theta + r - (L/2 - x * \tan\theta) * \sin\theta < 0 \qquad (6)$$

$$-L * \sin\theta - b/2 - a/\sqrt{1 - h^2/L^2} - r < 0 \qquad (7)$$

The structure size of the track wheel arm structure robot is as follows: L=350mm, a=50mm, b=170mm, r=30mm, track maximum deflection angle α= 75°, the distance between the body's center of gravity and the chassis is 100mm.

From the above constraints the optimal solution can be obtained for 9 =0.49 rad, h=160mm. So, the maximum obstacle height of crawler wheel arm robot is 160mm.

4. Design of control system

The robot control system consists of three single-chip processing terminal, defined as MCU1, MCU2 and MCU3. MCU1 uses the SPI standard serial port to connect to the PTR8000 full duplex wireless communication system so that it can communicate with the host computer for data exchange; MCU2 is mainly to control the four way power motor and three dimensional rotation platform; MCU3, connected with temperature, ultrasonic measurement distance and gas detection sensors, converts the analog signal into digital signals and uploads to MCU1; MCU3 controls the image acquisition and processing system and its signal is uploaded to the host computer by 2.4G microwave transmitter. The block diagram of the control system of the robot is shown in figure 8.

In order to obtain the environmental conditions in the cable tunnel, the robot is equipped with a variety of sensors.

(1) Temperature sensor

The robot can quickly and accurately measures the temperature of the surface of the robot by the non-contact infrared temperature measurement technology, which can be used for the measurement of hot, dangerous or difficult to contact and other objects.

(2) Attitude sensor

The robot uses Holzer sensor to monitor the rotation angle of the wheel arm. Put micro cylindrical magnets magnetized on the first stage wheel of gear box and detect the rotation number of the gear by Hall sensor. MCU records the turning number of driven wheel and converts it into wheel arm angle information so that it can achieve precise control.

(3) Gas sensor

Gas sensor selects MQ-7 sensor. The detector has high sensitivity to carbon monoxide and methane, and has a long service life and high stability.

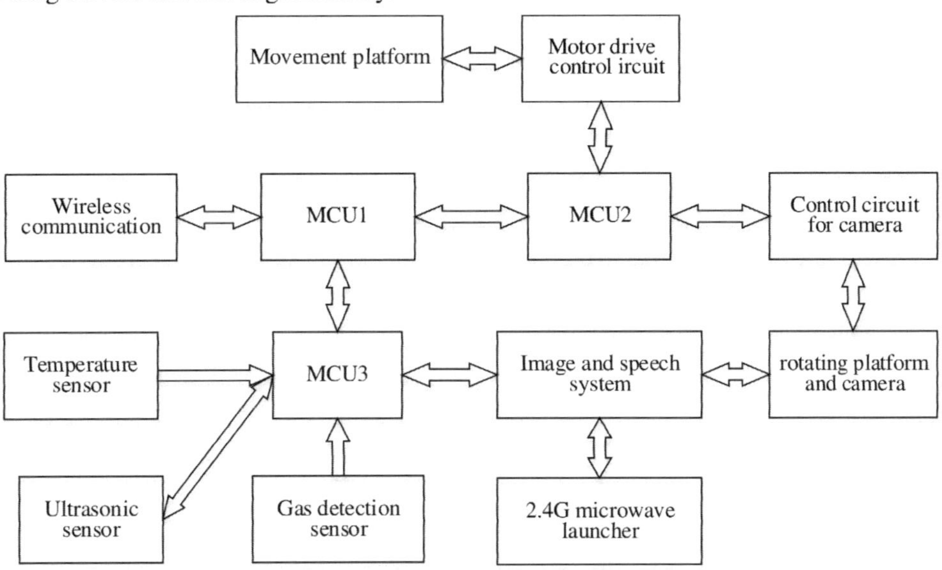

Figure 8. Robot terminal control system

(4) Camera

The wireless infrared camera uses 2.4G wireless transmitter ultra micro CCD chip, which has the advantages of signal stability and image anti interference. In addition, use the UTV330+ TV box to receive the image signal transmitted from the wireless camera.

(5) Motor drive circuit

The rotation and rotation direction of the DC motor is controlled by the driving circuit. Two driving schemes designed are as follows.

Scheme one:

Control four DC motor using H bridge, so each tube has a large tube pressure drop and the power consumption is large in each tube. Besides, when the PWM (pulse width modulation) there is a switching loss, so we are not able to adopt this plan.

Scheme two:

Use high power transistor to drive the DC motor, use the relay to control the rotation of motor. The scheme has the advantages of small power consumption, high efficiency and simple and practical design. This scheme is as shown in figure 9.

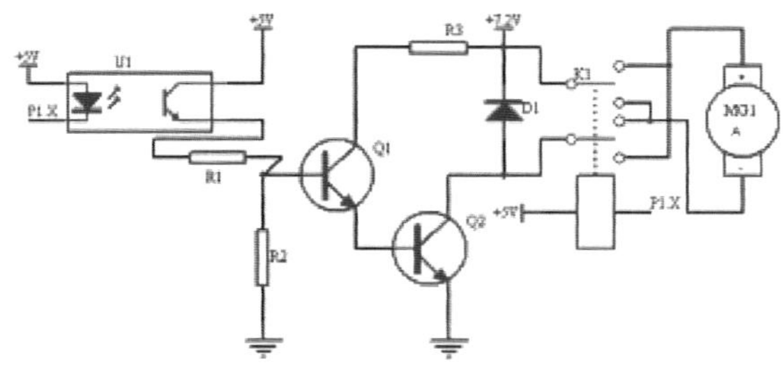

Figure 9. DC motor drive mode

(6) Steering gear control

The steering gear is composed of a DC motor, a reduction gear set, a sensor and a control circuit, which is used for controlling the angle and position. Control the rotation of the motor and output the torque by sending a pulse signal. At the same time, potentiometer on output shaft detects the rotation angle of the output shaft and converts it into electrical signal. The electric signal is compared with the electric signal of the input shaft angle, which can be used to control the DC motor to rotate feedback. Servo drive module uses high power MOS tube as the power element. The driving module is designed for DC motor speed control by PWM mode, which is suitable for small DC motor.

5. Conclusion

In this paper, based on the practical application and combined with the existing technical conditions, a design scheme of a micro tunnel inspection robot is proposed. Then introduce the main function, mechanical structure and control system of the robot from the whole angle. From the mechanical structure, the robot crawler wheel arm structure enriches the robot movement capability, which is more suitable for the complex terrain such as cable tunnel; From the control circuit, the circuit has the advantages of simple structure, strong anti-interference ability, stable operation of the circuit system and so on. The experimental results show that the crawler wheel arm robot not only is mobile, flexible and cost-effective, but also realizes the function of remote control robot and wireless data transmission.

6. References

[1] Bing Jiang; Sample, A.P.; Wistort, R.M.; Mamishev, A.V., "Autonomous robotic monitoring of underground cable systems," in Advanced Robotics, 2005. ICAR '05. Proceedings., 12th International Conference on , vol., no., pp.673-679, 18-20 July 2005

[2] Jiang Yun, Fu Zhang. Design of a Small Cable Tunnel Detection Robot. 2009.East China Electric Power ,01:95-97

[3] Lin Yongfeng. Research on Computer Integrated monitoring System for Power Cable Tunnel. Shanghai Jiao Tong University,2007

[4] Fu Ruhui. Research on a small track robot for tunnel detection. *Equipment Manufacturing Technology*, 2010, 03:21-22.

[5] Songyi Dian; Tao Liu; Yan Liang; Mengyu Liang; Wei Zhen, "A novel shrimp rover-based mobile robot for monitoring tunnel power cables," in Mechatronics and Automation (ICMA), 2011 International Conference on , vol., no., pp.887-892, 7-10 Aug. 2011

[6] Ma Jinmeng, Li Xiaofan, Yao Chen, Wang Zhong. Modeling and analysis of obstacle dynamics for ground mobile robot. *The Robot*, 2008, 03:273-278.

[7] Bin Li; Shugen Ma; Jinguo Liu; Yuechao Wang, "Development of a shape shifting robot for search and rescue," in Safety, Security and Rescue Robotics, Workshop, 2005 IEEE International , vol., no., pp.31-35, 6-9 June 2005.

[8] Rao Wei, Wang Jianzhong, Shi Jiadong. Analysis of obstacle performance of articulated tracked robot. *Machinery Design & Manufacture*, 2014, 06:39-42.

[9] Zheng Chao, Zhao Yanzheng, Fu Zhuang. Structure design and implementation of a small crawler robot. *Mechatronics*, 2007, 04:70-72.

[10] Ohno, K.; Morimura, S.; Tadokoro, S.; Koyanagi, E.; Yoshida, T., "Semi-autonomous control system of rescue crawler robot having flippers for getting Over unknown-Steps," in Intelligent Robots and Systems, 2007. IROS 2007. IEEE/RSJ International Conference on , vol., no., pp.3012-3018, Oct. 29 2007-Nov. 2 2007

[11] Li Yunwang, Ge Shirong, Zhu Hua, Liu Jian. The mechanism and ability of obstacle crossing for four track double swing arm robot . *The Robot*, 2010, 02:157-165.

[12] Chen Changzheng, Xiang Hongwei, Yang Kongshuo, Ye Changlong. Dynamic analysis of a deformable tracked robot over a step. *Journal of Shenyang University of Technology*, 2015, 02: 165-170.

[13] Liu J G, Wang Y C, Ma S G, et al. Analysis of stairs-climbing ability for a tracked reconfigurable modular robot. IEEE International Workshop on Safety, Security and Rescue Robotics. Piscataway, NJ, USA: IEEE, 2005: 36-4

IOP Conf. Series: Materials Science and Engineering **157** (2016) 012033 doi:10.1088/1757-899X/157/1/012033

Parametric and Generative Design Techniques for Digitalization in Building Industry: the Case Study of Glued-Laminated-Timber Industry

G Pasetti Monizza[1], D T Matt[1,2], C Benedetti[2]

[1] Faculty of Science and Technology, Free University of Bozen
[2] Fraunhofer Italia

Gabriele.Monizza@natec.unibz.it

Abstract. According to Wortmann classification, the Building Industry (BI) can be defined as engineer-to-order (ETO) industry: the engineering-process starts only when an order is acquired. This definition implies that every final product (building) is almost unique, and processes cannot be easily standardized or automated. Because of this, BI is one of the less efficient industries today, mostly leaded by craftsmanship. In the last years, several improvements in process efficiency have been made focusing on manufacturing and installation processes only. In order to improve the efficiency of design and engineering processes as well, the scientific community agrees that the most fruitful strategy should be Front-End Design (FED). Nevertheless, effective techniques and tools are missing. This paper discusses outcomes of a research activity that aims at highlighting whether Parametric and Generative Design techniques allow reducing wastes of resources and improving the overall efficiency of the BI, by pushing the Digitalization of design and engineering processes of products. Focusing on the Glued-Laminated-Timber industry, authors will show how Parametric and Generative Design techniques can be introduced in a standard supply-chain system, highlighting potentials and criticism on the supply-chain system as a whole.

1. Introduction

According to ISTAT surveys [1], Building Industry (BI) is one of the less efficient industries in Italy and the productivity gap with other industries is growing faster. This gap is due to two main reasons. Firstly, BI can be defined as an engineer-to-order (ETO) industry following Wortmann classification [2]. Secondly, BI identifies different responsible figures for each process usually structured in a serial workflow. Serial workflow reduces the efficiency of information management by enhancing the propagation of errors because of fragmented information flow among the responsible figures.

Today, the challenge for BI is to reduce the productivity gap with other industries by reducing the wastes of resources and improving the overall efficiency of the supply-chain system [3] [4]. Improving the overall efficiency in BI may generate huge benefits: reduce social and economic costs, reduce the environmental footprint and enhance the quality of final products (buildings).

1.1. Parametric and generative design techniques

According to Jabi´s definition [5], Parametric Design is "the processes based on algorithmic thinking that enables the expression of parameters and rules that, together, define, encode and clarify the relationship between design intent and design response". This relationship is expressed through a relationship between elements (components) that leads the manipulation and the generation of complex geometries and structures. Whether parametric algorithms use iteratively some components in order to identify the best solution to a specific design intent within design boundaries (parameters and rules), the process may be defined as Generative Design technique.

Nowadays, computers aid the algorithmic thinking and scripts are the formal expression of design algorithms. Parametric and Generative Design techniques have been adopted with effective result for design automation, form-finding strategies and digital fabrication strategies [6] [7]. Nevertheless, due to a lack of automation in standard manufacturing processes, digital fabrication strategies are far from a broad application in the BI.

Parametric and Generative Design techniques may be considered Knowledge-Based Engineering (KBE) techniques, according to its definition [8]. Nevertheless they lack in common libraries and dictionaries which have to be defined on purpose time by time.

1.2. Glued-Laminated-Timber technology

Glued-Laminated Timber (also known as GLT or glulam) is the most ancient engineered wood product. Since from Leonardo da Vinci's studies until the official patent by Hetzer in 1905, the main purposes of this product are to enhance the mechanical performances and to increase the dimension limits of the raw material (wood). In order to fulfil these purposes, it is manufactured by bonding together selected timber elements (laminates) and reducing the natural defects of the raw material.

Focusing on the case study of the glued-laminated-timber industry, the research presented in this paper aims at highlighting whether Parametric and Generative design techniques allow reducing wastes of resources and improving the overall efficiency of the supply-chain system as a whole, by pushing the Digitalization of design and engineering processes of products. Authors will discuss possible improvements to an ordinary production system of GLT in order to fulfil the minimum requirements of automation in order to make effective parametric and generative design techniques.

2. Methods

Research activities have been planned though two main phases:

- The first phase develops a specific parametric and generative algorithm that can be applied on design and engineering process of GLT products. The algorithm will be stress-tested by designing and engineering some prototypes, which have to satisfy different customer's need randomly generated. This test has to highlight the robustness and the flexibility of the algorithms against different designed geometries or different structural designs.
- The second phase aims at testing benefits and criticisms of parametric and generative algorithm on an ordinary production system of GLT. The test is performed using the Value-Stream-Mapping (VSM) technique. In order to measure effects of parametric and generative algorithms, a comparison between two value-steam maps (one referred to an ordinary production system and one using parametric and generative algorithms) will be developed focusing on both process timing and resource commitment.

The research is currently going to develop the second phase. A first version of the algorithm has been programmed and authors concluded a first survey of an ordinary production system of GLT. Authors are going to conclude the value-steam map of the ordinary production system of GLT and to start activities aiming at defining the value-steam map of a production system using parametric and generative algorithms.

Referring to case studies of KBE [9] [10], the approach proposed is more similar to KBE applications in automotive or aeronautical industry than applications in building industry: usually aiming at facilitating early design activities of entire buildings instead of single elements or products.

2.1. The programmed algorithm

The programmed algorithm develops two specific design intents (Figure 1 shows the algorithm's strategy): reduce the usage of unneeded high-quality raw material in final products and facilitate the manufacturing process. These design intents will be developed independently from the designed geometry and the overall structural design. In order to fulfil the first design intent, the algorithm has been designed combining the possible optimizations offered by standards and codes (EN 14080, the

German standard DIN 1052 and the European structural code for timber structures - Eurocode 5) [11] [12] [13]. Once inputs have been defined, the algorithm performs two different analysis:

- a mechanical stress analysis (through FEM simulations) to define the section areas of final products that are more stressed in order to use high-quality raw material only where needed (Figure 2-3);
- a curvature analysis to identify manufacturing strategies that should prevent possible reductions of mechanical performances of products.

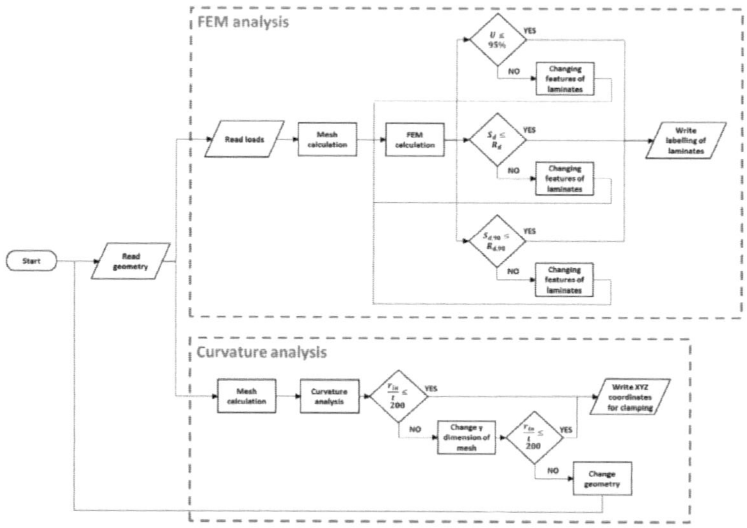

Figure 1. Preliminary flowchart of the designed strategy.

Whether optimizations are not performed the algorithm has to reduce mechanical performances of final products and iteratively change their geometry until the required performance is reached. For further details about the algorithm's strategy, readers may refer to previous publication by authors.

The outputs of the algorithm may be used for a cross-analysis with the ordinary supply-chain system of glued-laminated timber, highlighting benefits, criticism and possible improvements. The main intent of the authors during the discussion is to provide a case study, which may become a reference for transferring the adopted strategy to other building products.

3. Discussion

According to scientific literature [14] and to the first survey on a real production system, the production system of GLT is divided in five main stages:

- The process starts with a strength grading of sawn timber. Usually this is a machine-driven stage. It allows to grade timber elements by removing parts containing defects (such as large knots, irregular grains, etc.).
- After trimming out defects, according to the identified timber strength class, the required lamination length is obtained by finger jointing. This stage is machine-driven as well. At the end of these two stages selected laminates could be stored for a while for glue curing or, if allowed by a fast curing glue, they are sent to assembly area for the laying-up stages.
- Laminates strength for each layer are identified and manually selected for further stages. Just before starting the laying-up stages, regular laminates' thicknesses and depths are achieved using planers.
- Once laminates are covered by adhesive on selected faces, they are assembled and cramped together waiting for glue curing. Straight glued-laminated products are usually cramped using

automatic pressing machines. In case of complex geometries or curvatures, manual jigs are used for laying-up and cramping.
- Regular products' dimension are achieved using automatic or manual planers. At the end, after finishing products using coatings and paints, they are packed and labelled, ready for shipping.

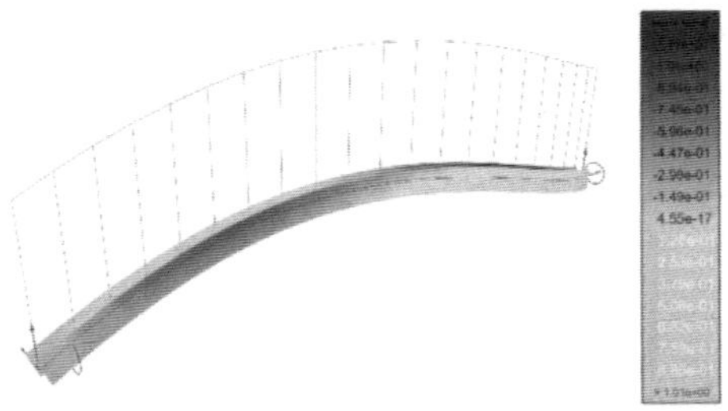

Figure 2. Example of FEM simulation of a double curved beam performed through the algorithm.

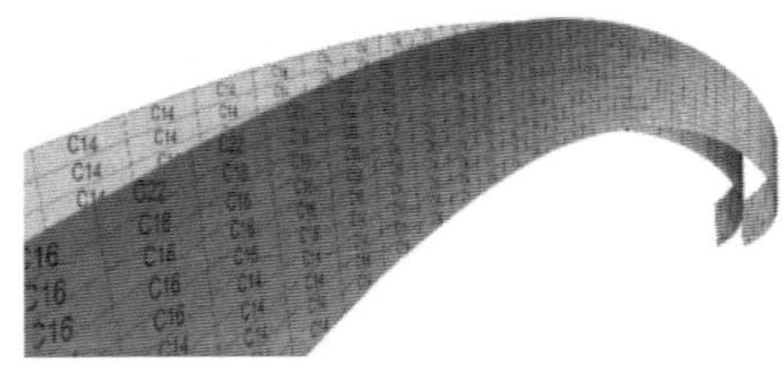

Figure 3. Labelling of laminates after FEM simulation.

The cross-analysis can be developed crossing the analysis performed by the programmed algorithm and manufacturing stages of the ordinary production system previously defined. From early evaluations, waiting for results from VSM analysis, authors identified some improvements of front-end activities of the supply-chain system such as design and engineering processes of GLT products. A uniform language (scripts) offered by parametric and generative algorithms avoids possible misunderstandings among responsible figures of design and engineering processes and enhance the efficiency of the overall communication management. Moreover, the automation of specific stages across design and engineering processes allows to overcome limitations in optimizations offered by the normative and to speeds up ordinary tasks as well as custom tasks in special commitments, such as the design and the engineering of double-curved GLT products.

Although parametric and generative algorithms can save 30% of high-quality raw material thanks to optimizations introduced across design and engineering processes, benefits on the manufacturing process of GLT are limited by lack of automation of specific stages across the production system. The strength grading and the finger jointing stages are arranged in a rigid serial workflow. This means that only homogeneous laminates can be manufactured. Parametric and generative outputs may become effective only whether machineries can manufacture inhomogeneous laminates. Companies could achieve this result without a huge effort. The sawn timber has to be arranged to provide different qualities of raw material just before the strength grading. This could be done by manually loading the strength-grading machinery or by adopting a machine that select the needed sawn timber. The strength-grading machine, interacting with the cutting machine and with the finger-jointing machine, has to acquire outputs from the algorithm in order to drive the length of each element across laminates and in order to request specific quality of raw material. Moreover, during laying-up and cramping stages, the curvature and the final geometry of products is achieved using manual jigs, this means that algorithm outputs may support operators setting-up jigs but this stage is still done manually. Whether companies would adopt an automatic pressing machine that can handle curvature in final products by a numerical control, the benefits of parametric and generative outputs could be stronger. Nevertheless, this kind of pressing machine is not available on the market nowadays: the installation would be a custom installation that needs huge investments.

4. Conclusions

Considering that results could be assumed only in a preliminary stage due to incomplete research phases and to the absence of a prototyping phase due to a lack of resources, it is possible to highlight that:

- Parametric and Generative Design techniques allow a strong improvement acquiring and handling design contents from responsible figures of design and engineering processes. This strongly increase the efficiency of the value-chain in the front-end processes mainly thanks to a more efficient communication management due to a uniform language.

- These approaches make optimization of production system easier thanks to a front-loading of manufacturing strategies and concerns during design activities.
- In order to produce effective benefits on manufacturing processes, these approaches require a strong automation across the production system. Referring to the case study, authors would like to remark that the interaction between parametric and generative outputs and machine-driven manufacturing stages could be not so easily achieved. Whether custom machinery has to be adopted, return of investments might be not sustainable in a short term for companies.

5. Acknowledgements

The authors would like to express their gratitude to Prof. Maurizio Piazza from University of Trento and to Eng. Giorgio Bignotti from Holzbau Sud company for their support and precious suggestions.

6. References

[1] ISTAT - Istituto Nazionale di Statistica - Italian Istitute of Statistics 2015 *Censimento dell'Industria e dei Servizi 2015*

[2] Wortmann J C 1983 *A Classification Scheme for Master Production Schedule. Efficiency of Manufacturing Systems* (New York: Plenum Press)

[3] Matt D T, Benedetti C, Krause, D and Paradisi, I 2011 *Proc. of 1st Int. Workshop on Design in Civil and Environmental Engineering* (Yuseong, Korea) pp 52-62

[4] Dallasega P, Matt D T and Krause D 2013 *Proc. of 2nd Int. Workshop on Design in Civil and Environmental Engineering* (Worcester, MA) pp 7-15

[5] Jabi W 2013 *Parametric Design for Architecture* (London: Laurence King)

[6] Sacks R, Eastman C M and Lee G 2004 *Parametric 3D Modeling in Building Construction with Examples From Precast Concrete* (Automation in Construction vol 13) pp 291-312

[7] Shea K., Aish R and Gourtovaia M 2005 *Towards Integrated Performance-Driven Generative Design Tools* (Automation in Construction vol 14) pp 253-264

[8] Prasad B., 1977 Capturing Life-Cycle Intent and KBE *Concurrent Engineering Fundamentals* (*Integrated Product and Process Development* vol 2), (New Jersey: Prentice Hall) Chapter 6

[9] Chapman C B and Pinfold M 2000 *The application of a knowledge based engineering approach to the rapid design and analisys of an automotive structure* (Advances in Engineering software vol 32) pp 903-912

[10] Whitney R S 1987 *Proc. of 2nd New Zeland Conf. of Expert Systems* (Auckland, New Zeland) pp 5-11

[11] UNI EN 14080 2013 *Strutture di Legno - Legno Lamellare Incollato e Legno Massiccio Incollato - Requisiti* (*Glued Laminated Timber - Requirements*) (UNI - Ente Italiano di Normazione)

[12] UNI EN 1995-1-1 2014 *Eurocodice 5 - Progettazione delle Strutture di Legno - Parte 1-1: Regole Generali - Regole Comuni e Regole per gli Edifici* (UNI - Ente Italiano di Normazione)

[13] DIN 1052 2008 Entwurf, Berechnung und Bemessung von Holzbauwerken - Allgemeine Bemessungsregeln und Bemessungsregeln für den Hochbau (Design of Timber Structures - General Rules and Rules for Buildings) (DIN Deutsches Institut für Normung e.V.)

[14] Piazza M, Tomasi R and Modena R 2005 *Strutture in Legno: Materiale, Calcolo e Progetto secondo le Nuove Normative Europee* (Milano: Biblioteca Tecnica Hoepli)

IOP Conf. Series: Materials Science and Engineering **157** (2016) 012034 doi:10.1088/1757-899X/157/1/012034

Type selection and design of hybrid propulsion system of ship

Nengqi Xiao, Riping Zhou, Xichen Lin

School of Energy and Power Engineering, Wuhan University of Technology, Wuhan 430063, China

E-mail: xiaonengqi@126.com, rpzhouwhut@126.com, 228128296@qq.com

Abstract: Hybrid propulsion system is a new type of dynamic form. It has the characteristic structural complexity and the diversity of operating conditions. Due to the different vessel functions, different sailing areas or different control performance requirements of the ship, types of hybrid propulsion systems are not the same. In this paper, 6000HP platform supply vessel is an example. Hybrid propulsion the system is selected by the fuzzy comprehensive evaluation method.

1. Introduction

Hybrid electric propulsion system is a new type of dynamic form. The hybrid propulsion system has been widely used in passenger ships, Marine work ship and public service and so on [1-3]. For different types of ships, the propulsion system of the host type and number, transmission mode, gear box device, the number of shaft and bearing and propeller device models and numbers are not the same.

Völker T [4] the feasibility of hybrid propulsion system for different types of ships is studied in the paper. The port tugs and ferries as the research object, team hybrid propulsion system for economic analysis.Shen Jing-kang [5] this paper analyzes the advantages and disadvantages of mechanical propulsion system, electric propulsion system and hybrid propulsion system, and introduces the configuration and type spectrum of HAVYARD 845 and A122 hybrid propulsion system. Simon Sortland [6] the configuration of the hybrid propulsion system of VS491CD three is studied, and the fuel consumption and exhaust emission of the system are compared with the traditional propulsion system. Sha Feng [7] the hybrid propulsion system is applied to the tuna fishing line, and the difference between the hybrid propulsion system and the traditional propulsion system is analyzed from the point of view of economy and reliability.

In this paper, the types and characteristics of hybrid propulsion system are studied. In this paper, the evaluation system of the hybrid propulsion system is established, and the reasonable pattern spectrum is selected based on the fuzzy comprehensive evaluation method.

2. Composition and type of hybrid propulsion system

Hybrid propulsion system is mainly composed of mechanical systems and electrical systems, as shown in figure 1 composition of hybrid propulsion system. The hybrid propulsion system has several typical operating conditions:

a) Power Take Off The main engine drives the propeller through a gearbox. A generator is connected at the gearbox that provides the electric power for the vessel auxiliary load, as shown in figure 2.

b) Boost Power Take In Diesel engine and electric motor delivers extra power through the gearbox to the propeller. Electric motor is provided by the diesel generators, as shown in figure 3.

c) Power Take Home In the process of ship navigation, diesel engine failure can't run. Propulsion power can still be provided by the electric motor, as shown in figure 4

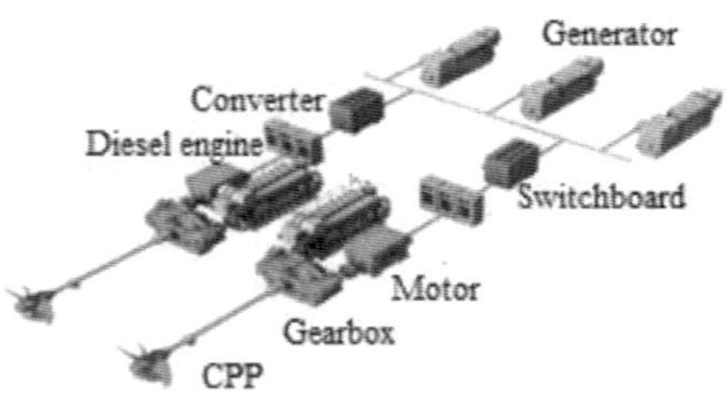

Figure 1. Composition of hybrid propulsion system

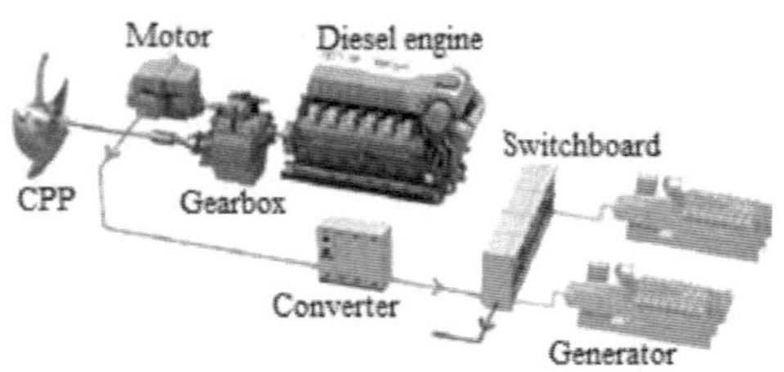

Figure 2.PTO of hybrid propulsion system

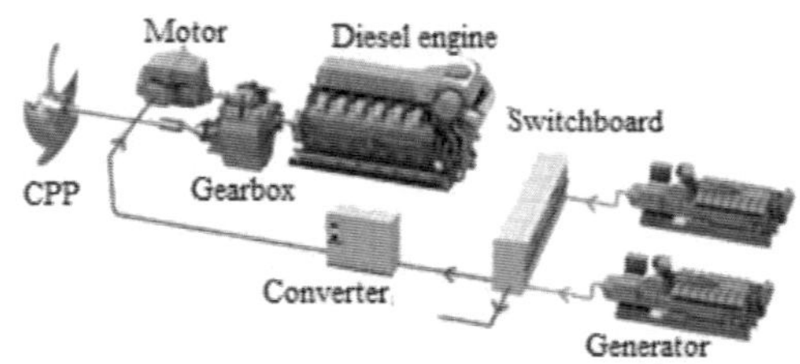

Figure 3.PTI model of hybrid propulsion system

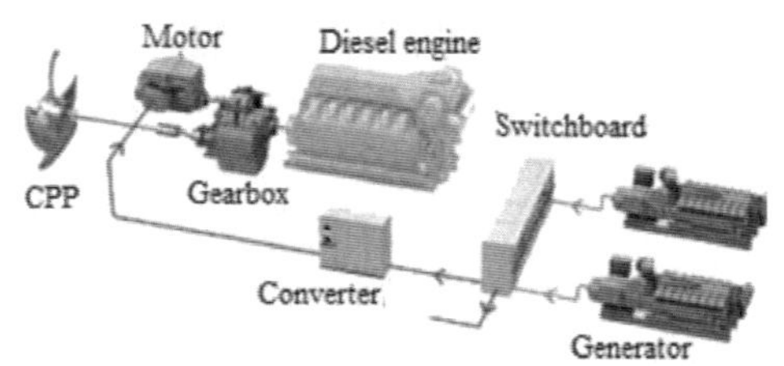

Figure 4.PTH model of hybrid propulsion system

After the collection and induction of the data of the domestic and international ship hybrid propulsion system type, we can get the propulsion system spectrum as shown in Table 1.

Table 1. Using hybrid propulsion system of the ship at home and abroad

No.	Ship name	Power system type	No.	Ship name	Power system type
1	VS4612 AHTS	twin machine one screw	8	A122	Twin screw diesel mechanical + PTO/PTI system
2	VS483M kII PSV	twin machine one screw	9	AX104	Diesel- electric hybrid propulsion
3	UT 730 WP	Diesel - electric hybrid	10	A101	Four machine twin screw system
4	UT 712 CD	Diesel - electric hybrid	11	Fujian 1000 tons of patrols	twin machine twin screw + PTO system
5	UT 759 ICE	Diesel - electric hybrid	12	Train ferry	Twin screw diesel

					mechanical
6	UT788CD	Double vehicle driver	13	Environmental protection ship	Four machine twin screw
7	(HAVYARD)845	Twin screw diesel mechanical + PTO/PTI system	14	Shandong 1000 tons of patrols	twin machine twin screw + PTO

3. Spectrum selection based on fuzzy comprehensive evaluation method

3.1 Establish the evaluation index system of the propulsion system spectrum

The evaluation index system needs to meet the principle of system optimization, comparability, scientific, stability, hierarchical, practicality and associability. According to the above principle, the evaluation index system of the hybrid propulsion system is established, see figure 5. Target layer: A = (hybrid propulsion system's spectrum).Factors layer: B_i= (B_1,B_2, B_3,B_4,B_5).The sub factors layer: C_i=(C_{11},C_{12}, C_{21},C_{31},C_{32},C_{41},C_{42}, ,C_{51}).

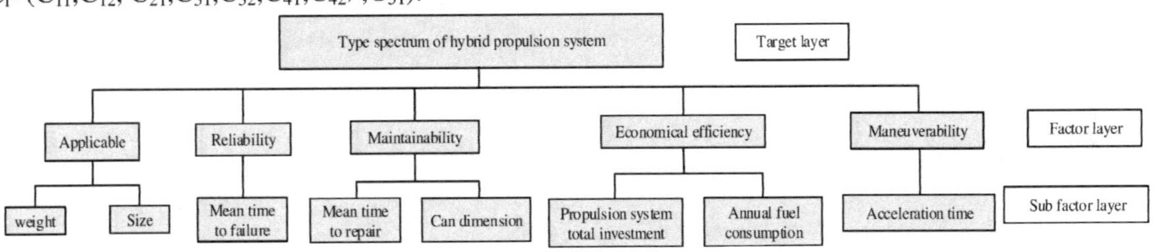

Figure 5. Evaluation index system of hybrid propulsion system based on spectrum

3.2 Determine factor weight vector

In the process of hybrid propulsion system selection, evaluation index system of hybrid propulsion system gives 5 factors. The weight of various factors is not the same in the evaluation system in Table 2. Factors layer B_i relative to the target layer A of the judgment matrix can be established in Table 3.

Table 2. Judgment matrix scale and its implication

Scale	Meaning
1	The two factors are of the same importance
3	The former is a little more important than the latter
5	The former is significantly more important than the latter
7	The former is more important than the latter
9	The former is more important than the latter
2/4/6/8	Intermediate value of the above judgment

Table 3. Factors B_i judgment matrix of the target A

Factors	B_1	B_2	B_3	B_4	B_5
B_1	a_{11}	a_{12}	a_{13}	a_{14}	a_{15}
B_2	a_{21}	a_{22}	a_{23}	a_{24}	a_{25}
B_3	a_{31}	a_{32}	a_{33}	a_{34}	a_{35}
B_4	a_{41}	a_{42}	a_{43}	a_{44}	a_{45}
B_5	a_{51}	a_{52}	a_{53}	a_{54}	a_{55}

By computing the judgment matrix, factor layer relative to the target layer of the weight vector U= (u_1, u_2, u_3, u_4, u_5) is obtained. Similarly according to the method described above, the sub factors layer relative to the factor layer of the weight vector A= (A_1, A_2, A_3, A_4, A_5) is obtained.

3.3 Establish the evaluation matrix of each factor

From design of ship to operation of ships, experts from different professional backgrounds give different evaluation opinion on factors and sub factors layer. According to the expert evaluation of the factors layer and sub-factors layer, evaluation set is established, S= (s_1, s_2, s_3, s_4, s_5, s_6) = (very good, good, relatively good, general, poor, very poor). The group of experts is composed of ship design experts, marine engineers, and maintenance management engineer and ship management. Using expert investigation method and probability statistics method, evaluation of sub factors layer are obtained. Because of the different factors layer has different sub factors, fuzzy evaluation matrix K of sub-factors are established. The evaluation matrix of factors layer M=AK. The evaluation matrix of hybrid propulsion system Q=AM.

4. Example analysis

6000HP platform supply vessel for the study, the ship in accordance with the American Classification Society (ABS) design, build and test the current specifications and rules ， as shown in Table 4 and figure 6.

Table 4. 6000HP platform supply ship basic parameters

Project	Unit	Values
overall length	m	79
LBP	m	71.4
beam	m	17.6
depth	m	7.7
Maximum draft	m	6.5
Maximum load	t	3800

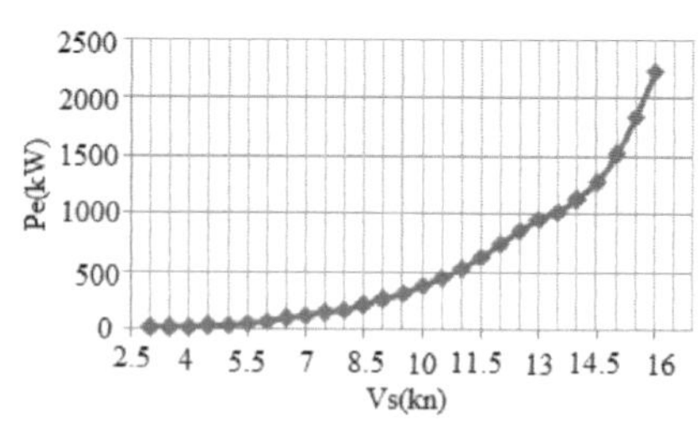

Figure 6. The relationship between speed and power curve

In order to realize the systematic comparative analysis of spectrum of different hybrid propulsion, double sculls and four sculls main configuration of the system is given, as shown in Table 5.

Table 5. 6000HP platform supply vessel system configuration

Name	**Double machine twin screw**		**Four machine twin screw**	
Project	Equipment details	Main parameters	Equipment details	Main parameters
Diesel	Rated power	1720kW	Rated power	960kW
	Rated speed	900r/min	Rated speed	720r/min
	number	2	number	4
	Type	MAN 8L21/31	Type	MAN 6L23/30A
Generator	number	3	number	3
	Rated power	180kW	Rated power	180kW
Emergency generators	number	1	number	1
	power	220kW	power	220kW
PTO	number	2	number	2
	power	1100kW	power	1100kW
CPP	number	2	number	2
	Type	CPP	Type	CPP

According to the theory of 3.2 section, the relevant theory is used to construct the factor Bi to the target A judgment matrix, as shown in Table 6.

Table 6. Factor B_i judgment matrix and weight vector of target A

Factors	**Applicability B_1**	**Reliability B_2**	**Maintainability B_3**	**Economy B_4**	**Mobility B_5**	**Factor weight U**
Applicability B_1	1	1/4	1/3	1/2	2	0.089
Reliability B_2	4	1	4	5	7	0.484
Maintainability B_3	3	1/4	1	6	4	0.260
Economy B_4	2	1/5	1/6	1	3	0.116
Mobility B_5	1/2	1/7	1/4	1/3	1	0.051
Maximum eigenvalue λ_{max} =5.432		CI=0.108	RI=1.12	CR=0.096	Meet the consistency check	

Where consistency index $CI = (\lambda_{max} - n)/(n-1)$; consistency ratio $CR = CI/RI$; If $CR < 0.1$, the judgment matrix is reasonable, and the random consistency index RI is given in Table 7.

Table 7. Random consistency index

n	**1**	**2**	**3**	**4**	**5**	**6**	**7**	**8**	**9**
RI	0	0	0.58	0.89	1.12	1.25	1.32	1.41	1.45

In the same way, due to the influence of each factor layer, it is necessary to construct the weight distribution set, as shown in Table 8 to Table 10.

Table 8. Judgment matrix and weight vector of C_{11} and C_{12} relative to B_1

Applicability	**Factors**	**Weight C_{11}**	**Size C_{12}**	**Factor weight A_1**
	Weight C_{11}	1	1/3	0.250
	Size C_{12}	3	1	0.750

Table 9. Judgment matrix and weight vector of B_{31} and C relative to B_3

Maintainability	Factors	MTTR C_{31}	Dimensions C_{32}	Factor weight A_3
	MTTR C_{31}	1	4	0.80
	dimensions C_{32}	1/4	1	0.20

Table 10. Judgment matrix and weight vector of B_{41} and B_{42} relative to B_4

Economy	**Factors**	**Total investment C_{41}**	**Fuel consumption C_{42}**	**Factor weight A_4**
	Total investment C_{41}	1	1	0.50
	Fuel consumption C_{42}	1	1	0.50

Selection of ship design experts, engineers, maintenance engineers, management of ship management experts such as establishing evaluation system of expert group. According to the evaluation opinions and suggestions given by each expert, the expert comments $S=(s_1, s_2, s_3, s_4, s_5)=$ (very good, good, relatively good, general, poor, very poor) set can be determined. The evaluation matrix of factor layer is calculated based on 3.2 section, as shown in Table 11 and Table 12.

Table 11. Evaluation matrix of sub factor layer and factor layer of twin screw propulsion system

Twin screw propulsion system			
Applicability factors evaluation matrix	$K_{11}=\begin{bmatrix}0.3 & 0.5 & 0.2 & 0.0 & 0.0\\ 0.3 & 0.3 & 0.3 & 0.1 & 0.0\end{bmatrix}$	$M_{11}=A_1K_{11}$	$[0.3 \quad 0.35 \quad 0.275 \quad 0.075 \quad 0]$
Reliability factors evaluation matrix	$K_{12}=[0.3 \quad 0.4 \quad 0.3 \quad 0.0 \quad 0.0]$	$M_{12}=A_2K_{12}$	$[0.3 \quad 0.4 \quad 0.3 \quad 0 \quad 0]$
Maintainability factors evaluation matrix	$K_{13}=\begin{bmatrix}0.2 & 0.4 & 0.4 & 0.0 & 0.0\\ 0.3 & 0.4 & 0.3 & 0.0 & 0.0\end{bmatrix}$	$M_{13}=A_3K_{13}$	$[0.22 \quad 0.4 \quad 0.38 \quad 0 \quad 0]$
Economic factors evaluation matrix	$K_{14}=\begin{bmatrix}0.3 & 0.4 & 0.3 & 0.0 & 0.0\\ 0.2 & 0.3 & 0.4 & 0.1 & 0.0\end{bmatrix}$	$M_{14}=A_4K_{14}$	$[0.25 \quad 0.35 \quad 0.35 \quad 0.05 \quad 0]$
Mobility factors evaluation matrix	$K_{15}=[0.3 \quad 0.4 \quad 0.3 \quad 0.0 \quad 0.0]$	$M_{15}=A_5K_{15}$	$[0.3 \quad 0.4 \quad 0.3 \quad 0 \quad 0]$

Table 12. Evaluation matrix of sub factor layer and factor layer of four machine twin screw propulsion system

Four machine twin screw propulsion system			
Applicability factors evaluation matrix	$K_{21}=\begin{bmatrix}0.2 & 0.4 & 0.4 & 0.0 & 0.0\\ 0.2 & 0.3 & 0.3 & 0.2 & 0.0\end{bmatrix}$	$M_{21}=A_1K_{21}$	$[0.3 \quad 0.325 \quad 0.375 \quad 0.15 \quad 0]$
Reliability factors evaluation matrix	$K_{22}=[0.5 \quad 0.4 \quad 0.1 \quad 0.0 \quad 0.0]$	$M_{22}=A_2K_{22}$	$[0.5 \quad 0.4 \quad 0.1 \quad 0 \quad 0]$
Maintainability factors evaluation matrix	$K_{23}=\begin{bmatrix}0.4 & 0.4 & 0.2 & 0.0 & 0.0\\ 0.5 & 0.4 & 0.1 & 0.0 & 0.0\end{bmatrix}$	$M_{23}=A_3K_{23}$	$[0.42 \quad 0.4 \quad 0.18 \quad 0 \quad 0]$
Economic factors evaluation matrix	$K_{24}=\begin{bmatrix}0.1 & 0.3 & 0.4 & 0.2 & 0.0\\ 0.4 & 0.5 & 0.1 & 0.0 & 0.0\end{bmatrix}$	$M_{24}=A_4K_{24}$	$[0.25 \quad 0.4 \quad 0.25 \quad 0.1 \quad 0]$

Mobility factors evaluation matrix	$K_{25} = [0.3 \quad 0.4 \quad 0.3 \quad 0.0 \quad 0.0]$	$M_{25} = A_5 K_{25}$	$[0.3 \quad 0.4 \quad 0.3 \quad 0 \quad 0]$

According to the 3 section, we can get the evaluation matrix Q_1 and Q_2 of the hybrid propulsion system spectrum of twin screw propulsion system and four machine twin screw propulsion system. The experts' comments S=(s_1, s_2, s_3, s_4, s_5)= (very good, good, relatively good, general, poor, very poor) are graded according to the percentage, S= (95, 80, 65, 45, 30). The hybrid propulsion system spectrum of twin screw propulsion system and four machine twin screw propulsion system scores in Table 13:

Table 13. Matrix and score of hybrid propulsion system

Type	Spectrum evaluation matrix	Score of hybrid propulsion system
Twin screw propulsion	$Q_1 = UM_1 = [0.273 \quad 0.39 \quad 0.324 \quad 0.012 \quad 0]$	$Q_1 = 0.273x95 + 0.39x80 + 0.324x65 + 0.012x45 + 0 = 79.578$
Four machine twin screw propulsion	$Q_2 = UM_2 = [0.413 \quad 0.393 \quad 0.168 \quad 0.025 \quad 0]$	$Q_2 = 0.273x95 + 0.39x80 + 0.324x65 + 0.012x45 + 0 = 83.587$

According to the score of the hybrid propulsion system can be known, 6000HP platform supply ship selection four machine twin screw hybrid propulsion system is better than two machine twin screw hybrid propulsion system.

5. Conclusion

Five aspects of applicability, reliability, maintainability, economy and mobility to establish a hybrid propulsion system based on spectrum evaluation system. Fuzzy comprehensive evaluation method is used to comprehensively consider the weight of various factors in the system, so that the results of hybrid propulsion system type spectrum selection is more scientific and accurate.

6. Reference

[1] Dedes E K, Hudson D A, Turnock S R. Assessing the potential of hybrid energy technology to reduce exhaust emissions from global shipping . *Energy Policy*, 2012, 40: 204-218.

[2] Alf Kare Adnanes. Reduction of fuel consumption and environmental footprint for AHTS and OSV using electric or hybrid propulsion . *World Ships& Boats*, 2008(7), 32-34.

[3] Reljić M, Matika D, Gržan M. Dynamic Position Of Offshore Anchor Handling Tug Supply (AHTS) Vessels (UT 788 CD Project) . *Brodogradnja*, 2014, 65(1): 55-64.

[4] Völker T. Hybrid propulsion concepts on ships. *Zeszyty Naukowe Akademii Morskiej w Gdyni*, 2013.

[5] Shen Jing-kang, Chen Wei-zhong. Prospect of hybrid power on multi purpose working ship in deep sea .*Jiangsu Ship*,2010.8:1-7

[6] Simon Sortland. Hybrid propulsion system for anchor handling tug supply vessels.*WÄRTSILÄ TECHNICAL JOURNAL*. 2008(01):45-48.

[7] Sha Feng, Wang Yong-ding,Ye-Shou-jian. Study and analysis on tuna longliner hybrid power system. *Journal of shanghai ocean university*, 2014, 23(2): 279-283.